Handbook of Surfactants

Handbook of Surfactants

Second edition

M. R. PORTER, BSc, PhD, CChem, MRSC
Maurice R. Porter & Associates
Consultants in Speciality Chemicals
Sully
South Wales

SPRINGER-SCIENCE+BUSINESS MEDIA, B.V

First edition 1991
Reprinted 1993
Second edition 1994

© 1994 Springer Science+Business Media Dordrecht
Originally published by Chapman & Hall in 1994
Softcover reprint of the hardcover 2nd edition 1994
Typeset in 10/12 Times by Photoprint, Torquay, Devon

ISBN 978-94-010-4580-3   ISBN 978-94-011-1332-8 (eBook)
DOI 10.1007/978-94-011-1332-8

Apart from any fair dealing for the purposes of research or private study, or criticism or review, as permitted under the UK Copyright Designs and Patents Act, 1988, this publication may not be reproduced, stored, or transmitted, in any form or by any means, without the prior permission in writing of the publishers, or in the case of reprographic reproduction only in accordance with the terms of the licences issued by the Copyright Licensing Agency in the UK, or in accordance with the terms of licences issued by the appropriate Reproduction Rights Organization outside the UK. Enquiries concerning reproduction outside the terms stated here should be sent to the publishers at the Glasgow address printed on this page.

The publisher makes no representation, express or implied, with regard to the accuracy of the information contained in this book and cannot accept any legal responsibility or liability for any errors or omissions that may be made.

A catalogue record for this book is available from the British Library
Library of Congress Catalog Card Number: 94-70709

∞ Printed on acid-free text paper, manufactured in accordance with ANSI/NISO Z39.48-1992 (Permanence of Paper)

Preface

In preparing the second edition of this book, I have revised, updated and extended the original material, with particular attention to two areas of the book where there has been considerable publication of new material. The chapters on the use of surfactant theory and polymeric surfactants have been completely rewritten. Surfactant theory has seen considerable progress in the 1980s, but it is only in the last few years that it has been simplified such that it can be used in helping to formulate compositions containing surfactants in different applications. It must be admitted that even now most applications utilise empirical methods of formulation but the results in many cases can be better interpreted. Wetting and micro-emulsions are now much better understood, but foams and defoamers still elude simplification. The use of theory in formulating compositions will probably very slowly be extended, but only if fairly simple rules, preferably non-mathematical, can be taught to industrial chemists. The concept of geometrical packing of surfactant molecules at an interface and the effect of the relative size of the head and tail has probably been the major advance in simplifying theory. This concept is now well recognised in the scientific literature but is not yet used widely by technologists using surfactants. I hope that this book will help in spreading the use of this simple concept. The chapter on polymeric surfactants was rewritten, due to the paucity of data available on polymeric surfactants during preparation of the first edition. Even now the amount of data published on well defined chemical structures is very limited. Manufacturers still do not publish the chemical structures of most polymeric surfactants, probably because many of the structures are still uncertain.

The opportunity has also been taken to extend the technical data on the basic surfactants themselves, in particular the alkyl polyglycosides and the narrow range ethoxylates, but there are many small additions to make the data more complete. The aim of the first edition was only to cover in depth those surfactants commercially available; as a result many very interesting surfactants were not described. This edition includes a few novel surfactants (bolaform, Gemini, polymerisable and labile surfactants), but only those which are probably on the point of commercialisation.

As in the first edition, many generalisations have been made in order to keep statements as simple as possible. These generalisations are based on my own experience, supported by other data. I make no apologies for these generalisations being not entirely accurate for every surfactant, if they have enabled readers to improve their knowledge of surfactants and surfactant technology and use it in practice.

I would like to take this opportunity to thank the very large number of people in the surfactant industry who have willingly supplied both data and many helpful comments and suggestions. In particular Eric Lomax, who checked the section on amphoterics and helped in discussions on surfactant theory and how it can be used in practical work.

Maurice R. Porter
Sully, South Wales

Contents

1 General introduction   **1**

2 General approach to using surfactants in formulations   **6**

2.1 Introduction	6
2.2 Systematic approach	8
2.3 Practical formulation	9
2.4 Understanding formulations and end effects	10
2.5 Properties of the hydrophilic and hydrophobic groups	11
2.5.1 The hydrophilic group	11
2.5.2 The hydrophobic group	11

3 Information sources   **14**

3.1 Introduction	14
3.2 Manufacturers' literature	15
3.3 Published books	17
3.4 Journals and periodicals	20
3.5 Patents	21
3.6 Symposia, meetings and courses	23
3.7 Government publications	23
3.8 Databases	24

4 Use of surfactant theory   **26**

4.1 Introduction	26
4.2 Adsorption and critical micelle concentration (CMC)	27
4.3 Micelles, vesicles, liposomes and lamellar structures	39
4.4 Solubility and liquid crystals	46
4.5 Solubilisation and microemulsions	57
4.6 Wetting	61
4.7 Foaming/defoaming	65
4.8 Macroemulsions and HLB	72
4.9 Dispersing	78
4.10 Detergency	84
4.11 Surfactant mixtures and interactions	89
References	92

5 Surfactants commercially available   **94**

Reference	98

6 Anionics   **99**

6.1 Soaps	100
6.2 Modified carboxylates	104
6.2.1 Ethoxy carboxylates	105
6.2.2 Ester carboxylates	107
6.2.3 Amide carboxylates	109
6.3 Isethionates (ester sulphonates)	111
6.4 Phosphate esters	113

viii HANDBOOK OF SURFACTANTS

6.5	Sulphates	116
	6.5.1 Alcohol sulphates	118
	6.5.2 Alcohol ether sulphates	121
	6.5.3 Sulphated alkanolamide ethoxylates	126
	6.5.4 Sulphated oils and glycerides	127
	6.5.5 Nonylphenol ether sulphates	129
6.6	Sulphonates — general	130
	6.6.1 Ethane sulphonates	134
	6.6.2 Paraffin sulphonates	135
	6.6.3 Alkyl benzene sulphonates	138
	6.6.4 Fatty acid and ester sulphonates	143
	6.6.5 Alkyl naphthalene sulphonates	148
	6.6.6 Olefin sulphonates	151
	6.6.7 Petroleum sulphonates	155
6.7	Sulphosuccinates and sulphosuccinamates	159
	6.7.1 Sulphosuccinates	159
	6.7.2 Sulphosuccinamates	164
6.8	Taurates (amide sulphonates)	166
	References	168

7 Nonionics 169

7.1	General introduction	169
	7.1.1 The chemistry of ethoxylation	171
	7.1.2 General properties of nonionics	175
	7.1.3 Surface-active properties of nonionics	180
7.2	Acetylenic surfactants	187
7.3	Alcohol ethoxylates	188
7.4	Alkanolamides	194
7.5	Amine oxides, phosphine oxides and sulphoxides	198
7.6	Surfactants derived from mono- and poly-saccharides	202
7.7	Ethoxylated alkanolamides	210
7.8	Ethoxylated long-chain amines	212
7.9	Ethylene oxide/propylene oxide (EO/PO) copolymers	215
7.10	Fatty acid ethoxylates	222
7.11	Sorbitan derivatives	226
7.12	Ethylene glycol, propylene glycol, glycerol and polyglyceryl esters plus their ethoxylated derivatives	231
7.13	Alkyl amines and alkyl imidazolines	236
7.14	Ethoxylated oils and fats	242
7.15	Alkyl phenol ethoxylates	243
	References	246

8 Cationics 248

8.1	Cationics (general)	248
8.2	Quaternary ammonium	249
8.3	Amine and imidazoline salts	254
	Reference	257

9 Amphoterics 258

9.1	Amphoterics (general)	258
9.2	Betaines	264
9.3	Glycinates	269
9.4	Amino propionates	272
	References	275

CONTENTS — ix

10 Speciality surfactants — 276

10.1 General — 276
10.2 Silicone surfactants — 277
10.3 Fluorocarbons — 282
10.4 Miscellaneous specialities — 285
    10.4.1 Bolaform surfactants — 285
    10.4.2 Gemini surfactants — 286
    10.4.3 Labile surfactants — 287
    10.4.4 Polymerisable surfactants — 288
References — 292

11 Polymeric surfactants — 293

References — 304

Appendix 1 Names of hydrophobes and average composition of fats and oils — 305

Appendix 2 Ecological and toxicity requirements — 306

Biodegradation — 306
Toxicity — 313
References — 317

Index — 318

Abbreviations

General

ABS	Branched alkylbenzene sulphonate
Ac	Acetyl group CH_3CO
AE	Alcohol ethoxylate
AES	Alcohol ether sulphate
AOS	Alpha-olefin sulphonate
APE	Alkyl phenol ethoxylate
APG	Alkyl polyglycoside
AS	Alcohol sulphate
BOD	Biological oxygen demand
CD	Coconut diethanolamide
CD-ROM	Compact disk read only memory
CMC	Critical micelle concentration
COD	Chemical oxygen demand
DEA	Diethanolamine
DOC	Dissolved organic carbon
EO	Ethylene oxide
EO/PO	Ethylene oxide/propylene oxide copolymer
Et	Ethyl—C_2H_5
EtOH	Ethanol (ethyl alcohol)—C_2H_5OH
FES	Fatty ester sulphonate
HLB	Hydrophilic lyophilic balance
LES	Lauryl ether sulphate
LABS	Linear alkyl benzene sulphonate
MBAS	Methylene blue active substances
Me	Methyl—CH_3
MEA	Monoethanolamine
MeOH	Methanol (methyl alcohol)—CH_3OH
NPE	Nonyl phenol ethoxylate
NTA	Sodium nitrilotetra-acetate
O/W	Oil in water emulsion
PE	Phosphate ester
PEG ESTERS	Polyoxyethyleneglycol esters of fatty acids
PO	Propylene oxide
pvc	Poly (vinyl chloride)
QAC	Quaternary ammonium compound
RT	Room temperature
SAS	Secondary alkane sulphonate (paraffin sulphonate)
SDS	Sodium dodecyl sulphate
SLS	Sodium lauryl sulphate
ST	Surface tension
SXS	Sodium xylene sulphonate
TEA	Triethanolamine
TPS	Tetrapropylene benzene sulphonate
USP	United States patent
W/O	Water in oil emulsion

HANDBOOK OF SURFACTANTS

Units

k	kilo = $\times 10^3$
da	deca = $\times 10$
d	deci = $\times 10^{-1}$
c	centi = $\times 10^{-2}$
m	milli = $\times 10^{-3}$
μ	micro = $\times 10^{-6}$
n	nano = $\times 10^{-9}$
g	gram
kg	kilogram
M	g mol/1
m	metre
nm	nanometre = 10^{-9} metre = 0.1 Ångström unit
mol	gram-molecule
s	second
°C	degrees Celsius
l	litre (normally written in full except with a prefix, e.g. ml)
mN/m	surface tension, numerically equal to dyne/cm

1 General introduction

The basic aim of the book is to give practical help to users of surfactants and those people who make formulations with surfactants. The choice of a cost-effective surfactant, although essential, is not easy, due to the bewildering number of surfactants available on the market, insufficient data usually provided by the manufacturers and poorly defined chemical structure–effect relationships. This book attempts to provide some practical help with these problems, but also explanations of the basic properties of particular types of commercially available surfactants.

The practical help consists of describing the various types of surfactants in terms of their chemical structures, principal physical properties, functional properties and principal end uses, with some comments on the requirements for quality control. There are chapters on general approaches to formulating, use of surfactant theory, how and where to find information concerning surfactants and their applications, and a broad description of test methods for biodegradation and toxicity. However, the bulk of the book is about the surfactants themselves, in a concise, consistent format.

Surfactants are generally described as anionic, nonionic, cationic or amphoteric. These are physical properties of particular chemical structures. To describe a surfactant in these terms gives some very general properties which may, but usually does not, help a user to choose a surfactant. The more important factors in the choice of a surfactant are:

- Is it commercially available?
- Does it perform the function required?
- Is it expensive to use?
- Is the physical form convenient for manufacture and use?
- Is the surfactant stable to storage in its end use?
- Is it safe in manufacture and transport?
- Is it safe to use in the required end use?
- Does it pose any dangers to the ecology?

The last three requirements are becoming the most important requirements. No matter how good or cheap the finished product, if the surfactant (or any other component for that matter) gives a danger, real or imagined, to the end user or the ecology then the product is not saleable. No safety data or toxicity data are given on the various types of surfactants described, the reason being that it is impossible to generalise on toxicity, as small

2 HANDBOOK OF SURFACTANTS

changes in the chemical constitution can affect the toxicity of products. On environmental grounds, the requirements change so rapidly that statements on environmental safety could be misleading. Factual data such as toxicity tests and biodegradation tests are not useful unless full details of the test protocols are given. Instead a short chapter on this subject has been included to give some background on the various tests quoted for those new to this area. At least readers may then have a better understanding of the data and statements made by the surfactant manufacturer or supplier, who should be the major source of data and advice.

When picking surfactants for a particular end use, a user must bear in mind all these requirements in order to make cost-effective saleable products or to use surfactants without problems inside the factory. It is only sensible to use those types of surfactants which will cover all the requirements above. There are hundreds of surfactants commercially available but only a small number of chemical types are produced in large volume commercially world-wide. It is obviously preferable to select one or more of these types for use if possible. However there can be problems in picking one chemical type. Most surfactants that are commercially available are not pure chemicals but consist of mixtures of chemicals. It is these mixtures that often give the end effect required. This is why there can be variations in performance of surfactants from one manufacturer to another. Although manufacturers may give as full a chemical description as possible, the relationships between chemical structure and performance are still not well documented or understood.

A very large amount of information is available on surfactants in published papers and conference proceedings. The surfactant manufacturer will also have a considerable amount of information on the properties, end use and safety of products. Most published scientific information relates to the properties of the surfactants, whereas details of the use of surfactants in formulations tends to be in the patent literature. Patent literature can often be confusing rather than helpful if one is seeking the reason behind the choice of a particular surfactant in a specific end use.

The present tendency to stricter quality control will emphasise the need for the routine analysis of the surfactant plus the analysis of the surfactant in the formulation. Users need to know which tests need to be carried out in order to identify significant variations in the composition of the surfactant.

There is now a strong move by consumer groups and some sections of the surfactant producers to move away from petrochemical-based feedstocks. Surfactants made from natural materials other than petroleum have been neglected but are now making a comeback. It is for this reason that possibly a disproportionate amount of space has been devoted to products whose major raw materials are derived from animal fats, vegetable oils or carbohydrates.

GENERAL INTRODUCTION

3

The use of surfactants is extremely widespread not only in industry but in the home. There are three major reasons for this:

- Water rather than organic solvents is increasingly used in industrial products
- Most mixtures and formulations are applied to solid substrates
- Cleaning is a very common requirement

Most industrial and domestic processes using chemicals involve contact between a liquid and a solid where the solid needs wetting. This is exactly the function of the surfactant. However, there are many products which can easily wet substrates better than water, such as alcohols, hydrocarbons, etc. It is the particular property of surfactants to decrease the surface tension of water using very low concentrations that is so valuable. In practical terms it means that most of the properties of water can be retained and the wetting improved at a very small additional cost. The function of cleaning is extremely common both in consumer use and in industry. There is no other end use of formulated products that approaches the volume and number of applications which involve cleaning — the scientific term is 'detergency'. Water is by far the most common medium and all aqueous based detergents (formulations for cleaning purposes) contain surfactants.

'Surfactant' is an abbreviation for surface active agent, which literally means active at a surface. The surface can be between solid and liquid, between air and liquid, or between a liquid and a different immiscible, liquid. The primary property of a surfactant in a solution is that the concentration of the surfactant is higher at the surface than in the bulk of the liquid. Thus the surfactant concentrates at the surface, and it is therefore no surprise to find that surfactants can be very economical in use, as they are concentrated at the point where they are performing a useful function. This may be an elementary concept, but it does explain the effectiveness of surfactants in many applications compared to other products that do not show any marked surface active properties. The user is always concerned with the economics of the product, which are often dependent upon the basic surface active properties.

What do we mean by a formulated product? Manufacturing industry makes and uses a wide variety of formulations for use within industry and also as the manufactured product for use in industry or for use in the home and institutions. Detergents, paints, inks, adhesives, cosmetics, dyes, weedkillers, insecticides and ice cream are all examples of formulations familiar to everyone. However, inside manufacturing industry there are many formulations that are not seen by the end consumer but are essential as processing aids. The textile industry uses scouring aids to clean fibres, aqueous based lubricants for spinning synthetic fibres, warp sizes to protect

4   HANDBOOK OF SURFACTANTS

fibres in weaving, defoamers to suppress foam during dyeing, and softening agents to treat fabrics in order to give a soft 'handle'. The paper industry uses defoamers, dispersing aids and release agents. The engineering industry uses lubricants, anticorrosive treatments and metal working lubricants. Practically all these formulations are composed of mixtures of chemicals. Surface active agents are present in all the formulations mentioned so far and in numerous others.

Although formulations differ from one application to another there are some factors which are common to every formulation:

1. An active ingredient which carries out the primary function desired by the end user
2. A medium by which the active ingredient is carried
3. At least one secondary function which will usually, but not always, be achieved by at least one other ingredient

It is possible to provide all functions with one chemical product. A good example is soap, which in the bar form provides the active ingredient which washes, the solid medium which is convenient for washing hands and the good dispersion of the dirt in water for easy disposal. Modern synthetic detergents now contain active washing ingredients which can wash at much lower temperatures, are liquid in form by virtue of their solubility in water (which is now the medium), and provide better dispersion of dirt, particularly in hard waters. Although new synthetic organic chemicals have been produced that can give all these improvements in one chemical species, they are very rarely used. The reason is that mixtures of chemicals are easier and cheaper to produce than new molecules, particularly on the large scale. Soap has not disappeared from the market place, but the soap tablet of today is a carefully formulated product, not one chemical.

Another example is lubricating oils, which were at one time a refined fraction of crude petroleum. The oil would provide the essential function of lubricating, it would be its own medium, being liquid and easily handled, and it had secondary functions such as dispersing solids and giving some corrosion resistance. However, modern lubricants are complex mixtures with chemical additives, some giving improved lubrication and others giving improved corrosion resistance and improved dispersability of solids. It is much easier and cheaper to provide improved products by mixing than by synthesising new chemical molecules with the desired properties. New synthetic organic chemicals (esters, synthetic hydrocarbons, phosphate esters) have been produced that are superior to petroleum-based lubricating oils as lubricants, but these synthetic products are now themselves being formulated by the addition of additives to improve their basic properties.

These two examples illustrate the practical effect of using mixtures of chemicals to solve problems. In the great majority of formulations used by

GENERAL INTRODUCTION

industry and in the home, a surfactant or a mixture of surfactants will be used.

The information presented in this book has been obtained from a combination of personal experience, manufacturers' technical information, and the patent and scientific literature. There are some references, but a considerable number of generalisations are made in order to make the book easy to read. If there is no reference it will mean that there will be at least two sources of information which agree, plus the author's own experience. If a reference is given it generally means that the statement is reasonable within the author's knowledge, but a secondary independent source cannot be found. The number of generalisations means that exceptions can probably be found, particularly those relating chemical structure and physical/chemical properties. The end uses given, with very few exceptions, have been positively identified as those actually used in practice and are not quotations from patent literature. There will be many uses of particular surfactants that are not mentioned, but the author believes that he has identified the major uses of particular types of surfactant. If he is wrong in statements on data or end uses he would be most interested to be provided with appropriate data confirming the error.

2 General approach to using surfactants in formulations

2.1 Introduction

There can be two quite different approaches to formulation because of the very different requirements of the market. There are basically two different market conditions:

1. Where there is a large volume market and the formulation will be sold for several years without significant change e.g. a household detergent
2. Where the market is small and subject to change

In the first situation the potential profit in the future can be very large on one single product and hence a detailed technical program can be initiated. Planned experiments on end effect, storage stability and environmental acceptability with comprehensive testing of various surfactants and of hundreds of formulations are feasible. Detailed examination of the properties of the surfactant is possible, new methods of analysis can be devised and more information is often obtained than that possessed by the supplier. The formulator then becomes to a large degree independent of technical help from the supplier.

The picture is very different in the second situation. The potential profit is so much smaller that technical work has to be limited. The formulation is always required quickly, if not by the customer then at least by one's own sales staff. The resources are generally very much more limited. The overall result is that the formulator becomes very dependent upon the supplier. The main contact with the supplier comes via a sales representative and the technical literature published by the company. The formulator will be hoping that someone can specify all the formulation required together with all safety data and environmental acceptability. However, in the case of surfactants the formulator's major problem is finding and choosing the supplier. As surfactants are often similar in effect, most suppliers will be promoting a particular surfactant. This product may well be the best product in that supplier's range, but is it the most cost effective product available on the market? In the first situation above, the technical department has the time and resources to search for the best surfactant. In the second situation, formulators are dependent upon their own (sometimes literally) knowledge, and in extreme circumstances have to make decisions in a matter of days.

GENERAL APPROACH TO USING SURFACTANTS IN FORMULATIONS 7

The information in this book will not enable formulations to be made up quickly, but should help the formulator in choosing the right family of surfactants and posing the right questions to suppliers in order to identify the best surfactant to use.

There will be a reason for a new or modified formulation. This reason should be firmly established with other members of the company, e.g. the marketing department, before commencing work, as this reason can and does restrict the choice of surfactant which can be used. The most common reasons are:

- Meeting a new market requirement in terms of a completely new product
- Changing the physical characteristics of the formulation
- Improving the functional efficiency of a product
- Reducing costs of a formulation to meet competition
- Avoiding problems of human toxicity
- Avoiding problems of environmental acceptability
- Avoiding a patent

However, the exact reason for the need of a new formulation may not be clear in detail or quantifiable. 'It doesn't work' or 'it's too expensive' are often the reasons given by the marketing department. They, however, have their problems and it is likely that the customer has been vague with them. There is no substitute for a meeting with the end user, even if the need is for a cheaper product, in order to identify the critical requirements of the product as seen by the user. Thus after the reason, i.e. the overall objective, is established, there is the need to establish the technical and economic target; the main factors being:

- The end effect (or function) desired and the conditions of use
- The costs to be met
- The physical form
- Restriction on safety in manufacture, handling, transport and use

The end effect and costs are generally related; a high cost product can be sold if it is very efficient, i.e. used in smaller quantities than the cheaper formulation.

In the case of changing an existing formulation, the situation can often arise that the original formulation has been unchanged for many years, the original formulator has retired and there are no detailed records of the development work leading to the formulation. This situation often arises where a company places great importance on the confidentiality of the formulations. Particularly where a formulation contains more than one surfactant, the functions performed by each individual surfactant may not be at all clear. There is now considerable evidence to show that mixtures of different surfactants do show synergistic properties and so, if mixtures are

8 HANDBOOK OF SURFACTANTS

present, not only must the properties of each surfactant be identified but the interaction between them as well. When there is a need to change a complex formula it may be simpler to start from basics rather than modify by trial and error.

2.2 Systematic approach

The first essential in a systematic approach is to draw up a detailed requirement for the product:

- Identify the end user's requirement in terms of the function of the finished formulation and how to test for that requirement (if possible)
- Identify the physical properties required by the product and user
- Identify toxicity and ecological requirements
- Identify cost limitations
- Identify time limitations, i.e. when is the new formulation required

Experienced formulators will find that this target requirement can often be determined very quickly but the author strongly urges that time be spent on this analysis in order to avoid wasted time and effort. It must also be realised that one or more of these requirements can change during the course of the development of a formulation. Therefore there is a need to update these requirements if there is a time delay in producing a new formulation. An update should be made at a maximum of three months.

These targets can be quickly identified but the problems arise in translating the properties required into the type of surfactant that will satisfy the targets. The following approach is suggested:

1. Consider the cost of the formulation and the quantity of surfactant in the formula; if there is a high proportion of surfactant in the formulation then this can often eliminate high priced surfactants.
2. Identify the physical properties required, e.g. solubility, viscosity, pH range stability, chemical stability, compatibility with other components, hard water tolerance. Again this can often eliminate many surfactants
3. Try to identify the basic functions provided by the surfactant, namely wetting, foaming, emulsifying, solubilisation, dispersing and cleaning. Cleaning is not a basic property but it is included in this category because it demands the right combination of wetting, foaming, emulsifying and dispersing properties. It is also a very common requirement of household and industrial formulations.

The chapters on the different classes of surfactants have been written in such a way that it is easy to determine the physical and functional properties of the various classes of surfactants.

GENERAL APPROACH TO USING SURFACTANTS IN FORMULATIONS 9

Cost and availability are probably the most important considerations, and the next step would be to see if properties specified so far would be satisfied with those surfactants which are produced in large volume worldwide. Examine whether the following chemical types can satisfy these criteria: soaps; linear alkylbenzene sulphonates (LABS); alcohol ethoxy sulphates (AES); alcohol sulphates (AS); alkane or paraffin sulphonate (SAS); alcohol ethoxylates (AE), There are many other surfactants available in volume but the above families probably represent the most common, cheapest and most widely available products world-wide. There is also the added benefit that safety information on these surfactants is available in great detail.

Once a detailed target is established and also some idea of the surfactant's requirements then it is much easier to search the technical literature and to put the right questions to potential suppliers. Of major help in creating a new formulation is to find internal reports or personal knowledge from someone who has formulated a similar kind of product. If one lacks that help, external sources must be used (see chapter 3). The five main sources of information are: internal reports or verbal help; manufacturers' literature or advice from their technical department; specialist books; *Chemical Abstracts* as an index to published articles; and patents.

2.3 Practical formulation

In nearly all cases, a meaningful test of the functional use of a formulation is the most difficult to devise. There are laboratory tests for detergents, wetting agents, lubricants, defoamers, dispersing agents, etc., but no one laboratory test can simulate the many different end users' requirements. Therefore it is preferable to eliminate as many surfactants/formulations as possible with simple, quick and cheap tests that do not attempt to test for functional use. Requirements such as flash point, viscosity, minimum solids content, etc., can be quickly checked. As described in section 2.2, many products can also be eliminated on cost and safety considerations.

With the use of this handbook, the chemical type of surfactant to give the required end effect can usually be identified. However, the exact choice of surfactant will need laboratory work in testing for the end use. Such testing is more easily carried out in the laboratory but the essential need is to correlate laboratory tests with actual practice. With many variables, statistically designed experiments are more efficient but rarely carried out in practice.

When a finished formulation is ready for outside testing it is wise to carry out some simple stability tests, because surfactants are usually in a state of semi-solution, giving separation of phases, thickening, thinning and sometimes loss of activity. Stability tests at higher (40°C) or lower (freezing)

10   HANDBOOK OF SURFACTANTS

temperatures can often quickly identify unsatisfactory formulations. Visual examination is generally adequate for shelf stability tests. When such tests have been carried out the products can be released for commercial evaluation. Often a considerable length of time will elapse before results are available from outside customers. Such time should not be wasted. Extended storage trials do not involve much extra effort in visual examination once a week. Additional information can be obtained from suppliers on methods of analysis for quality control which will be needed if the product is successful commercially.

The quantitative analysis of a surfactant in a formulation is often needed for quality control. Analysis of a single surfactant species is relatively easy and well documented, but analysis of small quantities of surfactants in mixtures is not. The manufacturer of the surfactant can usually supply methods of analysis.

There is a need to check on published toxicity and environmental information. Check that the surfactant is in EINECS and EPA regulations. All the large volume surfactants are registered in EINECS and the surfactant manufacturer should provide help in other cases.

At all stages of the work keep neat detailed records of all results, particularly negative results, because some day someone will need them. There is a need to keep such results confidential but accessible to future workers.

The problem of trying to improve a present formulation is quite different, as one way of improving performance will be identifying a synergistic effect of a mixture of surfactants or identifying a new surfactant to solve the problem. A detailed literature survey is advisable, and some theoretical basis often helps in designing the experiments, because the number of possible combinations of mixtures of surfactants will run into thousands.

2.4 Understanding formulations and end effects

To understand the interrelationship between surfactant structure and the end effects caused by the surfactant is the goal of every formulator. Unfortunately surfactant theory is not yet sufficiently advanced to give more than guidelines. Empirical data are comprehensive but scattered throughout the literature, or more often than not never published but filed away in industrial files or in someone's mind. This book attempts to give basic data on different surfactant types which can be used for reference. However, in addition it is hoped that the data are presented in such a way that the reader begins to build a relationship of the properties given by a particular chemical structure. What is so confusing is that small variations in hydrophobic and hydrophilic groups give so many different properties.

GENERAL APPROACH TO USING SURFACTANTS IN FORMULATIONS 11

To obtain a simplified view of these relationships it is best to separate in the mind the properties conferred by the hydrophobic group and by the hydrophilic group. Whilst chapters 6–12 give detailed data on the individual surfactants this section attempts to give an overall simple view of the main properties of the hydrophobic and hydrophilic groups.

As pointed out in chapter 1, any attempt to summarise the properties of surfactants will fail to be entirely accurate, and examples will be found that do not exactly agree with the statements below. Thus the general comments presented here must only be used as a guide.

2.5 Properties of the hydrophilic and hydrophobic groups

The major applications of surfactants are in aqueous media and this summary is confined to applications where water is the continuous phase.

2.5.1 The hydrophilic group

The data in the book have been organised on differences in the hydrophilic group, because this group determines the main differences between the majority of surfactants on the market. Thus it is important to have in mind the major properties, i.e. end effects of the main surfactant types, namely:

- Anionics   detergents, adsorption on polar surfaces
- Nonionics   stability in varying pH
- Cationics   adsorption on surfaces
- Sulphonates   stable in solution
- Sulphates   unstable in solution
- Ethoxylates   stable in hard water

Other generalisations on the hydrophilic groups are given in chapter 5.

2.5.2 The hydrophobic group

The hydrophobic group for 99% of surfactants is made up of hydrocarbon chains and the majority of these are linear due to the demands for biodegradability. As the major application of surfactants is detergency and as detergency is generally in the C10–C16 range, the majority of surfactants on the market have these groups. Also, detergency tends to be a combined adsorption, wetting and emulsifying action, so the majority of other surfactant application will either be in this region, i.e. C10–C16, or at least close to it. The hydrophobic group will constitute the largest part of the molecule except for high ethylene oxide nonionics and thus is the major cost of a surfactant molecule. The commercial history of surfactants is the availability of hydrophobic groups at costs which the application can carry.

12   HANDBOOK OF SURFACTANTS

A good example is the use of the silicone chain in place of the hydrocarbon chain as the hydrophobe. The silicone chain has some advantageous properties compared to the hydrocarbon chain but the cost is so much greater that silicone-based surfactants are only used where they have special properties and can carry a higher cost.

The hydrophobic group based on hydrocarbons is basically available from three sources:

1. Petrochemicals
2. Natural vegetable oils
3. Natural animal fats

It is important to appreciate that in every case the hydrophobic group exists as a mixture of chains of different length, whether manufactured or found in nature. It is fortunate for the surfactant users that mixtures of varying chain length are normally better surfactants in practice than pure compounds. If this were not the case then the separation and purification costs would be higher and the resulting prices would be higher. Even more important the choice of hydrophobes would be more limited.

2.5.2.1 Petrochemicals. There is a comprehensive literature on the chemistry and production of hydrocarbons as raw materials for surfactants, but this information is not very relevant. How these raw materials are converted into the surfactants is covered in chapters 6–10, which describe the manufacture of all the major types of surfactants. However, the hydrophobe of the finished surfactant will be entirely dependent upon the starting hydrocarbon in the majority of cases. Thus, if the hydrocarbon used in the surfactant manufacture has a distribution of chain lengths of 25% C10, 50% C12 and 25% C14 then the resulting surfactant will have exactly the same distribution of chain lengths in the hydrophobe. The resulting surfactant properties of solubility, viscosity, wetting, etc., will depend upon this carbon chain distribution. Thus the surfactant properties will be dependent upon the starting hydrocarbon chain distribution.

All this seems very obvious but in many cases the formulator will not know the carbon chain distribution because the surfactant manufacturer will not include this in the specification. However, demand for the specification of carbon chain distribution is becoming more common, particularly by the big detergent manufacturers. Specification of carbon chain distribution to a large degree can specify surfactant performance, but not entirely; chapters 6–9 give some details where specification of carbon chain distribution is desirable. Surfactant performance can be critically dependent upon carbon chain distribution, yet there are surfactant types where it can be less significant than the variations in the hydrophilic group (e.g. ethoxylates).

Hydrocarbons from petrochemicals can have the following variations:

GENERAL APPROACH TO USING SURFACTANTS IN FORMULATIONS 13

1. Length of the hydrocarbon chain
2. Degree of branching
3. Odd or even carbon atoms

Note that variations in alkyl benzenes are a special case and are described in section 6.6.3.

The odd or even carbon atoms arise due to the method of manufacture. Mixtures of odd and even carbon chains are obtained by 'cracking' higher hydrocarbons, whilst even-numbered chains are derived by building up chains from ethylene. Hydrocarbons from natural sources invariably contain only even-numbered hydrocarbon chains, and thus products derived from ethylene are said to match 'natural' products more closely than those derived by cracking. At carbon chain lengths of C10 and greater, the difference in surfactant properties between surfactants made from even carbon chain lengths and mixtures of odd and even chain lengths is minimal.

3 Information sources

3.1 Introduction

Information on the properties of surfactants and their use in formulations is not neatly collected together in any one type of publication. Although surface active agents are chemicals they are not pure and the products of commerce are mixtures of chemicals of very similar but not identical properties. Thus publications such as *Beilstein* and the Registry File of *Chemical Abstracts* are of very limited use. The major source of information on surfactants is the surfactant manufacturers. Their information is generally inadequate because they concentrate on that information relevant to the manufacture and safety of the surfactants rather than their use. The larger manufacturers will have available considerable information on the use of their products as a sales aid, but this will inevitably be limited to the larger end uses.

The surfactant manufacturer will have a reasonable idea of the composition of the raw materials (which are generally mixtures) and a reasonable knowledge of the chemistry used, but that is all unless considerable research is undertaken to find out more. The profit margin on most surfactants is such that research will not be carried out unless motivated by legal or economic reasons. The profit margin per tonne on speciality surfactants may be good but the volume is small, so again there is a reluctance to determine exactly what is being made and sold. All surfactants, without exception, contain by-products not described by the general chemical description of the particular surfactant. In the majority of products the by-products are not relevant to the end use of the surfactant and may be disregarded. Occasionally a by-product can become so undesirable in the finished formulations that great care must be taken in the choice of surfactant to minimise the presence of the by-product. A knowledge of the chemical constitution of the surfactant is becoming more important as consumer standards rise ever higher. Fortunately a lot of information has been published on the composition of surface active agents, but it is scattered throughout the technical literature.

In addition to the actual composition of the surfactant the effect of surfactant composition on the physical properties and functional characteristics will be of prime importance to the end user. Manufacturers provide considerable data in this respect but generally only on their own products. Independent information on the relationship between the chemical struc-

INFORMATION SOURCES

ture of the surfactant and its functional properties will be of significant help to a user in choosing the most cost-effective surfactant. To the author's knowledge, there is no one single source of information that gives an in-depth account of such relationships. Most of the information is given at scientific conferences and meetings, not all of which have their proceedings published or easily obtained.

Most surfactant users deal only with a limited number of surfactants and are interested only in a limited number of applications. They will undoubtedly be building a database on raw materials (surfactants plus other chemicals) and also the functional properties obtained by these raw materials. This private database will be the technology for their products and it is essential to have a systematic method of keeping and retrieving such information. Very few companies do this job very well and considerable knowledge resides in a few individuals and is lost if they leave for any reason. Computer-based information systems are becoming common, but for the type of information required a computer-based information system will not be satisfactory unless a paper-based filing system already exists. Computer-based systems will make the problem more difficult rather than simplifying unless a good paper system already exists.

The sources of information on surfactants and their applications in approximate order of importance are:

1. Manufacturers' literature
2. Specific meetings, symposia and training courses
3. Published books
4. Patents
5. Published scientific papers
6. Government publications

Each of these will be described in more detail. It is worth pointing out at this stage that computer online databases or *Chemical Abstracts* are not primary sources but are methods of searching the primary sources. Generally, but not always, they are the quickest method of searching sources and identifying the particular source.

3.2 Manufacturers' literature

The quality of manufacturers' literature ranges from the very good to the very bad. That published by the very large surfactant manufacturers is extremely good and represents very significant research and development work, particularly on the chemical properties, physical properties and safety data. Suggested formulations using the surfactants can be useful but such formulations may be out of date.

16 HANDBOOK OF SURFACTANTS

However, every surfactant manufacturer will have considerably more data than that printed in glossy brochures, and it is worth writing to or contacting the technical departments who have generated the data. Very often they are only too anxious to give such data when asked. If the primary contact is via a sales representative, it is not always realised that sales staff are sometimes not chemists, and therefore requests for data should be specific rather than vague. The information required from the manufacturer will probably fall into several categories:

1. *Suitability* of the surfactant to do the job. Surfactant manufacturers can only recommend a particular surfactant if they know its exact use. Often this is confidential, so one must then define the requirements in terms of detergency, foaming, wetting, etc. and of course price. Many manufacturers produce application booklets on specific applications.
2. *Physical properties* such as specific gravity, colour, flash point, etc., which are generally contained in sales leaflets on the product.
3. *Safety data*. Recommendations on safe handling and storage will be available from all surfactant manufacturers. In addition, most manufacturers will give simple toxicity data and biodegradation characteristics on the type of chemicals that they sell. Most of this information is obtained from the literature, and therefore the results are obtained not on the actual surfactant being sold but on a chemical that has a similar composition to that sold. Only a very few surfactant manufacturers can give toxicity and biodegradation data that have been obtained from testing the products as sold.
4. *Specification* for a check on quality. Most glossy data sheets will have a 'manufacturer's specification', which may well fully characterise the surfactant, but more often the measurements on the specification will give very little in the way of guarantee of reproducible material. In chapters 6–9 there is a short section for each surfactant which gives some suggested chemical or physical tests that are likely to reveal any batch-to-batch variation. It is not suggested that these tests are the only ones or even are required in every case for checking quality. Each end use of a surfactant can demand a different requirement, which should be agreed as a specification between the supplier and the user. The tests for specification in this book are suggestions as a basis for discussion between the supplier and user.

Collecting manufacturers' literature is easy, as such literature is freely available. Writing to the manufacturer and visiting trade fairs and exhibitions are the quickest ways. Filing the information by name of manufacturer is convenient, but cross references to type of surfactant and application are also needed if such a databank is to be useful. Names and addresses of manufacturers can be obtained from directories of surfactants; a list of directories is given in section 3.3

3.3 Published books

There are now many books on surfactants but by far the majority are concerned with academic research. The following books all contain useful information to the practical chemist using surfactants. A brief note of the principal features of each book is given.

General

Encyclopedia of Shampoo Ingredients, A.L. Hunting, Micelle Press 1983, 479 pp. Gives quite good descriptions of some surfactants and LD_{50} data.

Industrial Applications of Surfactants, ed. D.R. Karsa, Royal Society of Chemistry, Special Publication No. 59, 1987, 352 pp. Proceedings of a Symposium organised by the Industrial Division of the Royal Society of Chemistry, held at the University of Salford, 15–17 April 1986. Good review of industrial applications.

Industrial Applications of Surfactants II, ed. by D.R. Karsa, Royal Society of Chemistry, Special Publication, No. 77, 1990, 400 pp. Proceedings of a Symposium organised by the Industrial Division of the Royal Society of Chemistry, held at the University of Salford 19–20 April 1990. Excellent account of speciality surfactants.

Industrial Applications of Surfactants III, ed. by D.R. Karsa, Royal Society of Chemistry, Special Publication No. 107, 1991, 277 pp. Proceedings of a Symposium organised by the Industrial Division of the Royal Society of Chemistry, held at the University of Salford 16–18 September 1991. Excellent account of speciality surfactants.

Surface Active Agents: Their Chemistry and Technology, Vol. 1, A.M. Schwarz *et al.* (Wiley, 1949, reprinted Krieger, 1978), 592 pp. Rather old and academic but comprehensive.

Surfactants in Consumer Products – Theory, Technology and Applications, ed. J. Falbe, Springer-Verlag, 1987, 547 pp. Good but brief summary of commodity surfactants. Good review of publications on household detergents.

Surfactants and Interfacial Phenomena, 2nd edn, M.J. Rosen, Wiley, 1989, 448 pp. Good general account of surface active properties.

Anionic surfactants

Anionic Surfactants, ed. W.M. Linfield, Parts 1 2, Marcel Dekker, 1967, 376 pp. Probably the best book on anionics but out of date.

Anionic Surfactants Biochemistry, Toxicology and Dermatology, ed. C. Gloxhuber, Marcel Dekker, 1980, 456 pp.

18 HANDBOOK OF SURFACTANTS

Anionic Surfactants–Physical Chemistry of Surfactant Action, ed. E.H. Lucassen-Reynders, Marcel Dekker, 1981, 413 pp. Somewhat theoretical.

Nonionic surfactants

Nonionic Surfactants, ed. M.J. Schick, Marcel Dekker, 1966, 1120 pp. Out of date on applications but excellent for the chemistry of ethoxylation and the physical properties of nonionics.
Nonionic Surfactants: Physical Chemistry, ed. M.J. Schick, Marcel Dekker, 1987, 1160 pp.
Surface Active Ethylene Oxide Adducts, N. Schonfeldt, Pergamon, 1970, 964 pp. reprint made to order by Micelle Press. Same comments as *Nonionic Surfactants*, ed. M.J. Schick, above.

Cationic surfactants

Cationic Surfactants, E. Jungermann, Marcel Dekker, 1970, 672 pp. Good but out of date.
Cationic Surfactants – Physical Chemistry ed. N. Rubingh and P.M. Holland, Marcel Dekker, 1991, 527 pp. Good but more theoretical than practical.

Amphoteric surfactants

Amphoteric Surfactants, B.R. Buestein and C.L. Hilton, Marcel Dekker, 1982, 352 pp. Not a good book but the only one available. Amphoterics form a rapidly changing field.

Analysis

The Analysis of Detergents and Detergent Products, G.F. Longman, Wiley, 1975, 625 pp.
Analysis of Oils and Fats, R.J. Hamilton and J.B. Rossell, Elsevier Applied Science, 1986, 440 pp.
Anionic Surfactants–Chemical Analysis, J. Cross, Marcel Dekker, 1977, 272 pp.
Nonionic Surfactants: Chemical Analysis, J. Cross, Marcel Dekker, 1987, 432 pp.
Systematic Analysis of Surface Active Agents, M.J. Rosen and H.A. Goldsmith, Wiley, 1972.

INFORMATION SOURCES

Applications

Detergents and Textile Washing, G. Jacobi and A. Lohr, VCH Verlags-gesellschaft, 1987. Excellent, plus many non surfactant aspects.

Emulsions and Solubilisation, K. Shinoda and S. Friberg, Wiley, 1986, 172 pp. Quite a short book, good but more theoretical than practical.

Encyclopedia of Emulsion Technology, P. Becher, Marcel Dekker, Vol. 1 *Basic Theory*, 1983, 752 pp. Vol. 2, *Applications* 1985, 536 pp. Vol. 3 *Basic Theory/Measurements/Applications*.

Microemulsions – Theory and Practice, L.M. Prince, Academic Press, 1977, 173 pp.

Synthetic Detergents, A. Davidson and B.M. Milwidsky, 7th ed. W. Godwin, 1987, 228 pp. The best practical book concerned with detergents, but limited.

Environment and safety

Surfactant Biodegradation, 2nd edn, ed. R.D. Swisher, Marcel Dekker, 1987, 1120 pp.

Biodegradation of Surfactants, ed. D.R. Karsa and M.R. Porter, Blackie, 1994.

Theoretical

Structure/Performance Relationships in Surfactants, ed. M.J. Rosen American Chemical Society, 1984, 356 pp. Does not quite live up to title.

Surfactant Aggregation, J.H. Clint, Blackie, 1982, 283 pp. By far the best book on the physical chemistry of surfactants and some applications.

Directories of surfactant manufacturers

Surfactants Applications Directory, Directory of the applications of surface active agents available in Europe, ed. D.R. Karsa, J.M. Goode and P.J. Donnelly, Blackie, Glasgow UK, 1991.

Surfactants UK, Directory of surface active agents available in UK 1979, ed. G. L. Hollis, Tergo-Data, Darlington, UK, 1979.

Surfactants Europa, Directory of surface active agents available in Europe, ed. G. L. Hollis, Tergo-Data, Darlington, UK, 1989.

Surveys on Surfactants Commercialised in Europe, D.T.A., 3 rue Lavoisier, BP 72, 77330 Ozoir-la-Ferriere, France.

20 HANDBOOK OF SURFACTANTS

McCutcheon's Emulsifiers and Detergents, North American and International editions, published annually by the McCutcheon Division of MC Publishing Company, 175 Rock Road, Glen Rock, New Jersey, USA.

3.4 Journals and periodicals

A considerable amount of technology and scientific literature is published regularly on surfactants. Probably the best way to keep abreast of most of this literature is to read *CA Selects*. This is a regular (every two weeks) publication of *Chemical Abstracts* on specific subjects. The most relevant are:

- Number SVC 089 Detergents, Soaps and Surfactants
  Preparation, properties and uses of soaps and synthetic detergents
  Formulations
  Dry-cleaning solvents
  Use of surfactants in petroleum recovery
  Not routinely covered: detergent additives for fuels and lubricants
- Number SVC 041 Colloids (Applied Aspects)
  Emulsions, gels, latexes, micellar solutions, sols, other forms of colloidal dispersions
  Uses of these materials in cosmetics, foods, fuels, metals, other products
  Excludes: routine application of silica gel, emulsions and suspensions
- Number SVC 02F Emulsifiers and Demulsifiers
  Preparation, properties, uses of surface-active agents in formation, stabilisation and destabilisation of emulsions
  Aqueous and nonaqueous emulsions
  Applications in cosmetics, food, petroleum, polymer industries

Others of more specific interest which do contain some references to surfactants are:

- Number SVC 082 Adhesives
- Number SVC 042 Colloids (physicochemical aspects)
- Number SVC 03J Cosmetic Chemicals
- Number SVC 05M Drilling Muds
- Number SVC 02G Emulsion Polymerisation
- Number SVC 04M Enhanced Petroleum Recovery
- Number SVC 03N Fats and Oils
- Number SVC 04H Lubricants, Greases and Lubrication
- Number SVC 08L Oleochemicals Containing Nitrogen

INFORMATION SOURCES 21

- Number SVC 04R Paint Additives
- Number SVC 07D Quaternary Ammonium Compounds
- Number SVC 05T Water-based Coatings.

They can be obtained from the Chemical Abstracts Service, 2540 Olentangy Road, PO Box 3012, Columbus, Ohio 43210 USA.

For a formulator or surfactant chemist the other journals which are useful are:

Tenside Surfactants Detergents, Carl Hanser Verlag, 8 000 Munchen 86, Postfach 86 04 20, Germany. Although this is mainly in German, many articles are in English. Probably the best journal for surfactants.

Journal of the American Oil Chemists Society, American Oil Chemists Society, PO Box 3489, Champaign, IL 61826–3489, USA. Practical articles on surfactants and applications.

Soap, Cosmetics and Chemical Specialities (USA), MacNair-Dorland Co., 101 West 31st Street, New York, NY 10001, USA. Occasional good articles on surfactants and applications.

SPC (Soap, Perfumery and Cosmetics), Wilmington Publishing Ltd, Wilmington House, Cloud Hill, Wilmington, Dartford, Kent, UK.

Performance Chemicals, Reed Business Publishing, Quadrant House, The Quadrant, Sutton, Surrey SM2 5AS, UK. Occasional articles on surfactants.

Speciality Chemicals, FMJ International Publications Ltd, Queensway House, Redhill, Surrey RH1 1QS, UK. Occasional articles on surfactants.

Surfactant Update, quarterly, Speciality Training Ltd., 14 Staple Hill, Wellesbourne, Warwickshire CV35 9LH, UK. Specifically published for the surfactant manufacturer or user. The best technical review on surfactants for the industrial chemist.

3.5 Patents

Patents are an enormous source of information on surfactants and their applications. The types of information which can be found in patents are:

1. Processes for the manufacture of surfactants
2. Formulations for many different end uses with details of the surfactant which is used in the formulation; often comparisons are given between one surfactant and another in the formulation

To use patents as a source it is essential to have a clear objective, as reading patents randomly is probably the most efficient way of wasting

22　　　　　　　　　　HANDBOOK OF SURFACTANTS

time, money and effort. There are technical and business objectives in the use of patents. The business objectives of avoiding patents, licensing and taking out new patents will not be covered in this book. The technical objectives resolve themselves into either a broad objective of using patents to keep up to date with technology or a specific objective of obtaining information on a particular surfactant or a particular application.

Taking the broad objective, the best method is to have a librarian or information expert who is conversant with patents produce regular monthly abstracts of relevant patents. A better definition than 'surfactants' will, however, be needed. Two typical broad objectives would be:

1. Anionic surfactants – particularly sulphonated, especially sulphur trioxide sulphonation; methods of manufacture, details of plants and materials of construction.
2. Surface coating antifoams and/or defoamers – particularly for emulsion paints and water based industrial paints and details of formulations and test methods for measuring efficiency of the antifoam/ defoamer.

In the absence of a librarian there are several abstracts services for patents which can be expensive and time consuming to wade through, as you cannot specify the area of search, except in some on line databases (see below). The easiest, cheapest and most accessible is the *CA Selects* Number 089 (see section 3.4) which includes world-wide patents. The most comprehensive is the Derwent Patent Abstracts (Derwent Inc. USA), Suite 500, 6845 Elm Street, McLean, VA 22101, USA, or Derwent Publications Ltd., Rochdale House, 128 Theobolds Road, London WC1X 8RP).

Taking a narrow objective, the two best ways of identifying the right patents are *Chemical Abstracts* and online databases. The latter is the quickest but it is also very easy to miss the relevant patents (see online databases below). The *Chemical Abstracts* Collective Indexes in a library are really the most efficient and, in the author's experience, superior to online searching of *Chemical Abstracts*. Thus, go to the *Chemical Abstracts* Collective Indexes, find the abstract number, look that number up, read the abstract, decide if the patent is worth reading, obtain the patent and read it.

The only short cut in this procedure is to use an online database when you have the *Chemical Abstracts* abstract number as it is quick and easy to find and print out the abstract using an online database. The newer technology of CD-ROM will make this procedure easier but it is likely that searching the CD-ROM records will have similar limitations to online databases (see below). The reason for the superiority of using *Chemical Abstracts* indexes is that browsing is much easier. Browsing means noticing a word or phrase in the text which had not been visualised at the beginning

INFORMATION SOURCES 23

of the search. Browsing is possible using computer databases, but it is expensive and not as easy.

Online databases are very rapidly increasing in number and quality: each of the host databases (see below) has its own patent files.

Thus patents, whilst possibly the largest source of information, are not the easiest source to use, particularly for the unskilled.

3.6 Symposia, meetings and courses

Surfactants are well covered by meetings and symposia. The large symposia publish their proceedings and there are many useful papers given and published on new surfactants and new applications. There are three regular (in the past) symposia which attract all the surfactant industry and the main users, the detergent industry, and these are:

1. The CESIO (Comité Européen des Agents de Surface et leurs Intermediaires Organiques) conference every four years – the last one was in the UK in 1992.
2. The Soap and Detergent Association (USA) holds regular meetings.
3. Organizados por la Asociacion de Investigacion de la Industria Española de Detergentes, Tensioactivos y Afines (A.I.D.) y el Comite Español de La Detergencia (C.E.D) hold an annual meeting, generally in Barcelona in March. Some papers are in Spanish, but there are usually some interesting papers in English. The address for correspondence is Secretaria General, Asociacion de Investgacion de Detergentes (AID), Jorge Girona Salgado, 18–26, 08034 Barcelona, Spain.

In addition the British Association for Chemical Specialities has a Surfactants Section which holds regular meetings covering various applications of surfactants.

All the meetings are well publicised with the exception of the Spanish Conference.

In the last few years the British Association for Chemical Specialities initiated training courses in surfactant technology for industrial chemists. These are now organised and carried out by Speciality Training Ltd, 14 Staple Hill, Wellesbourne, Warwickshire CV35 9LH, UK. The courses are of one or two day duration, and consist of lectures and workshops in surfactant technology at all levels.

3.7 Government publications

Legislation concerning chemicals is quickly becoming more complex on issues such as labelling, packing and transport. However, there is some

24   HANDBOOK OF SURFACTANTS

specific legislation that gives considerable information on the application of surfactants. FDA Regulations of the USA give many specific references to the use of specific surfactants in applications which may come into contact with food, e.g. antifoam in sugar beet processing. UK Food Regulations cover emulsifiers permitted for use in food.

3.8 Databases

Online databases have been mentioned several times as they are a convenient source of technical information. Undoubtedly computer databases will slowly become the means of holding the technical information base of the future. However at the present time they certainly have not completely replaced paper. The output of a computer database of technical information is always a printout on paper because it is so much more convenient to read a piece of paper. What then are the advantages of online databases?

1. *Accessibility* – connection to any online system world-wide is possible if there is a public telephone line. It is best to use a specific connection for computers but that is not absolutely necessary. Good accessibility is a great advantage if you have no library of your own or are very distant from a big technical library.
2. *Low cost to set up* – a personal computer and a telephone line are all that are needed to access online databases. To set up a system with new equipment can cost well under £1 000. If a IBM-PC (or compatible) with printer is available then the cost can be as low as £200. Libraries are very costly to build, maintain and run and also costly to use if they are very distant.
3. *Rapid searching* – a typical search for a subject can be seconds. The printout of the information will only take minutes, so within a very short time of requesting the information you can be reading it, compared to going to a distant library which will take at least one full day.
4. *Up to date information* – all databases are continually updated, unlike books.

With these advantages, what are the disadvantages and why cannot we rely on such databases for our information?

1. *Abstracts only* – with few exceptions, technical databases only contain abstracts of the primary information which is in the form of printed material, e.g. a patent. Thus the online data search will only give a reference to the full information, which must then be obtained separately. However, a few technical journals have the full text

available and this is where computer databases are probably going to develop.

2. *Complex to use* – you need to know which database to use. There are now so many databases available it needs an expert to pick out the best one. The commands for searching are not uniform; the command language can be different from one database to another. The user must be a frequent user of online databases in order to be economic and efficient. Occasional users find the language difficult to remember, although there are now improved computer programs to help.

3. *Expensive to use* – once you get to a library the only cost is your own time. Online databases are expensive to use; you pay for the time used and/or the information received. It is very easy to spend £100–200 on a single query if you have difficulty in finding the right keywords.

4. *Browsing is difficult* – although some systems are now claiming this function.

The on-line data bases could well become obsolete if CD-ROM develops quickly and cheaply to take their place. However, for the present, online databases are a very valuable source of information and should not be neglected. If you have not had the experience then the best way to start is to contact a database host. This is a company that offers a number of databases which can be searched for payment to the host. This host company will give full details on setting up and tuition in searching and accessing the system. Unfortunately, there is no common command language between database hosts; the languages used are very similar but not identical. At the time of writing two useful database hosts for chemicals (e.g. surfactants) are:

1. STN International Headquarters: Postfach 2465, D-7 500 Karlsruhe 1, Germany. USA: 2540 Olentangy River Road, PO Box 02228, Columbus, OH 43202, USA.

2. Info Pro-Technologies/Orbit. Headquarters: 8000 Westpark Drive, McLean, VA 22102, USA. Europe: 18 Parkshot, Richmond, Surrey TW9 2RG, UK.

Undoubtedly there are others which are wider in scope, but these two database hosts have specialised in scientific and particularly chemical databases. Each host provides databases covering *Chemical Abstracts* and a number of patent databases.

4 Use of surfactant theory

4.1 Introduction

Formulations using surfactants, particularly mixtures of surfactants, have been devised by trial and error, and theory has followed to explain the observed results. The application of theory will not lead to quick, easy solutions of practical problems. However, there is a need to have a grasp of theory for a better understanding of the mode of action of a surfactant. This helps in the more efficient and rapid selection of the correct surfactant and possibly other components.

Although a feasible theoretical explanation of the mode of action of a surfactant must be shown to be quantitative, the approach presented in this chapter is to describe the theoretical basis of surfactant behaviour in a pictorial rather than mathematical form. The present state of the theory of dynamic systems such as wetting and foaming is such that the quantitative laws can seldom be used in practical formulation. In addition such diagrams and pictures are more easily remembered and used than mathematical formulae.

The principal properties which characterise surfactants are described:

- Adsorption
- Micelles
- Solubility
- Solubilisation
- Microemulsions
- Wetting
- Foaming/defoaming
- Macroemulsions
- Dispersing/aggregation of solids
- Detergency
- Surfactant interactions

Numerous studies have been published on the correlation between the chemical structure of a surfactant and its properties. The purpose of this chapter is to show how a particular chemical structure produces specific properties. In other chapters there will be frequent reference to empirical relationships between structure and properties. The good formulator gradually builds up a mental picture of such relationships by experimental trial and error over a period of time. The theoretical approach enables one

USE OF SURFACTANT THEORY

27

to understand and use such relationships in a more systematic manner, but it will not solve every problem or eliminate the need to do experimental work.

The theory of surfactant behaviour is now an extensive branch of physical chemistry and considerable basic research has been carried out in the last 20 years. The major reason for this work has been the possible shortage of oil in the 1970s leading to the use of surfactants in enhanced oil recovery. It was found that the theoretical knowledge of the day was inadequate to explain the complex behaviour of surfactants in microemulsions. In addition, the importance of the physical chemistry of surfactant behaviour is now being recognised in biochemistry and the study of cell behaviour. A further impetus to surfactant research has been the development of thin film technology in the production of microelectronics by the use of Langmuir–Blodgett films. Theoretical work on surfactants is still being published at a considerable volume, and this chapter attempts to summarise the present state of knowledge.

There are three basic concepts which need to be well understood in order explain the majority of observed phenomena; these are solubility, adsorption of a surfactant at a surface, and the formation of micelles in solution. These three phenomena differentiate a surfactant from other chemical entities. It is the abnormal solubility characteristics of surfactants that give adsorption and form micelles. It is adsorption at surfaces that gives the surface active effects of foaming, wetting, emulsification, dispersing of solids and detergency. It is the micellar properties that give the solution and bulk properties of surfactants such as viscosity and solubilisation, but also there is increasing evidence that the micellar properties are necessary in functional effects such as emulsification and detergency.

4.2 Adsorption and critical micelle concentration (CMC)

Many chemicals produce foams and wet surfaces but are not considered surfactants, e.g. methyl alcohol in aqueous solution. The major characteristic of a surfactant is that it is at a higher concentration at the surface than in the bulk of a liquid. This phenomenon is known as adsorption and occurs at a liquid/solid interface, at a liquid/liquid interface and at a air/liquid interface, as shown in Figure 4.1.

In Figure 4.1 the surfactant molecule is pictured as a long straight hydrophobic group and a small round hydrophilic group. The surfactant molecule can be oriented in various ways, as shown in Figure 4.1. As we shall see later (section 4.3), this is a gross over-simplification, and it is the relative sizes and shapes of the hydrophobic part and the hydrophilic part of the surfactant molecule that determine many of its properties. For now we will stay with the simple structure. The adsorption of a surfactant at an

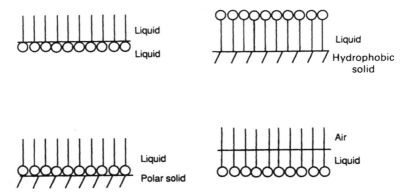

Figure 4.1 Adsorption at interfaces.

air/water surface will result in pronounced physical changes to the liquid; the more surfactant there is at the surface, up to complete coverage of the surface, the more the change.

A surfactant adsorbs because there are two groups in the molecule: a **hydrophobic** (water hating) group and a **hydrophilic** (water liking) group. A surfactant molecule can be shown as in Figure 4.2.

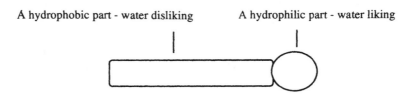

Figure 4.2 Basic structure of a surfactant.

If soap is taken as an example, with the hydrophobic group as a long-chain alkyl group $(CH_2)n$ and the hydrophilic group as a carboxylate ion COO^- neutralised with the sodium ion Na^+, we find that the practical surfactant properties (foaming, wetting, emulsifying, dispersing, detergency) are dependent upon the value of n:

If n is less than 8 the soap is quite soluble in water but its surfactant properties are minimal

In n is 10–18 then the soap is sparingly soluble and the surfactant properties are at a maximum

If n is greater than 18 the soap is practically insoluble while the surfactant properties are minimal.

Thus the solubility and practical surfactant properties are related. However, the relationship is not simple. The hydrophobic group, e.g. a long-chain alkyl group, is not repelled (as often stated) by water, as the attraction of the hydrocarbon chain for water is approximately the same as that for itself. In fact, at very low concentrations of surfactant in water the hydrocarbon chains will actually lie flat on the surface and thus the picture of a **solitary** surfactant molecule at a water/air interface as shown in Figure 4.3 is incorrect. It lies flat, as shown in Figure 4.4. It is only when there is a sufficient number of surfactant molecules at the surface that the surfactant molecule is oriented as shown in Figure 4.1.

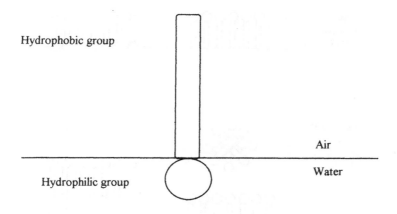

Figure 4.3 Adsorption.

The hydrophobic effect

The reason for the insolubility of the hydrocarbon chain in water is complex. The mechanism involves both enthalpic and entropic contributions and results from the unique multiple hydrogen-bonding capability of water. There is a restructuring or re-orientation of water around nonpolar solutes, which disrupts the existing water structure and imposes a new and more ordered structure on the surrounding water molecules, giving a decrease in entropy. For those interested in a more detailed account then see Clint (1992) and Israelachvili (1992).

The hydrophilic effect

While hydrophobic groups tend to increase the degree of order in water molecules around them, hydrophilic molecules are believed to have a disordering effect. Thus, ions in solution are hydrophilic (e.g. sulphonate, carboxylate, sulphate, phosphate, quaternary ammonium). Polar groups

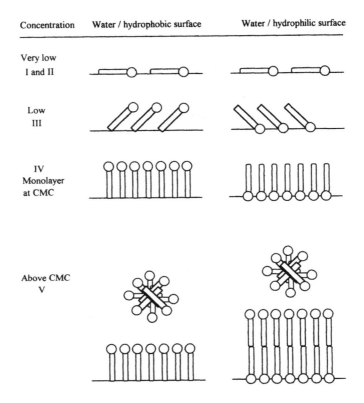

Figure 4.4 Adsorption and concentration.

with a highly electronegative character show strong electrophilic properties (e.g. primary amines, amine oxides, sulphoxides, phosphine oxide). Other nonpolar molecules can be hydrophilic if they contain electronegative atoms capable of associating with the hydrogen-bonding network in water. The oxygen atom in alcohols, ethers, aldehydes, amides, esters and ketones, and the nitrogen atom in amides, nitroalkanes and amines are such groups. When they are attached to a hydrophobic group their effect is diminished or even eliminated if the hydrophobic group is large enough. However, if a number of such nonpolar groups, e.g. polyoxyethylene groups, are attached to the hydrophobic group, then limited or entire water solubility can be achieved, depending upon the relative size of the hydrophobic effect and the number of hydrophilic groups.

The aqueous solubility of a lone surfactant molecule will depend upon the relative strengths of the hydrophobic and hydrophilic effects. They are not independent, since both rely on the structure of the hydrogen bonds around the hydrophilic and hydrophobic groups.

USE OF SURFACTANT THEORY

Adsorption and critical micelle concentration (CMC)

The adsorption of a surfactant from solution on to a surface depends upon the concentration, and Figure 4.4 shows the effect of increasing concentration. At very low concentrations (I and II), there is no orientation and the molecule lies flat on the surface. As the concentration increases (III), the number of surfactant molecules on the surface increases, there is not enough room for them to lie flat and so they begin to orient, the orientation depending upon the nature of the hydrophilic group and the surface. At concentration IV, the number of surfactant molecules available is now sufficient to form a unimolecular layer. This particular concentration is of importance and known as the **critical micelle concentration** (CMC). At concentration V (above IV), there is no apparent change in adsorption at hydrophobic surfaces, but at hydrophilic surfaces more than one layer of surfactant molecules can form ordered structures on the surface of the solid. In addition, the surfactant molecules in the solution will form an ordered structure known as a **micelle** so long as the concentration is above the CMC.

If adsorption is quantitatively measured and plotted against concentration the results are as shown in Figure 4.5. The adsorption moves in a series of steps with significant differences between the highly hydrophobic solid and the highly hydrophilic solid. In practice many solids are intermediate in hydrophilic/hydrophobic character, and the steps in the adsorption graph are smoothed out. Nevertheless in all cases **maximum adsorption** is reached, showing that some kind of ordered layer of surfactant molecules is on the surface.

Surface adsorption leads to pronounced physical changes to the solution, the effect on surface tension being particularly evident. Surface tension is the pull exerted by aqueous solutions and is shown most clearly where a free drop of water pulls itself into a sphere, minimum surface area for a given volume). The effect of surface tension against the overall concentration for the majority of surfactants is shown in Figure 4.6.

The surface tension falls to a minimum value, which will be at concentration IV Figure 4.4, i.e. where the surface is completely packed with a monolayer of surfactant molecules. The formation of multilayers (concentration V) has no significant effect, although in practice the surface tension does fall very slowly as the concentration increases.

There is a similar effect when a surfactant is present in a system of two liquids which do not mix. The type of adsorption is shown in Figure 4.7, where a hydrophobic liquid is in contact with water. The surfactant adsorbs at the interface and reduces the surface tension at the interface. This surface tension is known as the interfacial tension. If an interfacial tension is given then it is necessary to give a description of both liquids at the interface. The interfacial tension versus concentration gives a similar curve

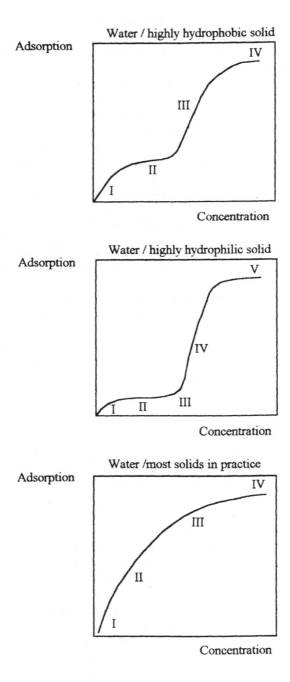

Figure 4.5 Adsorption *versus* concentration.

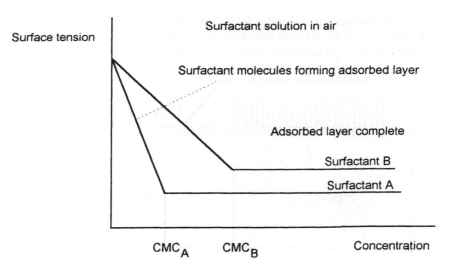

Figure 4.6 Surface tension *versus* concentration.

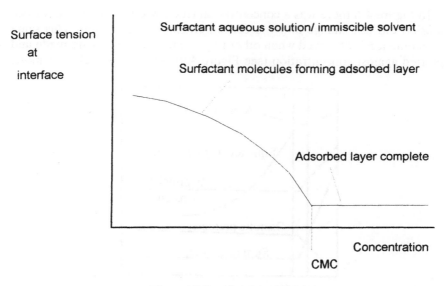

Figure 4.7 Interfacial tension.

to that of surface tension versus concentration (Figure 4.6) and shows a discontinuity, the CMC, which is characteristic of the surfactant.

At concentrations above IV (CMC), the molecules have no longer any sites available for adsorption. However, by the appropriate orientation of the hydrophobic parts of the molecules towards each other they can form ordered structures in solution (micelles), as shown in Figure 4.4 on hydrophilic surfaces. Exactly the same situation will exist in the adsorption of a surfactant at an air/water interface, as shown in Figure 4.8.

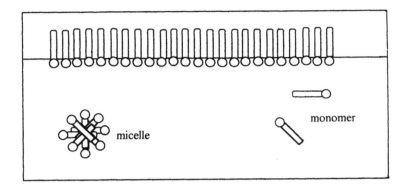

Figure 4.8 Micelles in solution.

In Figure 4.6 there was a concentration (the CMC) at which the surface tension did not change above a certain critical concentration. This same phenomenon is observed when other physical measurements are made and plotted against concentration (see Figure 4.9). Properties such as osmotic

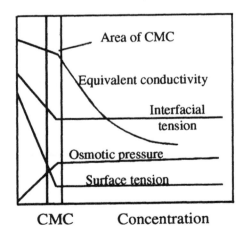

Figure 4.9 Critical micelle concentration (CMC).

pressure behave as if the number of molecules in solution has been reduced compared to what would be expected. This indicates aggregation and some form of structure.

The shape of micelles in solution

The actual structure and shape of a micelle depends upon the temperature, the type of surfactant and its concentration, other ions in solution and other water-soluble organic compounds, e.g. alcohols. It is important to understand that the structure and shape of the micelle can change and that the micelle is a dynamic entity.

Depending upon the conditions micelles can form spherical, rod-shaped or lamellar shapes (see Figure 4.10). The formation of micelles will be dealt with in greater detail in section 4.3. For the moment, consider micelles as simple spheres.

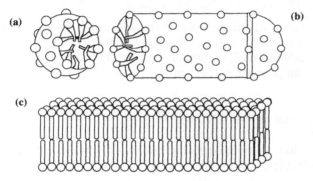

Figure 4.10 Shape of micelles: (a) spherical; (b) rod-shaped or cylindrical; (c) lamellar.

The number of surfactant molecules in a micelle is known as the *aggregation number* and some typical aggregation numbers (measured using light scattering) are shown in Table 4.1.

Table 4.1 Aggregation numbers

Surfactant	Temperature (°C)	Aggregation number
$C_{12}H_{25}SO_4Na$	23	71
$C_{12}H_{25}O(CH_2CH_2O)_6H$	25	400
"	35	1400
$C_{10}H_{21}N(CH_3)_3Br$	23	36

As shown in Table 4.1, nonionics generally have aggregation numbers (measured by light scattering) considerably higher than anionics or cationics. The aggregation number of nonionics is also very dependent upon temperature, unlike the ionic types, whose aggregation numbers are reduced by increasing temperature but not to any significant degree.

The CMC of a surfactant indicates the point at which monolayer adsorption is complete and the surface active properties are at an optimum. Hence there is considerable practical interest in the CMC as it

36   HANDBOOK OF SURFACTANTS

represents, in practice, the lowest concentration needed to get the maximum benefit. The CMC values for some typical surfactants are shown in Table 4.2. Note the much lower concentration at which nonionics form micelles compared to the ionic species.

The CMC is dependent upon the chemical structure of the surfactant, and some generalisations based on experimental data can be stated. For any homologous series of surfactants $CH_3(CH_2)_nX$ with the same hydrophilic group X then:

The CMC decreases as n increases
The CMC is at a minimum when X is at the end of a molecule
Where $X = (EO)_m$ the CMC decreases as m decreases

Aqueous solutions of surfactants invariably contain electrolytes (e.g. calcium ions from hard water). The CMC decreases significantly when electrolyte is added to **ionic** surfactants but hardly changes when it is added to **nonionic** or **amphoteric** surfactants.

Why do surfactants have a CMC?

The CMC is due to two opposing forces of interaction between the surfactant molecules:

Force 1. The polar groups in water if ionic will **repel** one another due to mutual charge repulsion. The larger this charge the greater the repulsion and the **less** tendency to form micelles. The hydrophilic groups may also have a strong affinity for water (e.g. polyoxyethylene chains) and there will be a tendency for them to be spaced out to allow as much water as possible to solvate the hydrophilic group.
Force 2. The hydrophobic groups act as if there is a bond **attracting** them together. The reason for this is complex and due to enthalpy and entropy changes when an alkyl group is transferred from a hydrocarbon environment to solution in water. This is basically the hydrophobic effect already described at the beginning of this section. A diagram illustrating these two forces is shown in Figure 4.11.

When the molecules are very far apart (very low concentration) then both forces above are weak. When the concentration increases, i.e. the surfactant molecules get closer, the two interactions described above will increase. If Force 1 exerted by the hydrophilic group is very much greater than Force 2 exerted by the hydrophobic group, then the molecules will probably not aggregate; they will remain monodisperse in solution at high concentration. This is the situation when the hydrophobic effect is very small and the molecule will be very soluble, e.g. sodium benzene sulphonate.

Table 4.2 CMC values

Surfactant	Temperature (°C)	CMC (mol l^{-1})	CMC (%)
Anionic			
$C_{12}H_{25}SO_4Na$	40	8.6×10^{-3}	0.2
$C_{12}H_{25}C_6H_4SO_3Na$	60	1.2×10^{-3}	0.04
Nonionic			
$C_{12}H_{25}O(CH_2CH_2O)_4H$	25	4×10^{-5}	0.0014
$C_{12}H_{25}O(CH_2CH_2O)_7H$	25	5×10^{-5}	0.0025
Cationic			
$C_{12}H_{25}N(CH_3)_3Br$	25	2×10^{-2}	0.46
Amphoteric			
$C_{12}H_{25}NHCH_2COOH$	27	1.3×10^{-3}	0.033

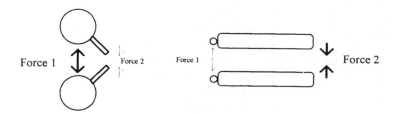

Figure 4.11 Forces between surfactant molecules in solution.

If Force 1 exerted by the hydrophilic group is very much smaller than Force 2 exerted by the hydrophobic group, then the molecules will aggregate together (Force 2 > Force 1) at very low concentrations. This is the situation when the hydrophobic effect is very large; aggregation is easy and the molecule is practically insoluble, e.g. $C_{26}H_{54}OH$. There will probably be a CMC, but it so low and the surface active properties are so weak that they will be difficult to detect.

The **relative** strength of Force 1 and Force 2 determines the CMC. At a particular concentration Force 1 will equal Force 2 and the molecules will aggregate. For this to occur the hydrophobic and hydrophilic effects must be of similar orders of magnitude. In comparing two surfactants, the one with the larger hydrophilic effect will have a higher CMC than the other.

The effect of various factors on the CMC

1. *The effective size of the polar or hydrophilic group.* As this gets larger, Force 1 increases and the CMC goes up. So strongly ionised polar groups, e.g. sulphates and quaternaries, will have higher CMCs than

less ionised groups. It does not matter whether the polar groups are positive (cationics) or negative (anionics); the effect is the same. With nonionics which have a very low polar charge, the CMC would be expected to be very small, as is shown in practice. In the case of amphoterics with both positive and negative groups, which may nominally have zero charge, the CMC is dependent upon the separation of the two charges in the molecule. The further apart the two charges, the higher the CMC, showing that the effective polar size increases as the two charges move apart.

2. *The effect of counterions.* If the counterions reduce the charge density on the polar group then Force 1 decreases and the CMC should reduce. Thus, calcium salts of sulphonates will have lower CMCs than the sodium salts.

3. *The addition of electrolyte to ionic surfactants.* This will reduce the charge density; Force 1 decreases and thus will lower the CMC. The reason for this difference is that the electrolyte can reduce the electrostatic repulsion between the ionic hydrophilic groups leading to a **increased** tendency to form micelles. The reduction in repulsion between the ionic groups also favours the change from spherical micelles to more complex aggregates, e.g. cylindrical or disk-shaped micelles. For nonionic and amphoteric surfactants this effect is considerably less.

4. *Effect of pH.* If the charge density is increased then Force 1 increases and the CMC will rise. Thus increasing the pH of weak acids, e.g. soap, will increase the ionisation and hence increase the CMC. The effect on amphoterics should be quite marked as the change in pH can make the polar groups go from high positive charge (high Force 1) to zero charge (low Force 1) to high negative charge (high Force 1). Thus the amphoterics would be expected to have a low CMC at the isoelectric region and high CMCs outside that region.

5. *Effect of the chain length of the hydrophobe (an alkyl group).* Force 2 (Figure 4.11) will be proportional to the length of the hydrophobic chain; the more methylene groups, the larger Force 2. A long alkyl chain will have a higher Force 2 than a short alkyl chain and therefore a lower CMC as long as Force 1 is constant.

6. *Effect of temperature on nonionics.* The increase in temperature will reduce the hydrogen bonds between the poly(EO) groups and water and thus reduce the effective size and reduce Force 1, giving a low CMC.

7. *Effect of temperature on anionics and cationics.* It is not readily apparent how the size of Forces 1 and 2 are affected. In fact there is usually a temperature at which the CMC is a minimum. This can be explained by thermodynamics but not easily by the simple picture proposed above.

USE OF SURFACTANT THEORY 39

Thus the simple concept shown in Figure 4.11 qualitatively predicts a number of experimental relationships between CMC, the molecular structure of a surfactant and the effect of some factors. It is extremely useful in practice and can be used in surfactant technology until we have better quantitative theories that can be applied in practice to predict performance and behaviour. However, it must be emphasised that it is a simplified picture; Force 2 as such does not really exist and therefore the concept must be used with caution.

The micelles are not fixed structures but must have a transient nature as the mechanism of aggregation should be reversible, as it is shown to be in practice, i.e. one can approach the CMC from either above or below the CMC. Surfactant molecules very rapidly join and leave micelles; the rate varies with molecular structure, and ionic surfactant micelles will have an average lifetime of milliseconds. The time scale of a monomer surfactant molecule entering a micelle is of the order of 10^{-6}s, while monovalent counterions at the micelle surface will exchange at rates of 10^{-8}s with the similar counterions in the solution around the micelle. The bound water molecules are also very mobile and typical lifetimes of water in the micelle are approximately 10^{-8}s (see Clint, 1992).

If micellar properties are averaged over periods of seconds then their average aggregation number and degree of counterion binding are well defined. This is only the case for the smallest micelles, and when we consider assemblies of micelles (see section 4.4) different orders of magnitude of the micellar structure are evident.

4.3 Micelles, vesicles, liposomes and lamellar structures

In 1913 McBain proposed that soap could form spherical or lamellar micelles as shown in Figure 4.10. Hartley in 1936 put forward a model of a spherical micelle of an ionic surfactant where the hydrophobic groups formed a hydrocarbon core and the polar groups formed a charged surface of a sphere with some of the counterions bound to the micelle surface. The diameter of the micelle is approximately twice the length of the hydro-phobic group and there is no water in the interior of the micelle. If a geometrical model is made, then to accommodate the alkyl chains in a sphere the aggregate number must be in the region of 50–100. Modern techniques of measuring micellar sizes have shown this picture to be correct and the aggregate numbers are in the region of 50–100 for most ionic surfactants.

However, many surfactants, particularly nonionics have aggregation numbers far in excess of 100 and molecular sizes of orders of magnitude greater than the length of the hydrophobic group. Cylindrical, disk-shaped and lamellar micelles have all been identified. These various shapes have

40   HANDBOOK OF SURFACTANTS

all been simply explained by the geometrical packing of the hydrophilic and hydrophobic groups as proposed by Israelachvili *et al.* (1976). Their ideas are an extension of the concept in Figure 4.11 where the head groups were repelling one another and the hydrocarbon groups were attracting one another. If they are to pack together to take up a geometrical structure then the relative **sizes** of the head group and hydrophobic group will determine not only the shape but the size. Note that in the case of an ionic group it is the **effective** size, which will take also account the electrostatic repulsion between the ions and any possible effect of counterions.

Ionic group size = actual size + electrostatic forces
Nonionic groups head-group size = actual size + bound water

The area occupied by the hydrophobic group and the hydrophilic group was defined by:

Hydrophilic group area = a_0
Hydrophobic group area = v/l_c where v is the alkyl chain volume and l_c its maximum length

Then, on geometrical grounds:

Spherical micelles are formed where	$v/a_0l_c < \frac{1}{3}$
Non spherical micelles where	$\frac{1}{3} < v/a_0l_c < \frac{1}{2}$
Bilayers where	$\frac{1}{2} < v/a_0l_c < 1$
Inverted micelles	$1 < v/a_0l_c$

A simple pictorial method of illustrating this feature was devised by Israelachvili and shown in Figure 4.12.

When the ratio of the hydrophilic head group area to the hydrophobic head group area becomes less than 2 then non-spherical shapes are preferred. By calculation it can be shown that a wide variety of shapes can satisfy these conditions, but the most likely shapes are cylindrical-like micelles, where the cross-section of the cylinder is similar to the cross-section of a spherical micelle with the ends rounded and similar to a hemisphere. This has already been shown in Figure 4.10. Measurements have been made on the sizes of these cylindrical micelles. The cross-sectional area is as would be expected for a spherical micelle but the lengths in some cases are surprisingly long. Balzer (1991) examined alkylpolyglycosides with alkyl groups in the range 12–14 and found cylindrical micelles of 3 nm diameter but 219 nm length. Lin *et al.* (1992) examined an ethoxylated C16 alcohol with 6 mol of ethylene oxide by electron microscopy and showed long, thin thread-like objects which were curved, bent or even looped with a diameter of 3–8 nm but lengths of the order of 700–1000 nm.

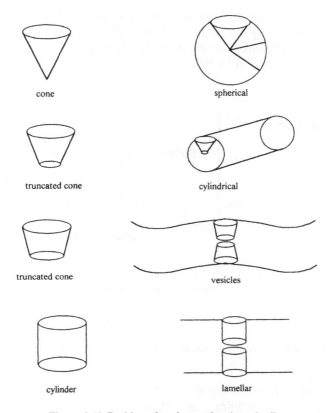

Figure 4.12 Packing of surfactant heads and tails.

Cylindrical-shaped micelles have unusual properties. Whereas the size of spherical-shaped micelles is not very dependent upon concentration it can be shown that the size of cylindrical micelles is dependent upon concentration, and this has been confirmed by measurements. The size is also sensitive to chain length of the hydrophobic group, temperature and (for ionic surfactants) the ionic strength of the solution. In addition, whereas the spherical micelles are monodisperse (all the same size), cylindrical micelles show a wide variation of aggregate number, i.e. they are polydisperse. The practical effects of such changes is that the transition to cylindrical micelles can give rise to increased viscosity. These viscosity effects are more common in more concentrated solutions.

The concepts of geometrical packing can be used to explain these phenomena. Any change which reduces the **effective** size of the hydrophilic head group will increase the micellar size, as more molecules can pack into a sphere. Also as the hydrophilic head group size is reduced, there will a certain point at which spheres cannot be formed, and cylindrical micelles, or other non-spherical shapes, are preferred. This is strikingly illustrated by the effect of adding sodium chloride to a solution of sodium dodecyl sulphate and measuring the aggregation number, as shown in Figure 4.13 from Clint (1992).

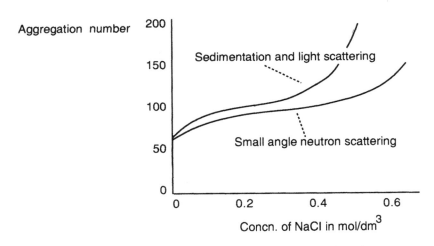

Figure 4.13 Addition of sodium chloride to sodium lauryl sulphate.

There is an increase in the aggregation number on addition of sodium chloride up to 0.2 mol/dm^3, and then on further addition of sodium chloride no large change in aggregation number ocurs until the sodium chloride concentration is about 0.6 mol/dm^3. Separate measurements show that the micelles change from spheres to cylinders and there is also a large increase in viscosity.

Cylindrical micelles can be formed in a number of ways:

1. *The counterion*. The addition of electrolyte e.g. sodium chloride in Figure 4.13) reduces the effective hydrophilic area for ionic surfactants because of the increase in counterions which will reduce the repulsion between the polar head groups (Force 1, Figure 4.11). The size of the counterion also contributes to the overall head group size and therefore changing from Na$^+$ to Mg^{2+} decreases the effective head area because only half the number of Mg^{2+}ions are needed by the micelle.
2. *Addition of cosurfactant with a compact head group*. Mixed micelles can be formed if the head group of the cosurfactant is of similar character to that of the surfactant but smaller in size.
3. *Change in the hydrophilic nature of the head groups*. A good example

is ethoxylated nonionics, where the polyoxyethylene hydrophilic group is effectively reduced in size with increasing temperature.
4. *Change in pH of amphoterics*. Changing the pH of amphoterics will change the degree of protonation and thus the effective head size.

Lamellar micelles

When the effective areas of the hydrophilic group and the hydrophobic group are nearly equal, the micelles can take up planar or lamellar structures. In addition the surfactants become practically insoluble in water although still dispersible. There is such a strong tendency of the lamellar micelles to form multilayers that they form two-phase regions at very low concentrations, which consist of a lamellar liquid crystal structure (see later) dispersed in a solution of surfactant monomers. This is exactly the same as the cloud point of nonionics. In addition, vesicles and liposomes can be formed under the right conditions. A simple pictorial view of these is given in Figure 4.14.

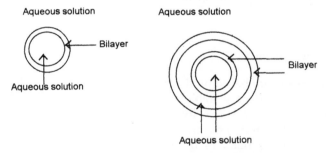

Figure 4.14 Simple view of vesicle (left) and (right) liposome.

Vesicles can be imagined as lamellar micelles bent around and joined up in a sphere, leaving an aqueous solution inside the sphere and one outside. Liposomes are concentric spheres of vesicles. There has been considerable confusion in the naming of these structures and 'vesicle, has sometimes been used as the generic name for both single spheres and concentric spheres. The term **unilamellar vesicle** is often used to make it clear that the structure represents only one sphere. For this book term **vesicle** and **liposome** will be used as shown in Figure 4.14.

Both vesicles and liposomes are well known, and vesicles in particular have been used by biologists as models for cell membranes. The most common application known to the surfactant chemist are cationic softeners, where the solutions of quaternary ammonium compounds have been shown to form vesicles and liposomes. The structure of the surfactant needed to form such structures is one where the size of the hydrophobic

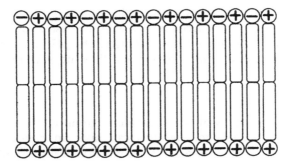

Figure 4.15 Lamellar structure of mixed anionic/cationic surfactants.

group is nearly the same size as the effective size of the hydrophilic group. In quaternary softeners this means that two C18 alkyl chains are needed on one nitrogen atom to equal the effective size of the quaternary group. The existence of such structures goes back to Bangham *et al.* (1965), who first showed that phospholipid molecules arrange themselves into microscopic vesicles.

Why are vesicles formed?

The ideas following are based on the proposals put forward by Kahler *et al.* (1992), who produced very stable vesicles by simply mixing anionic and cationic surfactants with single hydrophobic groups in nearly equimolar concentration. At equimolar concentrations vesicles are not formed.

Four major factors need to be explained for the formation of these vesicles:

1. Why do mixed surfactant systems of anionics and cationics of similar hydrophobic size form vesicles when the component surfactants form micelles?
2. Why do they form vesicles at nearly equimolar concentration but not at equimolar concentration?
3. What is different about the interaction of the bilayers in vesicles which are unilamellar, compared to those in liposomes, which are multilamellar?
4. What factors control the size of the vesicles and their polydispersity?

The first question can be answered using the geometric packing models of Tanford and Israelachvili and co-workers. The individual anionic and cationic surfactants normally form spherical and rod-like micelles with values of $v/a_0 lc$ around 1/3. When the two surfactants are mixed they probably form a double tailed salt with (a) a larger hydrophobic group, and

(b) a smaller effective charge density as the anionic and cationic hydrophilic groups neutralise one another.

The overall effect is that the value of v/a_0l_c is made larger and moves nearer to 1, and hence promotes the formation of vesicles. This explanation may even account for the fact that equimolar mixtures of cationic and anionic do *not* form vesicles because the charge densities have moved v/a_0l_c too far towards 1. This explanation is very satisfactory in showing the formation of vesicles but does not explain their considerable stability. The detailed structure of the vesicle would be something like Figure 4.15. The other questions (2, 3 and 4 above) are related, as the size of the vesicles is related to their stability.

The curvature of a **monolayer** of surfactant is related to the value of v/a_0l_c. As it approaches unity the monolayer will be flat, the curvature infinite, and bilayers can form easily. But, due to the lamellae being flat, vesicles of infinite size would be formed. Only if this bilayer is flexible and can bend to form a sphere will vesicles be formed. This must involve a **bending modulus**. Such a modulus has been measured in monolayer films in microemulsions, but not so far in bilayers and vesicles. It would seem that flat bilayer membranes could be formed, but they are never found in practice as they probably aggregate together to form packed multilayer micelles.

The reason that the bilayers form spheres is that the edge of a bilayer sheet with the hydrophobic groups inside will have their hydrophobic groups exposed to the water. The only way that the hydrophobic groups can avoid the water is by forming into a sphere. This concept is discussed in greater detail in the paper by Kahler *et al.*, (1992) with quantitative estimates which give reasonable results in terms of the vesicle size. The more stable vesicles are found in either cationic or anionic rich systems and the presence of these excess surfactants may well be affecting the edges to reduce the energy at the bilayer edges, as shown in Figure 4.16.

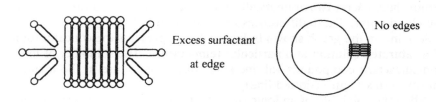

Figure 4.16 Bilayer edges.

An alternative theory of mixed surfactant systems with different sized head groups would allow the inner and outer **monolayers** to have different curvatures. The smaller head group would then have to be in the inside and

46   HANDBOOK OF SURFACTANTS

the larger head group on the outside. This simple picture looks attractive but has theoretical problems. The stability of the vesicles as opposed to the multilamellar liposomes depends primarily on the nature of the bilayer–bilayer interactions especially as the concentration is increased. It is proposed that the bilayers have **repulsive** forces between them, unlike phospholipid bilayers which have a net attraction. This repulsion may be due to the fact that the stable vesicles always have an excess of one surfactant, and consequently the bilayers must possess a net charge. But it has been shown that these vesicles, with anionic surfactant in excess, are stable in 1M sodium chloride. Thus electrostatic repulsion is not likely to give this surprising stability. The more likely explanation is hydration repulsion.

4.4 Solubility and liquid crystals

Solubility is one of the most important properties of a surfactant. Adsorption and hence surface active properties are high when solubility is poor. However, in practice formulations need good solubility particularly in concentrates that need dilution. Practical formulae are a compromise between surface active properties and solubility.

The surfactant molecule consists of a water-soluble group and a water-insoluble group (the hydrophobe). As the hydrophobic group increases in molecular weight (for the same hydrophilic group), the surfactant becomes less soluble in water. Similarly, for the same hydrophobic group, the surfactant becomes more water soluble as the hydrophilic group becomes more water soluble.

Surfactants show abnormal solubility characteristics compared to most other organic chemical compounds. The solubility of most common chemical compounds in water increases as the temperature rises. However, the solubility of ionic surfactants **increases** dramatically above a certain temperature known as the Krafft point. The solubility of nonionic surfactants (ethoxylates) **decreases** dramatically above a certain temperature known as the cloud point. These differences are illustrated in Figure 4.17.

A point on the graph shown in Figure 4.17 represents the concentration of a saturated solution at a particular temperature. The solubility of most non-surfactant compounds will increase with temperature in the manner shown in this Figure (dotted line)

When the solubility of an ionic surfactant is plotted against temperature a complex graph is obtained. The solubility slowly increases as the temperature rises up to a certain temperature (the Krafft temperature), after which there is a very rapid rise in solubility with practically no further increase in temperature. The Krafft temperature is where micelles are formed. Below the Krafft temperature no micelles are formed **and** the

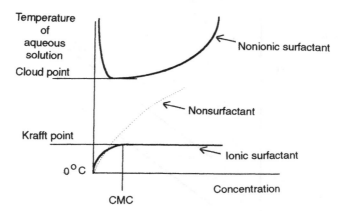

Figure 4.17 Solubility of surfactants. Reprinted from Gu, T. and Sjoblom, J. (1992) *Colloids and Surfaces*, **64**(1), 39–46, by permission.

solubility is limited. Above the Krafft point micelles are formed and the solubility is appreciably increased. Thus it is desirable to make formulations (particularly concentrated ones) at above the Krafft temperature if complete solubility is required. Below the Krafft temperature the surfactants exist as hydrated crystals. Also, the concentration at which micelles are formed at the Krafft temperature is the critical micelle concentration (CMC).

The solubility behaviour of ethoxylated nonionics appears to be quite different. At sufficiently high ethoxylate levels, nonionics are quite soluble at 0°C, but on heating they will come out of solution at a certain temperature (the cloud point).

Thus, knowing the Krafft point and CMC of a surfactant gives practical information about the conditions under which a surfactant operates, and thus helps in the choice of a particular surfactant for a particular application. The relationship between Krafft point and CMC and the chemical structure of surfactants is therefore of considerable interest in helping to choose surfactants and even to tailor-make new surfactants.

The solubility of ionic surfactants

The Krafft point of ionic surfactants is influenced by chain length of the hydrophobe (see Figure 4.18). Note that both series are the sodium salts. There is a large change in Krafft point if the type of salt is varied, or if an electrolyte is present.

The Krafft point plotted against the logarithm of the CMC gives a straight line with the higher Krafft point products (longer hydrophobic groups) having smaller CMCs (see Figure 4.19). The addition of electrolytes increases the Krafft point at the CMC. At higher concentrations of surfactant the crystalline surfactant can be dissolved in the micelles. Soap

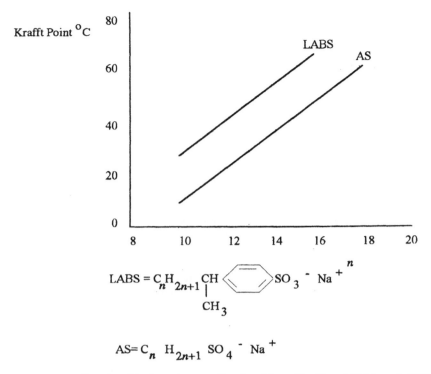

Figure 4.18 Krafft point of ionic surfactants. Reprinted from Gu, T. and Sjoblom, J. (1992) *Colloids and Surfaces*, **64**(1), 39–46, by permission.

will form a precipitate in hard water due to the calcium salt being formed, which gives an increase in the Krafft point (see Table 4.3). Addition of excess soap will cause the soap curds to redissolve.

The effect of added electrolyte is complex. On the addition of high concentrations of monovalent anions the activity coefficient of divalent ions (e.g. Ca^{2+}) falls dramatically and the overall result is a **lowering** of the Krafft temperature.

There are different values of Krafft points reported in the literature, e.g. for C12 sulphate values of 7–17 °C. The probable reason is that pure products usually give higher values than impure (with varying alkyl groups).

High Krafft temperatures are associated with surfactants which will form stable monomer crystal lattices with water, i.e. the region below the Krafft point in Figure 4.17. Thus surfactants with a high Krafft temperature have ionic head groups or compact highly polar head groups and long straight alkyl chains. Lower Krafft temperatures can be obtained with branched alkyl chains or bulkier hydrophobic groups (e.g. two alkyl chains). However, surface activity is generally shown by those structural features which show higher Krafft points, and thus in practice a compromise must

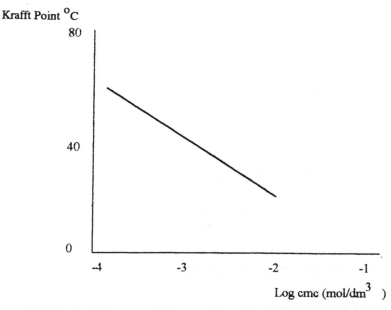

Figure 4.19 Relationship between Krafft temperature and CMC for alkane sulphonates (as defined in Fig. 4.18). Reprinted from Gu, T. and Sjoblom, J. (1992) *Colloids and Surfaces*, **64**(1), 39–46, by permission.

be reached. Table 4.3 shows the influence of incorporating EO in the molecule, showing why ether sulphates are so useful.

Table 4.3 The effect of EO and the counterion on the Krafft point

Surfactant	Krafft point of sodium salt	Krafft point of calcium salt
C12 sulphate	8	48
C12 alcohol (EO) sulphate	5	15
C12 alcohol (EO)$_3$ sulphate	<0	<0

It is tempting to conclude that the factors controlling CMC are the factors controlling the Krafft point. However this is not so although the Krafft temperature still depends upon the structure of the surfactant. The main factor controlling the Krafft temperature is the formation of the hydrated crystal structure.

Nonionic surfactants

As can be seen in Figure 4.20, the plot of Cloud Point against logarithm of the number of moles of EO is linear, and increases with increasing EO content. There is a different relationship depending upon the surfactant structure.

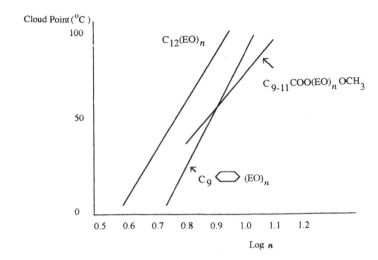

Figure 4.20 Cloud point and degree of ethoxylation. Reprinted from Gu, T. and Sjoblom, J. (1992) *Colloids and Surfaces*, **64**(1), 39–46, by permission.

Figure 4.21 shows that:

1. The cloud point of ethoxylated alcohols decreases as the chain length of the hydrophobe increases.
2. The cloud point of ethoxylated alcohols increases as the ethylene oxide (EO) content increases at a constant chain length of hydrophobe.

In fairly dilute solutions (0.1–5%) the cloud point is independent of concentration and is a characteristic of a particular nonionic surfactant.

The cloud point of nonionic surfactants arises because the solubility of the polyethylene oxide entity is due to hydrogen bonding. Materials relying solely on hydrogen bonding for solubilisation in aqueous solution are commonly found to exhibit an inverse temperature/solubility relationship. Above the cloud point, a second swollen phase appears, which has a micellar structure with a size appreciably larger than that of a normal micelle.

The liquid crystalline region

The simple curves shown in Figure 4.17 are much more complex in practice. If the temperature/concentration diagram is taken as a type of phase diagram then a typical type of phase diagram for anionic surfactants is shown in Figure 4.22.

All ionic surfactants with a single hydrophobic chain show a similar phase diagram; contrast this with the curve for the ionic surfactant in

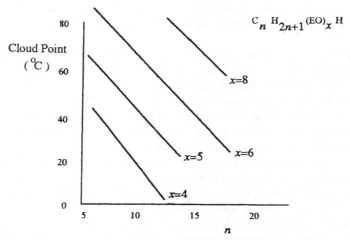

Figure 4.21 Cloud point and hydrophobic chain length.

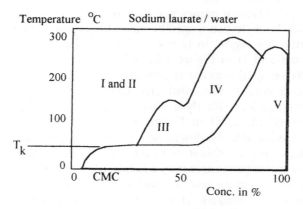

Figure 4.22 Phase diagram for concentrated solutions of surfactants.

Figure 4.17, which is basically similar, but Figure 4.22 has more phases present (phases III and IV). The phase diagram is somewhat simplified and there are more phases and structures than indicated (e.g. a cubic face-centred arrangement of spherical micelles has been detected. Nevertheless, the ones shown can explain most of the characteristics of aqueous surfactant solutions.

The more dilute part of the diagram is as shown in Figure 4.17, and below the Krafft temperature no micelles are formed, only surfactant monomer/solvent crystals in equilibrium with a saturated solution of surfactant at a concentration equal to the CMC. The Krafft temperature is connected with the surfactant/solvent interactions and not the surfactant/

52 HANDBOOK OF SURFACTANTS

surfactant interactions, which control the micelle formation. This is why there will be different effects with electrolytes on the CMC and Krafft temperature.

To the left of the dotted line, again no micelles are formed because the concentration is below the CMC. However, there is no solid surfactant present and therefore the solution will be entirely surfactant monomer.

Above the Krafft temperature and the CMC (Region I and II), the surfactant will be in the form of micelles plus surfactant monomer at a concentration equal to that of the CMC. The micelles can be spherical or cylindrical/disk shaped, as already described.

At higher concentrations two new phases, III and IV (in practice a much larger number), can be identified, particularly by their appearance under polarised light. They appear to have a crystalline structure although still liquid. They have been given various names, lyotropic mesomorphs, lyotropic mesophases, liquid crystalline phases or lyotropic liquid crystals. The word lyotropic means 'solvent induced'. There have also been attempts to give various parts of the phase diagram particular designations such as L_1, H_1, and L_α but this can cause confusion as different authors and publications use different names and symbols. Therefore there will be no attempt in this book to use a specific name for a liquid crystalline region but to describe the region in terms of the molecular structures present. Thus, Region I and II in Figure 4.22 will be described as the spherical/cylindrical/disk-shaped micellar region instead of L_1 (commonly used).

There are generally two distinctly characterised liquid phases, at medium concentrations. Firstly a a viscous phase often called a 'middle phase' is formed (area III in Figure 4.22). This phase consists of the cylindrical micelles already described, which have packed together in a hexagonal structure. It is this structural arrangement of the micelles which gives the anisotropic properties (e.g. rotation of polarised light). This has a two-dimensional structure and a considerably higher viscosity than the cylinders randomly arranged. This structure is shown in Figure 4.23.

The other region (IV) is often called the 'neat phase' or 'lamellar phase', and exists at a higher concentration. It is considerably less viscous and consists of extended regions of surfactant double layers with water between the layers. This again will be the structure formed by lamellar micelles (see Figure 4.24).

The phase diagram for a nonionic such as a C16 alcohol + 8EO of limited solubility shows that the liquid crystalline phases are similar to those present in ionic surfactants (see Figure 4.25).

In addition, optically isotropic liquid crystalline phases have also been observed near the other two anisotropic phases. These have been shown to be cubic face-centred or cubic body-centred structures of spherical micelles. A good review description of these phases has been given by Fontell (1992).

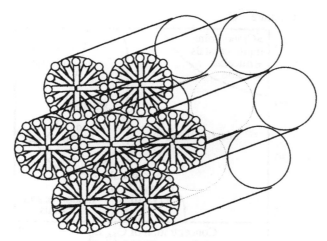

Figure 4.23 Hexagonal structure.

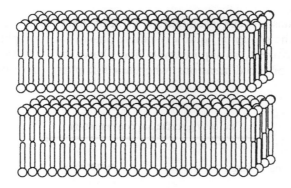

Figure 4.24 Lamellar structure.

Cubic liquid crystalline phases occur both in technical processes and in nature. Single membrane lipid species, mixtures of them, and membrane liquid extracts may form cubic phases together with water. These phases are now considered to have a functional role in cell processes. Cubic phases are not easily detected as a separate phase, but they give rise to startling increases in rheology which are often a nuisance in processing surfactant solutions, but can also be used in cosmetic and pharmaceutical preparations to form gels or **isogels**. The existence of cubic phases as separate phases was unnoticed for a long time, as they are optically isotropic, and thus are not detected in the polarising microscope. When they were first noticed they were termed 'viscous isotropic phases' or 'non-equilibrium isogels'. They occur interspersed between the other phases where liquid-

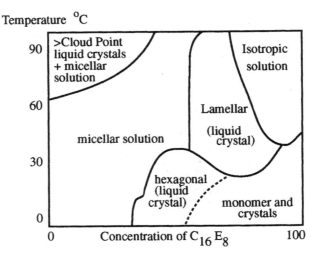

Figure 4.25 Phase diagram for C16 alcohol + 8 EO.

crystalline phases occur (see Figure 4.25). For that reason cubic phases are assigned to the group of liquid crystalline phases (or mesophases). This is not strictly correct, as liquid crystalline phases by definition only have order in one or two dimensions, whereas cubic phases have order in three dimensions.

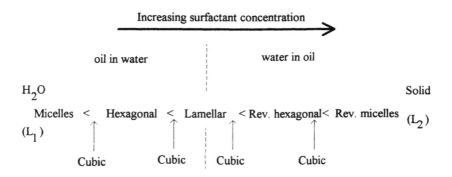

Figure 4.26 Possible locations of cubic phases.

Figure 4.26 (see Fontell, 1992) shows the ideal sequence of phases. When the surfactant concentration is low, micelles are first formed and then, on increasing concentration, hexagonal and then lamellar phases form. In between the changes from one phase to another, cubic phases can be detected. When the surfactant becomes more concentrated, then reverse hexagonal phases and eventually reverse micelles are formed, and there is again the opportunity for cubic phases to be formed when one phase changes into another. This diagram is an ideal situation; in practice, one or more of the phases may be missing, and also there are more cubic phases formed than indicated in Figure 4.26. Typical examples of cubic phases are shown on Figure 4.27, from Sjoblom *et al.* (1987).

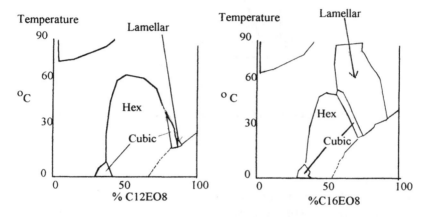

Figure 4.27 Cubic phases in nonionics.

These two examples show the very large difference which can exist between the cubic phases and, particularly, the lamellar phases with two surfactants that are very similar in chemical structure, differing only in the size of the hydrophobe. These data were obtained from pure compounds, so commercial samples with the same nominal chemical constitution will behave very differently. This can explain many observations of large changes in viscosity and gel behaviour of concentrated ethoxylated nonionics with only, apparently, small changes in constitution. Experimental methods for studying cubic structures are not easy to carry out. X-ray diffraction is the main tool, but the diffraction patterns obtained are not easy to interpret. Electron microscopy, NMR, infrared and differential scanning calorimetry are also used. The presence of a third component, such as a hydrocarbon, in emulsions and microemulsions can produce more cubic structures than are found in the two-component surfactant–water systems so far described. Such structures could well play a role in stabilising emulsions.

All surfactants show similar phase diagrams but, depending upon the molecular structure, not all phases may be present or some may be in a concentration range that may not seem appropriate. If there are two large hydrophobic groups, then the phase diagram is similar but the spherical micelles have disappeared and the lamellar phase appears at very low concentrations (see Figure 4.28).

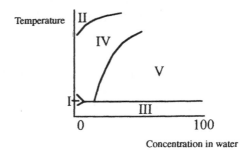

Figure 4.28 Phase diagram of a surfactant with two hydrophobic groups.

Viscosity

The viscosity behaviour of surfactants has in the past been very difficult to interpret and predict. High viscosities and gels are often encountered. In many cases the high viscosities and gels can be a nuisance, particularly in trying to dilute high concentrations of surfactants in aqueous solution. In some cases the gel-like behaviour has been used to advantage, e.g. in the manufacture of 'ringing gels' for hand cleaners and thickening bleach solutions. The use of surfactants as thickeners is often more cost-effective than water-soluble polymers. The reason for the high viscosities encountered has already been mentioned in passing but Table 4.4 sums the effect of micellar structure on viscosity.

Table 4.4 Viscosity and micellar structure

Type of micellar structure	Viscosity characteristics	Effect of shear
Spherical	Low viscosity	Independent of shear
Non spherical micelles (cylinders)	Moderate viscosity	Dependent upon shear rate
Hexagonal packing of cylinders	High viscosity	Highly shear dependent
Lamellar packing at high concentration	High viscosity	Often gel-like but shears easily
Cubic phases	Highest viscosity often in the form of gels	Difficult to shear

An example of a viscosity/concentration diagram is shown in Figure 4.29, which has a striking similarity to the phase diagrams in Figures 4.22

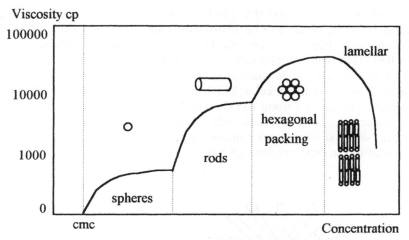

Figure 4.29 Viscosity/concentration data for an ether sulphate.

and 4.25. The viscosity at low concentrations is low due to the spherical micelles (Region I). There is a viscosity increase when the micelles change shape to rods, cylinders or disks (Region II), but an even more marked viscosity increase as the rod-liked micelles take up hexagonal packing (Region III). When the concentration reaches the lamellar stage (Region IV), the viscosity falls as the lamellar structure can shear much easier than the hexagonal packed array (Region III).

The effect of various variables on the viscosity of surfactants can now be better understood:

1. *Concentration.* Increasing the concentration of any product, particularly water-soluble polymers, will increase the viscosity of a solution. A surfactant will give very high increases in viscosity over a narrow concentration range. It is now obvious that there are changes in micellar structure occurring — either changes from spheres to rods which, will give moderate increase in viscosity, or changes in the aggregation behaviour of the micelles themselves.
2. *The salt effect.* For many years, shampoos have been thickened by the addition of salt. Figure 4.13 has shown that addition of electrolyte can change spherical micelles into cylindrical micelles, but spectacular increases in viscosity will be achieved if the spheres pack into cubic structures or the cylinders pack into hexagonal packed structures.

4.5 Solubilisation and microemulsions

Organic substances that are nearly insoluble in water may be dissolved in aqueous solution of surfactants up to certain levels. The surfactant solution must contain micelles and the solutions formed are clear and stable. In a similar manner, water can be dissolved in surfactant solutions of apolar solvents (e.g. heptane). The surfactant solution must contain reverse micelles. This phenomenon is known as **solubilisation**.

The reason for the formation of a stable solution is that the organic liquids 'dissolve' in the micelle. The location of the solubilised molecules in a micelle will depend upon the chemical nature of the organic liquid and the surfactant. Hydrocarbons will be associated with the core of the micelle, while slightly polar compounds (fatty acids, alcohols, esters) can be found in the outer regions of the micelle (see Figure 4.30).

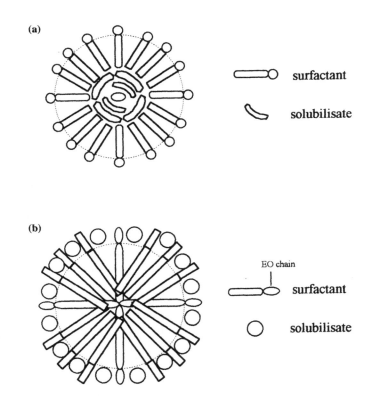

Figure 4.30 (a) Hydrocarbon solubilisation and (b) polar compounds solubilisation.

The extent of solubilisation is limited by the size and number of micelles present and therefore the surfactant concentration must always be well in excess of the material being solubilised. In aqueous solution the amount of organic substance capable of being solubilised is small (see Table 4.5, which gives data from Satsuki *et al.* (1992)).

Table 4.5 Solubilisation of oleic acid in anionic surfactants

Surfactant	Micellar size (nm)	Amount of oleic acid solubilised (%)	Micellar size (nm) after solubilisation
FES	12.8	0.2	28
LABS	12.5	0.1	19
AS	7.5	0.1	14

However, under certain conditions a very much larger amount of organic compound can apparently 'dissolve' in an aqueous solution of surfactant, or a large quantity of water can dissolve in an apolar solvent. The resulting 'solutions' are optically clear and indefinitely stable. They are known as O/W microemulsions and W/O microemulsions respectively, but they are of a different order of size to macroemulsions (see Figure 4.31).

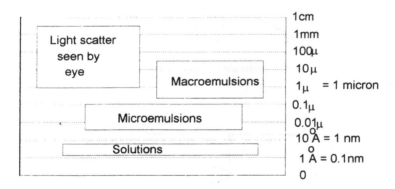

Figure 4.31 Size ranges.

Microemulsions

If an oil is added to an aqueous solution containing a surfactant, then low concentrations of oil can be 'solubilised' in the micelle, as described above. The swollen micelle is of such a size that the solutions remain completely transparent and stable. The micelle must increase in size, but this will be resisted by the interfacial tension between the oil and water. If this interfacial tension can be reduced then a larger 'swollen' micelle is possible which can take up considerably larger amounts of oil and still remain a transparent solution. At this point the oil can be looked upon as emulsified by the surfactant rather than solubilised, but the distinction is not clearcut.

Microemulsions have been manufactured and used for many years for purposes such as liquid polishes, enhanced oil recovery, etc. However, it is only in the last few years that it has been shown that the factors controlling the production of a very low interfacial tension are the same factors as those controlling the shape of the micelles, namely the packing of the surfactants at the interface. In Figure 4.12 a surfactant forming a spherical micelle can change to a cylindrical micelle and then to a lamellar micelle by decrease of the size of the head group relative to the hydrocarbon chain. If the same model is used for the packing of the surfactants at an oil/water interface, as the head group size and tail group size become similar then the O/W interface becomes more planar, and hence a spherical micelle will become larger and capable of taking up more oil. If the surfactant had head and tail of exactly equal size then a planar surface would be obtained with zero interfacial tension. If the head group becomes greater in size than the tail group, then the microemulsion inverts and there is a W/O interface (see Figure 4.32).

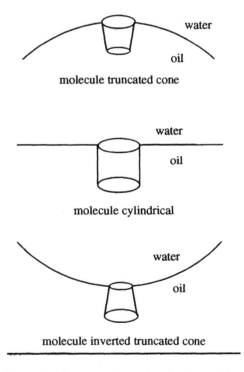

Figure 4.32 O/W interface and molecular packing.

However, the interfacial tension is an equilibrium concept, and in practice the surfactant molecules are in dynamic equilibrium. If the interface is subject to mechanical disturbance in any way, then the

USE OF SURFACTANT THEORY 61

reattainment of that equilibrium can be complex, depending upon the diffusion rate of the surfactant in solution.

Using the concept that the relative size of the head group and the tail group can determine the curvature of the surface, to change a system from O/W to W/O the polar head group in the water must be larger than the hydrophobic tail group in the oil. To invert a microemulsion from O/W to W/O, the size of the hydrophilic group must be decreased and/or the size of the hydrophobic group must be increased. Some possible methods are as follows:

1. *All surfactants.* Increase the chain length of the hydrophobic tail. Decrease the chain length of the oil. Add a cosurfactant with a very small head group
2. *Nonionic surfactants.* Increase the temperature (decreases solubility of EO chain). Decrease the number of EO units in the head group.
3. *Ionic surfactants* Add electrolyte. Change counterion to one that is less hydrated.

4.6 Wetting

When a drop of water is placed on a surface it can either spread over the surface, i.e. it 'wets', or it forms a drop which is stable, i.e. it does not 'wet'.

Reduction in surface tension of water by a surfactant can make a non-wetting solution into a wetting solution on particular substrates. The ability to wet or not to wet depends upon the surface tension of the solution and the **critical solid tension** (CST) of the solid. The CST is the surface tension of a liquid which will form a contact angle of zero measured through the film on that solid, i.e. the liquid spreads over the solid.

Glass, with a CST of more than 70 dynes/cm, can be wet with water. Polypropylene, with a CST of 28, cannot be wet with water but can by aqueous solutions of surfactants. Polytetrafluoroethylene, with a CST of 18, cannot be wet with most common surfactants, although there are now silicone and fluorochemical surfactants which can achieve this very low surface tension.

Although wetting is often described in terms of the spreading coefficient and Young's equation, such mathematical relations are true only at equilibrium. In most practical applications of wetting, e.g. a detergent, or removing water from a metal surface, the process is only in one direction and it is the kinetics of the wetting process which is more important. In addition, very few materials have smooth surfaces (e.g. textiles, powders and chalk do not), and there are considerable capillary effects. Most practical tests (e.g. the Draves wetting test for fabrics using tape; see Shapiro, 1950) are dynamic rather than at equilibrium.

An important concept is the **rate** at which the surface tension decreases. Imagine a drop of liquid placed on a surface with a zero contact angle. In order to spread, the surface of the liquid must expand; a new fresh surface is formed and the surface tension increases unless the surfactant is immediately adsorbed at the new surface at the same concentration as at the original surface. However, surfactant molecules will diffuse from the interior back to the surface in order to lower the surface tension. Thus, if the diffusion is slow then the rate of wetting is poor (see Figure 4.33).

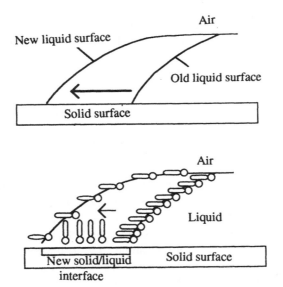

Figure 4.33 Diffusion and rate of wetting.

Correlation between the rate at which the surface tension is lowered and wetting has been observed (see Gruntfest, 1951), but it is not the only factor. Nevertheless, it gives an easily understood phenomenon that the faster a surfactant molecule diffuses to a freshly formed surface then the better (faster) the wetting. This very general rule is borne out in practice, as smaller surfactants are generally better wetters than large molecules due to the smaller molecules diffusing faster through the solution than larger molecules.

There are methods of measuring surface tension of liquids in time periods of hundredths of a second by blowing bubbles in the test solution, measuring the pressure developed and recording the output to a computer. The surface tension can be calculated from the pressure. In testing surfactants a typical diagram of surface tension with time is shown in Figure 4.34 from Hua and Rosen (1988). This diagram is characterised by four different regions:

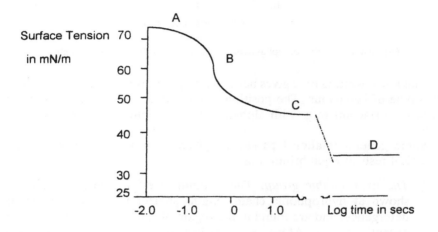

Figure 4.34 Measuring surface tension with time.

A. The induction region
B. The rapid fall region, which usually occurs in fractions of a second
C. The meso-equilibrium region where the rate of ST fall slows considerably
D. The equilibrium region

For most practical processes it will be the difference in surface tension and time between C and D that will be the most important. However, for very high speed processes, e.g. wetting yarn in high-speed melt spin yarn manufacture, then regions A and B could be relevant. The equipment and procedures for measuring the time to equilibrium — known as the meso-equilibrium — and the dynamic surface tension are more complex than those for than measuring Draves wetting times, but do give more information which may be more relevant to the application. Practical tests (Draves) have shown that wetting goes through a maximum as the chain length of the hydrophobic group is varied (see Figure 4.35).

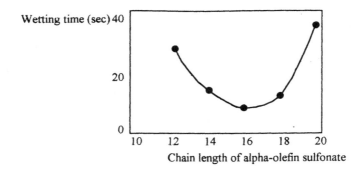

Figure 4.35 Wetting of alpha-olefin sulphonates of different chain lengths.

The lower wetting time gives better wetting. Many surfactant series show this type of behaviour. The position of the minima will depend upon the type of surfactant (e.g. with alpha-olefin sulphonates the minimum is at C16).

Some general relationships can be given on wetting as measured by practical tests at equilibrium, e.g. Draves):

1. *The hydrophobic group.* The optimum wetting characteristics are shown at hydrophobic chain lengths shorter than for foaming or detergency, and are found in the region of C9–12, although there are exceptions (e.g. AOS). Symmetrical located internal head group substitution and ortho-substituted alkyl benzene sulphonates are better wetters than straight-chain or para-substituted (see Gray *et al.*, 1965).
2. *The hydrophilic group.* Additional polar groups in the molecules (ester, amide, EO) usually result in loss of wetting power. Draves wetting times increase with each added EO group in ether sulphates.
3. *EO content of ethoxylated nonionics.* Ethoxylated nonionics will pass through a minimum of wetting performance as the EO content increases. This minimum is where the cloud point is just above the test temperature, (see Komor and Beiswanger, 1966). Ethoxylated fatty alcohols are better wetters than similar ethoxylated fatty acids (see Wrigley *et al.* 1957).
4. *Effect of temperature.* Increase in temperature reduces wetting power due to the better solubility and reduced adsorption, but not with ethoxylated nonionics.
5. *Addition of electrolyte.* If addition of an electrolyte causes a reduction in surface tension then wetting will improve.
6. *Addition of cosurfactants.* Addition of long-chain alcohols and non-ionic cosurfactants improves the wetting properties of anionics (see Biswas and Mukherji 1960)

7. *Effect of pH.* pH is important when weak basic and or acidic groups are present. Sulphocarboxylic acids show better wetting at low pH where the carboxyl group is not ionised.

4.7 Foaming/defoaming

Most surfactants give rise to foam, which can be desirable or undesirable depending upon the application. Whether needing copious foam, little foam or no foam, the user will wish to control the level of foam and to know what outside factors can affect the foam (e.g. foam increase or foam collapse). As a general rule, a foam is not generated in a pure liquid phase. A surfactant that strongly adsorbs at the air interface is necessary in order to produce a foam in aqueous solution.

There are several basic problems on foams which stand in the way of understanding how a particular surfactant performs. The biggest problem in practically all applications is that foam is simultaneously being generated and collapsing and that there can be different mechanisms operating for foam generation and foam stabilisation. The observed performance of the surfactant is the net effect. Another significant problem is the presence in many practical applications of large quantities of finely divided solid and/or liquids, the surfaces of which are in competition for the surfactant.

It is probably these two problems that cause fundamental studies on the foam behaviour of surfactants to give few guidelines in solving practical problems of the technology of foaming and defoaming. Most practical problems are still solved by empirical methods.

Foams consist of a thermodynamically unstable two-phase system of gas bubbles in a liquid. Foam is generated by air forming spheres in the liquid (spherical foam) but this then forms honeycomb foam with relatively thick lamellae between the cells (see Figure 4.36).

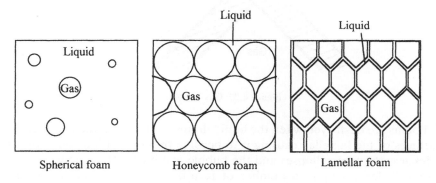

Figure 4.36 Different kinds of foam.

The formation of a foam from a bulk liquid involves the expansion of the surface area due to the work acting upon the system. As surface tension is the work involved in creating a new system, then the amount of new area formed (i.e. the foam) will be greater the lower the surface tension. Thus the reduction of surface tension by the surfactant is the primary requirement of foam formation.

The liquid in the lamellae drains into the junctions. This drainage is not only due to gravity but also due to the pressure difference across a curved liquid surface. The pressure difference (dP) quantified by Laplace is as follows:

$$dP = ST(1/R_1 + 1/R_2)$$

where ST = surface tension and, R_1 and R_2 are the radii of curvature of the liquid surface.

Thus the pressure difference at the junctions of the cells must be greater than in the walls. But the actual pressure inside the bubble must be the same at the wall and the junction, and therefore the pressure inside the junction must be less than in the walls. The liquid drains into the junctions between the cells leaving very thin lamellae (see Figure 4.37). The pressure difference is also proportional to the surface tension. The lower the surface tension the lower the pressure difference and the lower the drainage, i.e. a more stable foam.

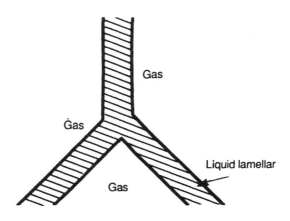

Figure 4.37 Gas bubble.

Foams are destroyed when the liquid drains out between the two parallel surfaces of the lamellae causing it to become thinner. At a certain critical thickness the film collapses and the bubble will burst. The stability of the film will depend upon a number of factors such as type of surfactant, concentration of surfactant, temperature, presence of electrolyte, the presence of other organic materials (foam stabilisers), particularly if polymeric and water-soluble, and the presence of finely divided solids which can be hydrophobic or hydrophilic.

Gibbs film elasticity

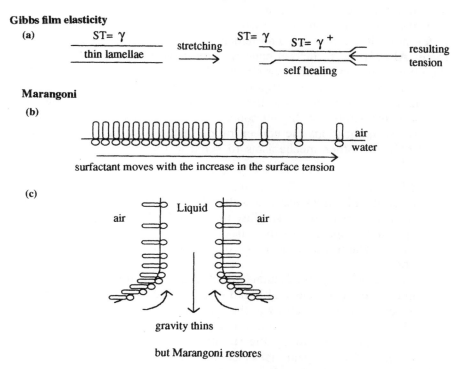

Figure 4.38 Gibbs film elasticity and the Marangoni effect.

The Gibbs surface elasticity and the Marangoni effect are often put forward to account for foam stability. Both effects postulate that film elasticity is due to the local increase in surface tension when a film is stretched (see Figure 4.38). The increase in surface tension causes liquid to flow from the thick section to the thinner section, thus 'healing' the thin spot. The two theories are mathematical in nature but both give similar conclusions in that the effect is dependent on concentration for maximum foaming but only operates in dilute solutions. This is borne out in practice, where there have been many studies to show that foam is at a maximum at about the CMC.

The basic difference between these two theories is that the Gibbs theory is at equilibrium whilst the Marangoni effect takes into account the diffusion rate of the surfactant and hence is time dependent. For a more detailed discussion see Molysa *et al.* (1981). Both effects work in foam because gravity causes liquid to drain out of the thin lamellae (Figure

68 HANDBOOK OF SURFACTANTS

4.38c), and hence the surface tension increases (there are fewer surfactant molecules) in the thin lamellae. There is now a restoring force from the Gibbs elasticity and the Marangoni effect bringing surfactant molecules back into the region of high surface tension.

If the concentration of surfactant is very low the surface tension difference will be very low and hence there will be little or no restoring force. Also, if the surfactant concentration is very high above the CMC, then the change in surface tension with the increase in the area of the film will be too small to prevent rupture of the film. Such theories can explain the stability of some foams at low concentrations of surfactant (e.g. in a bubble bath) whereas at higher concentrations they have lower stability.

In many applications the concentration of the surfactant is well above the CMC and other mechanisms to stabilise the foams have been put forward:

1. *Electrostatic effects*. These are only applicable to ionic surfactants when the film thickness is less than 1000 Å. The charged monolayers repulse each other and stabilise the foam. Such films are very thin and appear transparent so that they are of very little interest in applications where the appearance of the foam is important, e.g. shampoos.
2. *Surface viscosity*. Surfactants oriented as layers or packed together in multilayers (micelles and liquid crystals) can give very different and generally higher viscosity at the interface between the air and liquid. The interactions between alkanolamides and anionics has been suggested as explaining the stabilisation of anionic foams and the different foam appearances (see Wingrave, 1981).
3. *Bulk viscosity*. Polymeric additives enhance foam stability by increasing the bulk viscosity and slowing down the drainage.
4. *A second liquid or solid phase*. When an immiscible phase is present then the stability of the foam can be changed significantly, either increasing or decreasing the stability. Possibly the most important effect will be the change in the distribution of the stabilising surfactant molecules, which can be adsorbed at the new liquid/liquid or solid/liquid interfaces and lost to the air/water interface. Soil is such a possible interface and therefore during shampooing or cleaning of clothes the solution loses surfactant to the soil and destabilises the foam. Also, the effect of added particles can induce the Marangoni effect (see below).

Bearing in mind the very wide variations in foam performance, some empirical generalisations can be made relating chemical structure to foam performance. There are however distinct differences between concentrations at or close to the CMC and those considerably in excess of the CMC.

USE OF SURFACTANT THEORY 69

1. There is no direct relationship between the ability to produce foam and the ability to stabilise foam.
2. Foam volume formed increases with an increase in the concentration of the surfactant up to the CMC. Above that the amount of foam is relatively constant. There is no similar rule with respect to the stability of the foam.
3. Ionic surfactants produce more foam and more stable foam than nonionics.
4. Straight-chain hydrophobes show better foaming than the corresponding similar length branched-chain hydrophobes.
5. The amount of foam goes through a maximum as the chain length of the hydrophobe increases in a homologous series. The longer the hydrophobe the lower the surface tension and hence more foam, but as the chain length grows the solubility of the surfactant rapidly decreases.
6. If the hydrophilic group is moved from a terminal position to an internal position along the chain, higher foam heights but lower foam stability is the general rule (at concentrations above the CMC).
7. The effect of temperature on foaming ability will be similar to the effect of temperature on solubility. Thus, ionic surfactants will foam better with increasing temperature, whilst nonionics will either show a decrease or go through a maximum in foam production with increasing temperature.
8. Polar organic additives (other surfactants) which lower the CMC of a surfactant can improve the stability of foam. The most effective have a similar chain length to that of the surfactant. Examples are alkyl alcohols and fatty alkanolamides.
9. For ethoxylated nonionics, the foam efficiency and stability go through a maximum with increasing EO content. Replacement of the hydrogen atom on the terminal OH group with an alkyl group gives products which produce very little foam and very low foam stability.

Antifoams, defoamers and foam control

The easiest way to control foam is by the choice of a suitable surfactant, and chapters 6–9 give a considerable amount of data on the foam obtained by different surfactants in aqueous solutions. However, in many practical applications there will be severe limitations on the choice of the surfactant with relation to its foaming ability. The primary reason for using a surfactant, e.g. detergency, may give high foaming products; the ideal surfactant may be very costly; interaction with other components may change the foam performance. Therefore there can be a need to control the foam by addition of a foam control agent. Such foam control agents are called antifoams or defoamers. The two terms are not synonymous, as an antifoam is an additive which will prevent foam being formed whilst a defoamer will cause the collapse of foam which is already formed. As foam is simultaneously being formed and collapsing in most applications, the distinction between antifoams and defoamers is not easy to detect.

Antifoams are widely used in controlling the foam in systems which have surfactants that have not been added deliberately but are part of the system; a good example is paper pulping which has many natural surfactants present from the fatty acids and rosin acids in the wood. Another example would be an emulsion paint, which will contain surfactants used in the emulsion polymerisation of the binder. The need in such systems is to obtain no foam at all. It has been found that the antifoams that work in practice are highly specific, i.e. an antifoam which works with one type of paper pulp may be relatively ineffective with a different pulp. There are many such antifoams (see Porter, 1992). The most effective consist of a hydrophobic oil or compound with a surfactant or surfactant mixture that has very limited solubility in the aqueous media. The mechanism for defoaming is not yet well established. The most likely explanations are a combination of the following:

1. The hydrophobic oil causes lenses which cause a reduction in film elasticity to very low values.
2. The formation of an oil/water interface instead of an air/water interface will give a reduced surface tension and hence a reduced Marangoni effect as the surface tension gradient is less.
3. The surfactant in the defoamer can function as a dispersing agent for the hydrophobic oil. In addition, the surfactant may replace the surfactant causing the foam at the interface (see Figure 4.39).

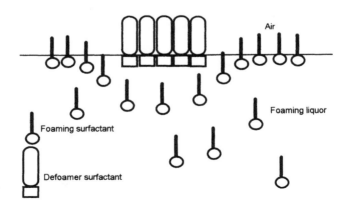

Figure 4.39 Replacement of foaming surfactant by defoamer surfactant.

4. Possible interaction between the foaming surfactant and the defoamer surfactant to form mixed micelles and/or mixed monolayers. Mixed micelles and mixed monolayers (see section 4.10) have the characteristics of being formed at specific ratios of the two component surfactants, with specific surfactant structures giving high synergism in surfactant properties. The highly specific nature of defoamers in paper and paint applications would suggest that mixed micelles and/or mixed surface layers are being formed.

The addition of hydrophobic finely divided solids in hydrophobic oils to high concentrations of high-foaming surfactants gives excellent foam control. In these applications the concentration of defoamer surfactant is very much smaller than the concentration of foaming surfactant. Surfactant interaction is still possible, but is likely to be of low significance. The theory developed by Garrett (1992) has been used to explain the high efficiency of finely divided hydrophobic solids which are used with hydrophobic oils. The overall mechanism proposed by Garrett is shown in Figure 4.40.

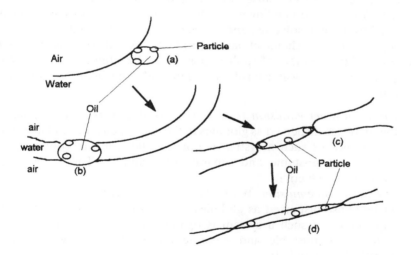

Figure 4.40 Process of film rupture by hydrophobic particle–oil mixtures. Reprinted from Garrett (1992) p.112 by courtesy of Marcel Dekker, Inc.

The particles rupture the air–water–oil film and the particles are at the surface of the oil droplets (a). The oil droplets adhere to the bubble surface and an oil lens is formed (b). The oil lens then elongates to form a bridge between the air–water–air foam films (c). Capillary pressure in the vicinity of the bridging lens increases the rate of drainage of the foam film (d).

72   HANDBOOK OF SURFACTANTS

There is also a large class of defoamers composed of hydrophobic oils, low-solubility surfactants and finely divided solids which are used in applications where the concentration of foaming surfactants can vary from low to high. The mechanism of defoaming will almost certainly involve the formation of mixed adsorption layers and mixed micelles, which could account for the high specificity of such systems.

4.8 Macroemulsions and HLB

One of the widest applications of surfactants is to solubilise or disperse water-insoluble substances (generally organic compounds) in water. Paint, adhesives, textile finishes, paper coatings, leather finishes, hand cleaners, etc. have in the past been formulated with organic compounds which would not dissolve in water. In recent years there is an accelerating trend to replace other solvents entirely or partly with water. The main reason for this change has been the increasing realisation of the toxicity of many of the cheap solvents employed. In addition, water-based products are favoured by customers because of the ease of handling and of cleaning up spillages. A further compelling reason has been the fact that water is cheap compared to organic solvents, and not subject to the fluctuations of price and supply of petrochemical-based products. All these reasons have resulted in a large number of applications where the intention is to dissolve or disperse in water a water-insoluble compound. Emulsions are generally found in two forms:

1. *An oil-in-water emulsion (O/W)*. This consists of two liquid phases, the oil phase dispersed as globules in the continuous water phase. The appearance of such a system is opaque and white in colour. Such systems are unstable and separate on standing. They show high electrical conductivity.
2. *A water-in-oil emulsion (W/O)*. This consists of two liquid phases, the water phase dispersed as globules in the continuous oil phase. The appearance of such a system is opaque and white in colour. Such systems are unstable and separate on standing. They show low electrical conductivity.

The relative sizes of macroemulsions are shown in Figure 4.31 (section 4.5). More complex emulsions, such as water in oil in water, where the continuous phase is water in which there is a W/O emulsion dispersed, also exist.

The most stable emulsions are found when the surfactant is more soluble in the continuous phase, i.e. for W/O emulsions a water-soluble surfactant is most efficient, and for O/W emulsions an oil-soluble surfactant is most efficient.

USE OF SURFACTANT THEORY 73

In practice it is found that a mixture of surfactants with differing solubility properties will produce emulsions with enhanced stability.

Many attempts have been made to correlate surfactant structures with their effectiveness as emulsifiers. The most successful method, still used, is the hydrophilic/lipophilic balance (HLB) first developed by Griffin (1949). This has proved very successful with alkoxylated nonionic surfactants but less successful with ionic surfactants. Griffin proposed to calculate the HLB number from its chemical structure:

HLB = % of the hydrophilic group (molar) divided by 5.

Thus the maximum HLB number was 20 and represented a product composed entirely of ethylene oxide with no hydrophobic group. HLB zero represented a completely water-insoluble product with no ethylene oxide. An approximate HLB number can be obtained by adding a small quantity of a surfactant to water and shaking. The appearance of the resulting solution is shown in Table 4.6.

Table 4.6 HLB and emulsifying properties

HLB number	Appearance on adding surfactant to water	Emulsion
1–4	Insoluble	W/O
4–7	Poor dispersion unstable	W/O
7–9	Stable opaque dispersion	—
10–13	Hazy solution	O/W
13–	Clear solution	O/W

Temperature has a big effect on emulsion systems made with nonionic (ethoxylated) surfactants. Increase in temperature will bring about a phase inversion from O/W to W/O due to the nonionic surfactant becoming less water-soluble as the temperature increases. The temperature at which the inversion takes place is known as the phase inversion temperature (PIT). The PIT can be used as a method for emulsion preparation. An emulsion is prepared near its PIT, where the minimum droplet size is obtained. Then it is cooled to its normal use or storage temperature.

Although both the HLB and PIT concepts are empirical they are extremely useful in practice. In most general applications the HLB system is the most useful in guiding a formulator to the choice of surfactant. There is a large amount of empirical information on the HLB requirement of oils in order to make O/W and W/O. In practice a considerable amount of trial and error is needed, as there can be many different mixtures which will possess a single HLB number or a single PIT.

The stability of an emulsion

In the same way as forming a foam, the formation of an emulsion from a liquid and water implies an increase in surface area of the liquid/water interface. The interfacial tension is a measure of the work performed on the system, and the lower the interfacial tension the lower work needed to make the emulsion. If the interfacial tension is very near zero then a microemulsion can be formed (see section 4.5). In the case of macroemulsions, the interfacial tension is not zero, and the emulsions are thermodynamically unstable and will separate. However, the separation can occur in more than one way.

1. *Creaming and sedimentation.* Creaming is the familiar phenomenon of creaming of milk, where the oil phase is lighter than the water phase, separates and collects on the top. Sedimentation occurs if the oil phase is heavier than the water phase and the separation is on the bottom. Relatively mild agitation can redisperse the oil to give the original emulsion. The oil droplets have retained their original size (see Figure 4.41). Very small droplets will cream more slowly because the Brownian motion will behave as a mild agitation to prevent creaming.

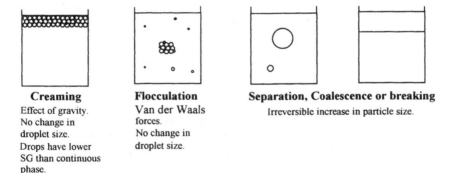

Figure 4.41 Instability of emulsions.

2. *Flocculation.* This refers to the mutual attachment of individual drops to form loose assemblies but in which the identity of each drop is still maintained. They can usually be separated by relatively mild agitation.
3. *Separation, coalescence or breaking.* The appearance is often similar to creaming but soon the oil is seen as a clear layer on the top (or bottom if it has a higher specific gravity than water). The oil droplets have coalesced, and shaking or agitation will not redisperse the oil droplets.

USE OF SURFACTANT THEORY

The rate of coalescence will depend upon the factors that bring the droplets close together, e.g. concentration, and then on the balance of forces that stabilise and disrupt the interface. The tendency for the drops to coalesce will be the van der Waals forces when the lamellae are thin enough, and the restoring forces will be the Gibbs–Marangoni effect (see Figure 4.42). The Gibbs–Marangoni effect will operate due to the distortion and increase in surface area of the drops as they get close together.

4. *Ostwald ripening.* Molecules in a curved surface are more exposed than molecules in a flat surface and have a lower net attractive force towards the bulk centre of the droplet. Smaller droplets therefore can lose molecules more easily than large droplets. If the molecules of the oil droplet have any solubility in the continuous phase, then molecules will transfer from the small droplets to the large droplets. The overall effect is that the large drops grow at the expense of the small drops and the emulsion becomes more polydisperse with time (see Figure 4.43).

The whole process is a dynamic one, with molecules passing in both directions. The process can only operate if the oil phase has some solubility in the water phase. In practice this is nearly always the case. In addition, micelles of surfactant are present in the aqueous phase, which can solubilise small quantities of the oil.

The stabilisation of an emulsion

All emulsions are unstable and will separate by coalescence back into their constituent oil and water phases unless prevented. Creaming can be prevented by making the specific gravity of the oil and water phase the same and by increasing the viscosity of the continuous phase, e.g. by the addition of a thickener. Ostwald ripening cannot be stopped but can be slowed if the drops are not too small and are of uniform size. The major problem is the prevention of coalescence, which is irreversible. The major effect will be the surfactant that can stabilise the droplets by electrostatic repulsion (see Figure 4.44).

Nonionics are equally good, often better, stabilisers and they operate by way of steric stabilisation (see Figure 4.45).

The previous figures have presented an idealised view of the surfactant interface as a curved plane separating the two phases. It is tempting to look upon a macroemulsion as having similar features to the microemulsion already described (see section 4.5). However, the interface in macroemulsions is more complicated and the phase boundary is not a simple curve but a more diffuse structure, as shown in Figure 4.46.

The adsorption of a surfactant at complex surfaces will be aided if there is a mixture of small and large surfactants, which will be able to pack more

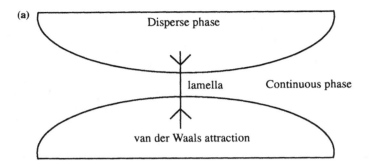

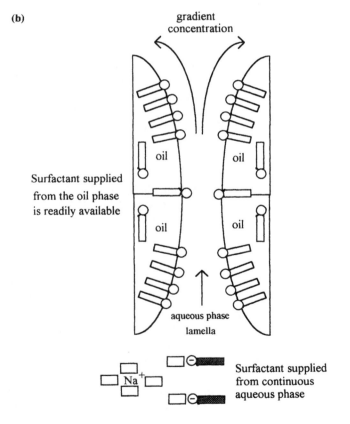

Figure 4.42 (a) The van der Waals forces can run across thin lamellae. (b) The Gibbs–Marangoni effect.

USE OF SURFACTANT THEORY

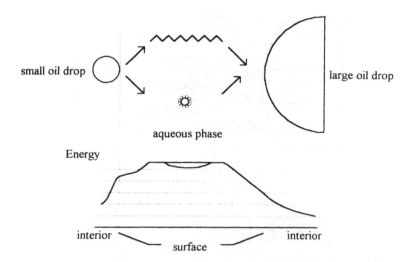

Figure 4.43 Ostwald ripening.

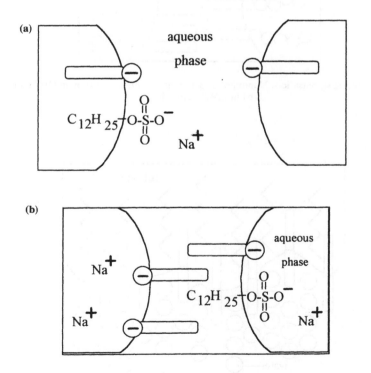

Figure 4.44 Electrostatic stabilisation. Anionic surfactant at the interface (a) in an O/W emulsion, (b) in a W/O emulsion.

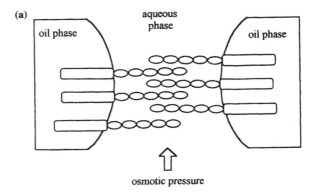

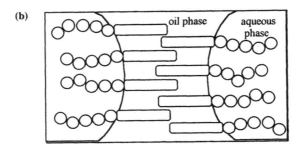

Figure 4.45 Steric stabilisation. Nonionic surfactant at the interface (a) in an O/W emulsion, (b) in a W/O emulsion.

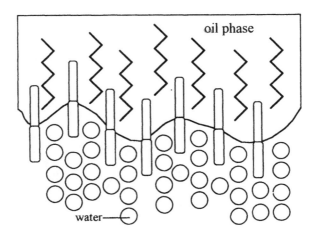

Figure 4.46 The diffuse interface.

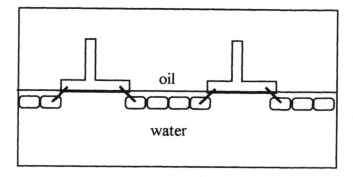

Figure 4.47 Polymeric stabilisers.

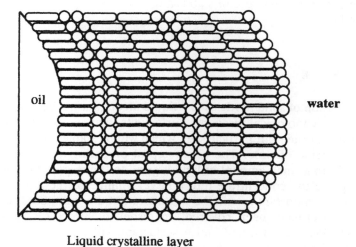

Figure 4.48 Liquid crystal stabilisation.

efficiently than a single uniform species. Also, surfactants which have more than one point of attachment at the interface (polymeric surfactants) will make them more resistant to lateral motion, since the single segments cannot move independent of the other segments (see Figure 4.47).

In Figure 4.47 the surfactant had a number of hydrophilic and hydrophobic groups on the same molecules which restricted the loss of surfactant when the surface area increased during distortion of the droplet. A similar situation will exist where the surfactants are not bound together by covalent bonds but exist in some form of liquid crystal structure. Such structures possess a degree of order which gives high viscosities. The exact

80 HANDBOOK OF SURFACTANTS

structures are not known but a suggested one is shown in Figure 4.48. Examples of polymeric surfactants are the EO/PO block copolymers (see section 7.9).

Multiple emulsions

These are emulsions composed of droplets of one liquid dispersed in droplets of another liquid which are then dispersed in a continuous phase (see Figure 4.49). The emulsification, stability and stabilisation are controlled by exactly the same factors as already described. However, as there are simultaneously both O/W and W/O interfaces present, the choice of surfactant is more limited and there are greater possibilities of instability. Polymeric surfactants have been found to be particularly suitable in the manufacture of stabilisation of these type of emulsions (see chapter 11).

4.9 Dispersing

The suspension of solid particles in a liquid media, particularly water, is an important technological process. Surfactants play a role in preparing suspensions of the right particle size, which will be stable on storage for an extended period of time. The usual step of preparing a suspension is to add a solid to a small amount of liquid, grind to the required particle size, then disperse the concentrate into a larger volume of the liquid. Surfactants play an important role in the wetting out of the solid, but the wetting process has already been covered. In dispersion processes, new solid/liquid interfaces are formed and a surfactant will reduce the interfacial energy at the solid/liquid interface and thereby facilitate the formation of new interfaces. The surfactant must also stabilise the particles against aggregation and/or sedimentation. In practice such surfactants are very different from those giving good wetting. Instead of being small molecules, one finds that the surfactants that stabilise solid suspensions are large.

The stability of particles in solution

A dispersion of particles in solution will be stable if any attractive forces can be countered by repulsive forces between the particles. The most common attractive forces are the dispersion forces, also known as London forces, charge-fluctuation forces and electrodynamic forces, from the fluctuation of electron distribution of the molecules which make up the

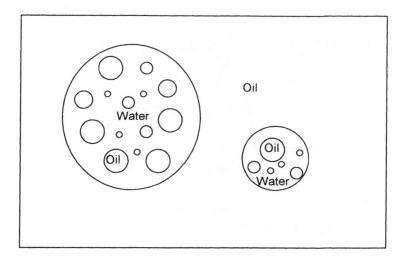

Figure 4.49 Multiple emulsions.

particles. They are always present, and are long-range forces effective from 10nm down to interatomic spacings (0.2nm). The repulsive forces can arise either from dispersion forces at very short range, electrostatic interaction of charged ions or steric interaction, all of which will depend upon the solid and the medium. All these forces can be calculated for simple molecules, but the theoretical treatment for complex molecules is as yet inadequate. For a fuller description of these forces, see Israelachvili (1992).

Lyophobic solids

The best known theory explaining stabilisation is the DLVO (Derjaguin, Landau, Verwy and Overbeek) theory, which covers only lyophobic solids. This theory takes into account the van der Waals attractive forces (this covers all attractive forces, the main component being the dispersion forces) and the electrostatic repulsion of similar charged particles. The net effect is that there is an energy barrier to overcome (see Figure 4.50).

The particles will be stable and not coagulate if there is a net repulsion. The net repulsion will obviously be larger if the electrostatic repulsion is larger. The influence of the surfactant is that adsorption of the lyophobic tail on the solid causes the solid to acquire a charge which will repel similarly charged particles by increasing the electrostatic force (see Figure 4.51).

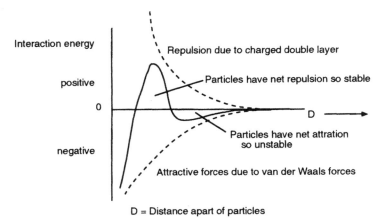

Figure 4.50 The DLVO theory for lyophobic solids.

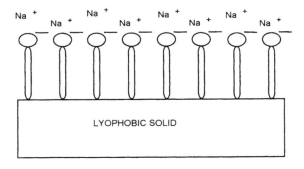

Figure 4.51 Increase in electrostatic force by adsorption of surfactant.

When adsorbed the surfactant then produces an electrostatic barrier (the negative ions) to prevent reaggregation of the particles. The type of surfactant that will give efficient dispersing properties will depend upon the nature of the solid to be dispersed. The adsorption is due to van der Waals interactions between the hydrophobic group (usually a hydrocarbon chain) and the solid surface. The driving forces for adsorption at a water/solid interface and a water/air interface are the same, as already described in section 4.2.

However, steric effects can also act in stabilising hydrophobic particles. EO/PO copolymers are good at stabilising hydrophobic particles in water. The polyoxypropylene chain will adsorb on the hydrophobic surface and allow the polyoxyethylene chain to form the steric barrier, because the highly hydrated polyoxyethylene chain extends into the aqueous phase in the form of coils which give an effective barrier against aggregation so long as the chains are long enough (see Figure 4.52).

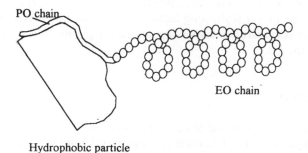

Figure 4.52 EO/PO copolymers as stabilisers.

The polyoxyethylene chain is usually more than 20 units long to obtain efficient stabilisation. There are two principal differences between the adsorption of ionic surfactants and non-ionic surfactants (e.g. EO/PO copolymers) on to hydrophobic solids. First the presence of electrolyte will change the adsorption characteristics of ionic surfactants whilst nonionics are hardly effected. Secondly an increase in temperature will decrease the adsorption of ionic surfactants whilst that of polyoxyethylene derivatives are increased due to the products becoming more hydrophobic (see cloud point in section 4.4).

Lyophilic solids

The theoretical basis for the stabilisation of charged particles will be basically the same as that already described for lyophobic colloids, but it is more difficult to apply a theoretical treatment. Lyophilic solids in water already have a charge. If a surfactant of the opposite charge is used then flocculation may occur. If a surfactant of the same charge is used then there is no improvement until a relatively high concentration of surfactant is added so that adsorption is sufficiently high to stabilise the particle. As the majority of polar solids are negatively charged, one finds that anionic surfactants are the most efficient stabilisers. Such surfactants are those of high molecular weight with a multiplicity of ionic groups; such products are now becoming known as polymeric surfactants. The multiple ionic groups serve a number of purposes. They can also adsorb and yet still give a electrical barrier on oppositely charged particles. They can give a steric barrier to coalescence.

Nonionic surfactants will have a lower degree of adsorption than that of ionic surfactants of opposite charge to that of the solid.

Examples of commonly used dispersing agents are naphthalene sulphonates condensed with formaldehyde and low-molecular-weight polyacrylic acid; also see chapter 11, Polymeric Surfactants.

In organic solvents the electrostatic effects will be minimal and the steric effect will be the major stabilising factor. This is why polymeric surfactants are proving so successful as dispersing agents for pigments and dyestuffs in nonaqueous media.

4.10 Detergency

Detergency is the most important application of surfactants. A detergent is a formulated product which is used to remove soil from surfaces. There is a very wide variety of soils, media and substrates; e.g. in normal household washes there will be a combination of mineral soils and liquid organic soils on both natural and synthetic fibres. It has been shown that the cleaning mechanism can differ depending on which soil and which fibre is being studied. The following therefore represents a **very** simplified picture of detergency.

In practice detergency involves essentially three processes: wetting the soiled substrate (e.g. fabric), removal of the soil from the substrate, and suspension of the soil and prevention of its redeposition. Thus detergency involves the surface-active processes of wetting, adsorption, emulsification, solubilisation and dispersion, all of which have already been described. The wetting of the substrate is a similar process to that already discussed under 'wetting' (section 4.6), but complicated by the fact that there is now an additional phase present. For instance, with an oily soil there will be an oil drop O surrounded by an aqueous surfactant solution S. The oil drop wets the surface of the solid R. The oil drop will have an angle of contact with the solid of θ (see Figure 4.53).

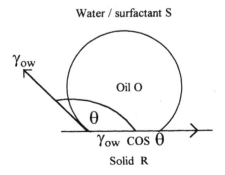

Figure 4.53 Oil drop on a solid surrounded by water.

If there is a greater tendency of the aqueous solution to wet the surface than the oil to wet the surface, then the aqueous solution will replace the oil drop and the oil will be released from the solid surface. This 'tendency' is measured by:

Wetting tension ('tendency') = Interfacial tension$_{ow}$ cos θ

The lower the interfacial tension the lower the wetting tension needed to release the oil drop.

The process of wetting and then subsequent removal of oily soil is known as the 'roll-back' mechanism. The contact angle between soil and substrate increases to either just below or just above 90°. In both cases the oily layer or drops are rolled back and then released. In the case where the contact angle is below 90°, a neck is formed and a small amount of the oil remains behind. Liquid soils can be skin fats, fatty acids, vegetable oils, liquid components of cosmetics, etc. Some solid soils such as animal fats will become liquid at the temperature used for the washing process.

The literature states that nonionics are more efficient than anionics in the roll-back mechanism because of their better solubility in oils. However, the synergism (see Section 11) which exists in many anionic/nonionic systems might give very low interfacial tensions. The 'roll-back' mechanism is shown in Figure 4.54. Where the contact angle < 90°, some work must be done to separate the soil, whereas where the contact angle is > 90° soil separates spontaneously.

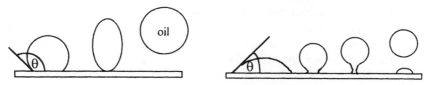

If θ >90° soil separates spontaneously If θ <90° work must be done to separate soil

Figure 4.54 The roll back mechanism.

The removal of particle (polar) soil is due to the fact that there is no electrical double layer at the point of contact of soil and substrate. With adsorption of surfactant an electrical double layer is formed by the diffusion of surfactant into the contact point between soil and substrate. The resultant repulsion thus releases the soil. Note that one wishes to increase the electrical charge. As most fibres are negatively charged, the adsorption of anionic surfactants is necessary to increase the negative charge. Nonionic surfactants have little or no effect on surface charge and thus are not efficient in particle removal. Figure 4.55 shows the removal of particulate soil. This mechanism is difficult to observe directly and detect experimentally. However the presence of Ca^{2+} and Mg^{2+} ions has been

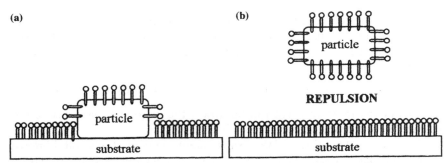

Figure 4.55 Removal of particulate soil in solution with (a) no surfactant, (b) surfactant.

shown to compress any electrical double layers and thus decrease any repulsion. This is borne out in practice, where hard water reduces detergency.

Once the soil is removed then it must be kept in suspension and prevented from redeposition. There are probably several mechanisms: solubilisation of organic materials in the micelles, stabilisation of the inorganic particles by dispersion forces, formation of macroemulsions of the liquid components, and possibly the influence of micellar aggregates (see later).

The presence of polyvalent cations, mainly calcium, in hard waters is detrimental to detergency because the adsorption of such cations onto negatively charged substrates (most fabrics are negatively charged when wetted with water) and on to the soil can reduce electrostatic repulsion and increase redeposition. Also, the corresponding calcium salts of anionic surfactants tend to be insoluble in water, precipitate out and become ineffective. Builders will eliminate or reduce the effect of hard water by sequestration or precipitation of polyvalent cations by sodium tripolyphosphate or zeolites.

Probably the main effect of inorganic builders is to increase the negative charges (known as the zeta potential) on the soil and substrate. The effects of various types of builders are complex. The major effect of inorganic builders on all surfactants is that they are all electrolytes and will reduce the solubility of the surfactant agents present in the detergent. This will increase the degree of adsorption on to substrate and soil and thereby increase their efficiency as soil dispersants.

A less obvious way in which divalent ions can cause a reduction in the efficiency of anionic surfactants is to change the orientation of anionic surfactants to negatively charged soil. In the absence of divalent ions, the anionic surfactant will adsorb to negatively charged soil with the hydrophilic head oriented away from the soil. In the presence of divalent ions these can act as a link between the surfactant head group and the soil, thus orienting the hydrophobic tail away from the soil. This makes the soil more hydrophobic and more difficult to wet, and increases the interfacial tensions between soil and detergent and between substrate and detergent. Thus, complexing the calcium ions with sequestering agents can improve the surfactant properties, and this effect is shown in Figure 4.56.

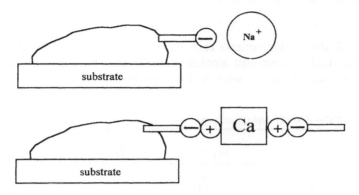

Figure 4.56 Effect of divalent ions.

Practical detergents have a number of other components: foam control agents, silicates to prevent corrosion, anti-redeposition agents such as carboxymethyl cellulose, colouring, perfume, and softeners, all of which can effect the properties of the surfactant. A detergent is a very complex formulation and it can be misleading to try and judge the performance of a surfactant in a 'built' detergent on data obtained in the absence of builders and other additives.

The theory of surfactant action in detergents is still evolving, and the picture presented so far must only be taken as a simplification. For instance, the ideas proposed so far could assume that the surfactant action is by the monomeric molecule and therefore would be expected to be at a maximum at the CMC. However, evidence is accumulating that the presence of a liquid-cystalline phase (cylinders, disks, or lamellar micelles) can have a pronounced effect on washing performance in certain systems. Nonionics seem to perform better above their cloud point; the maximum detergency of polyoxyethylene nonionics is 15–30°C above the cloud point, where particles of surfactant rich phases are present.

Raney (1991) has shown that the removal of hydrocarbons from fabrics could be correlated with the formation of a microemulsion phase of the surfactant, the hydrocarbon removed and water. Of key importance is the phase inversion temperature (PIT) of nonionic surfactant/oil/water systems. For a system with equal quantities of oil and water and less than 10% of a nonionic surfactant, then the effects shown in Table 4.7 occur.

HANDBOOK OF SURFACTANTS

Table 4.7 PIT and microemulsions

Temperature	Microemulsion
Below PIT	O/W microemulsion in equilibrium with an excess oil phase.
At PIT	3-phase system of water phase an oil phase and a microemulsion containing both oil and water.
Above PIT	W/O microemulsion in equil. with an excess of a water phase.

Table 4.8 shows the optimal removal of cetane from polyester/cotton fabric with single-component alcohol ethoxylates. The PIT is approximately 20°C above the cloud point of the nonionic.

Table 4.8 PIT and detergency

Ethoxylate	PIT (°C)	Optimum detergency temp.(°C)
C12E5	52	50
C12E4	31	30
C12E3	< 20	< 20

Similar results have been found with the removal of triglycerides, fatty alcohols, fatty acids and sebum-like hydrocarbon/polar oil blends. In the last-named case, the PIT was below the cloud point but there was still an excellent correlation with the optimum temperature for detergency. The importance of the PIT is that lamellar liquid-crystalline structures are shown experimentally to be formed as the temperature reaches the PIT. The same surfactant structures which produce lamellar liquid crystals also produce microemulsions; that is, the size of the hydrophilic group in the surfactant is equal to the size of the hydrophobic group, resulting in a flat interface.

These conclusions are based on the use of nonionics on their own. It has been shown by years of practical empirical experiments that optimum detergency is obtained by using mixtures of anionics and nonionics. The addition of nonionics to anionics will lower the CMC, PIT and phase behaviour. The addition of a hydrocarbon oil will effect the PIT considerably.

Direct evidence (see Raney, 1991) for low interfacial tension has been obtained and shown to be at a minimum (as low as 0.005 dyne/cm) at the same ratio of nonionic to anionic surfactants and the same optimum temperatures as optimum detergency. The low interfacial tension was only obtained very slowly and took up to 60 min to fall to the minimum value.

USE OF SURFACTANT THEORY 89

4.11 Surfactant mixtures and interactions

Surfactants are rarely used on their own; in practice they are invariably used as mixtures. All commercial surfactants are mixtures (see chapters 5–11), and their properties will deviate to some degree from those of a surfactant that has only one molecular species present. Many academic studies of surfactants use purified products and the conclusions based on pure surfactants may not always be applicable to commercial mixtures. Also, most practical formulations will consist of more than one type of surfacant, each of which will be a mixture.

CMC and surfactant mixtures

The thermodynamics of micelle formation can be used using a phase separation model which assumes that the micelles are a separate phase from the surfactant monomers in solution. For an ideal two-component mixture, which is above the CMC of both surfactants, the monomer concentration is proportional to the mole fraction of surfactant present if the activity coefficient of the free surfactant monomers is assumed to be 1. Then Clint (1975) showed that:

$$C_A = x \, CMC_A = \alpha \, CMC \tag{4.1}$$

$$C_B = (1\text{-}x) \, CMC_B = (1\text{-}\alpha) \, CMC \tag{4.2}$$

where C_A and C_B are the monomer concentrations of components A and B, x is the mole fraction of component A in the mixed micelles, and α is the mole fraction of component A in the mixture.

Hence, by rearrangement:

$$\frac{1}{CMC} = \frac{\alpha}{CMC_A} + \frac{(1\text{-}\alpha)}{CMC_B} \tag{4.3}$$

The surfactant with the lower CMC will have more influence in determining the mixed CMC.

Mixed systems of surfactants that are similar in chemical constitution obey this relationship. The surfactant with the lower CMC will have more influence in determining the mixed CMC, and therefore mixed systems of similar constitution (i.e. commercial surfactants) will have their surfactant properties influenced more by the longer hydrophobic chains than by the shorter chains.

Many surfactant systems do not obey this relationship and such systems are known as 'non-ideal'. In ideal systems the activity coefficient of the free surfactant monomer is assumed to be 1. Corkill in 1974 (see Clint, 1992) proposed introducing an activity coefficient into equation (4.1) as follows:

90 HANDBOOK OF SURFACTANTS

$$C_A = xf_A \, \text{CMC}_A = \alpha \, \text{CMC} \qquad (4.4)$$

$$C_B = (1\text{-}x)\, f_B \, \text{CMC}_B = (1\text{-}\alpha)\, \text{CMC} \qquad (4.5)$$

The activity coefficient is defined as:

$$f_A = \exp[\beta(1-x)^2] \qquad (4.6)$$

β is known as the molecular interaction parameter (MIP)

In most systems, experimental evidence finds that β has the same value across the concentration range and therefore can be calculated from the CMC of the two components and one other CMC of a mixture of known composition.

If β is zero then the mixtures behave ideally and there is no interaction of the two surfactants. If β is negative then the CMC will be lower than for an ideal system, i.e. it can be called synergistic. If β is positive then the CMC will be higher than for an ideal system. The value of β is then a measure of the deviation from 'ideality'. In practice β is generally negative. Typical values of β are shown in Table 4.9.

Table 4.9 Typical values of β for surfactant mixtures

	Anionic	Nonionic
Anionic	0 to -1	—
Nonionic	-1 to -5	<-1
Amphoteric	-5 to -15	<-1
Cationic	-15 to -25	-1 to -5

The physical significance of β is the hypothesis the two surfactants will form mixed micelles or mixed adsorption layers, and that there is an 'interaction' between the two types of molecules (see Figures 4.57 and 4.58). Figure 4.57 illustrates a mixed nonionic and anionic adsorbed layer.

Figure 4.58 illustrates a mixed micelle formed from an anionic and an amphoteric. In this example it is assumed that the two hydrophobic chains are equal in size. The positive charge on the amphoteric will reduce the repulsion of the anionic groups for one another and vice versa. The result will be a micelle of increased size (see section 4.2) and lower CMC.

Although there is no direct experimental evidence of mixed micelles, there is no doubt that significant changes in micellar size take place when two dissimilar surfactants are mixed together. A particular example is when sodium dodecyl sulphate (SDS) is mixed with a lauryl dimethyl-betaine (LB) and myrisityl dimethyl betaine (MB), the CMC of a mixture of a specific ratio is lower than the CMC of either surfactant see Figure 4.59, from Iwasaki et al. (1991).

If the aggregation numbers of the same mixtures are measured then the increased size of the micelles follows the fall in CMC, reaching a maximum

USE OF SURFACTANT THEORY 91

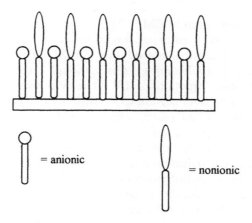

Figure 4.57 Mixed surface adsorption.

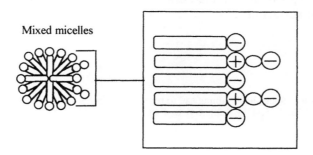

Figure 4.58 Mixed micelles.

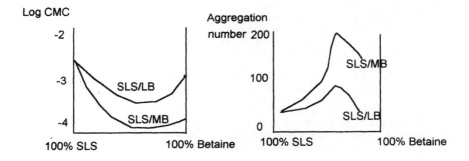

Figure 4.59 CMC and aggregation number of SDS mixed with LB or MB. Reprinted with permission from Iwasaki *et al.* (1991). Copyright (1991) American Chemical Society.

HANDBOOK OF SURFACTANTS

where there is a minimum in CMC. The lowest value of CMC, and hence free monomer, is at a ratio of 0.6 mole betaine to 0.4 mole SDS. Measurements of viscosity showed that the SDS/LB mixture displayed Newtonian viscosity at all compositions, although there was an increase in viscosity at 0.6 mole fraction of LB. The MB mixtures showed non-Newtonian behaviour over certain compositions, suggesting that sphere to rod transitions had taken place, which would explain the increases in aggregation numbers measured.

This reduction in CMC has a practical application in reducing eye irritation, because eye irritation is due to adsorption of charged surfactants at the eye protein–phospholipid complex. The adsorption is proportional to the free surfactant monomer, and therefore reduction in CMC of the mixed anionic and betaine will reduce the free anionic monomer concentration and reduce eye irritation.

There is now considerable theoretical work on mixed surfactant adsorption and mixed micelles. It is hoped that improvements in the theoretical understanding will help in the design of practical formulations made up of mixtures of surfactants.

Note

The permission of Speciality Training Ltd to reproduce Figures 4.4 to 4.12, 4.15, 4.16, 4.22 to 4.24, 4.27 to 4.33, 4.35, 4.36, 4.38, 4.41 to 4.48 and 4.52 to 4.57 is gratefully acknowledged.

References

Balzer, D. (1991) *Tenside, Surfactants, Detergents* **28**(6), 419–27.
Bangham, A.D., Standish, M.M. and Watkins, J.C. (1965) *J. Mol. Biol.* **13**, 238.
Biswas, A.K. and Mukherji B.K. (1960) *J. Appl. Chem.* **10**, 73.
Clint, J.H. (1975), *J. Chem. Soc. Faraday Trans. I.* **71**, 946.
Clint, J.H. (1992) *Surfactant Aggegation*, Blackie, Glasgow and London.
Fontell, K. (1992) *Adv. Colloid Interface Sci.* **41**, 127–47.
Garrett, P.R. (1992) In *Defoaming Theory and Industrial Applications*, ed. Garrett, P.R. Marcel Dekker Inc., New York, pp. 1–119.
Gray F.W., Krems I.G. and Gerecht J.F. J.A. (1965) *J. Am. Oil Chem Soc.* **42**, 298.
Griffin W.C. (1949). *J. Soc. Cosm. Chem.* **1**, 311
Gruntfest I.J. (1951). *Textile Research J.* **21**, 861–966.
Hartley, G.S. (1936) *Aqueous Solutions of Paraffin Chain Salts*. Hermann et Cie, Paris.
Hua, H. and Rosen, M.J. (1988) *J. Colloid and Interface Science* **124**(2), 652–659.
Israelachvili J., *et al.* (1976) *J. Chem. Soc. Faraday Trans. I.* **72**, 1525.
Israelachvili, J. (1992) *Intermolecular and Surface Forces*. Academic Press, London, pp. 128 *et seq.*
Iwasaki, T. *et al.* (1991) *Langmuir*, **7**, 30–55.
Kahler, E.W. *et al.* (1992) *J. Phys. Chem.* **96**(16), 6698–707.
Komor, J.A. and Beiswanger, J.P.G. (1966) *J. Am. Oil Chem. Soc.* **43**, 435.
Lin, Z. *et al.* (1992) *Langmuir* **8**(9), 2200–5.

McBain, J.W. (1913) *Trans. Faraday Soc.* **9**, 99.
Molysa, K. *et al.* (1981) *Colloids and Surfaces* **3**, 329–338.
Porter, M.R. (1992) In *Defoaming, Theory and Industrial Applications*, ed. Garrett, P.R. Marcel Dekker, New York, pp. 269–299.
Raney, K.H. (1991) *J. Am. Oil Chem. Soc.* **68**(7), 525–531.
Satsuhi, T. *et al.* (1992) *J. Am. Oil Chem. Soc.* **69**(7), 672–7.
Shapiro L. (1950) *Am. Dyestuff Rept* **39**, no. 2; *Proc. Am. Assoc. Textile Chemists Colorists* pp. 38–45, p. 62.
Sjoblom, *et al.* (1987) In *Nonionic Surfactants*, ed. Shick, M., Marcel Dekker, New York, pp. 369–434.
Wingrave J.A. (1981) *Soap, Cosmetics and Chemical Specialities* pp. 33–40.
Wrigley, A.N., Smith F.D. and Stirton, A. (1957) *J. Am. Oil Chem. Soc.* **34**, 39.

5 Surfactants commercially available

The first requirement of any practical user of surfactants is to find out if a particular surfactant is available commercially in the required quantities in the right price range. Directories do exist, giving details of which specific products are available. Chapters 6–11 do not attempt to replace these sources of information. Most directories will have considerable detail on where to find products, trade names, approximate compositions and suggested applications, but lack details of the composition and properties of different classes of surfactant.

Chapters 6–11 attempt to classify products, describe their composition, and give a description of their physical and chemical properties, the applications in which they are used and some suggested tests on which to base a specification for quality and reproducibility. There are speciality books dealing with all these aspects, but these chapters summarise these features to enable the user to obtain an initial choice of the surfactant. It should also give a quick and easy summary of the principal properties of those surfactant types which are unfamiliar. All the types of surfactants described are commercially available, even if only in development quantities, at the time of writing.

Although considerable work and care has been taken in drawing up this information, the reader must beware of the limitations in summarising information on surfactants. These are:

- Quantitative data is difficult to provide in every case
- Generalisations will always have exceptions
- Commercial surfactants are not pure compounds, and therefore samples of apparently the same product from different manufacturers may have slightly different properties

Nevertheless, the type of information given is that which accumulates in a formulator's mind over a period of time. These summaries try to give the beginner the benefit of experience.

Every statement should be checked against practical experience, and if it differs then a note to that effect should be made. However, do not let one exception lead to a new generalisation that is equally misleading.

The information presented does not generally have references, as it is a combination of the writer's experience, information from trade literature and information from published sources. Giving numerous references to

SURFACTANTS COMMERCIALLY AVAILABLE

justify every statement would not be helpful. However, there are occasions where information is available that is useful and seems logical, but the writer has not been able to find corroboratory evidence or has no personal experience. In this case a reference is given to the original information.

The information presented is in a standard format. The standard sections are:

1. *Nomenclature*. The various technical names and abbreviations are given for the appropriate surfactant. Common names and chemical names are given, but not trade names except where these have achieved 'common name' status. The names given are under two headings:

   Generic — describes the class of surfactants covered by the section
   Examples — gives specific examples of the most common surfactants in that generic class. In some cases a chemical formula is given, but readers must keep in mind that practically no commercial surfactant is a pure product.

2. *Description*. Identifies the major chemical constituents in the named class. All surfactants are mixtures, but some are more mixed than others and it is important to realise this fact when comparing one end effect between one product and another.

   A very brief description of the method of manufacture of the surfactant type is given. This is often useful in appreciating the composition of the surfactant and how products can differ in composition and hence properties.

   Chemical impurities are described when these might be important. Impurities are defined as chemicals that are not normally described by the chemical name of the class. This feature is becoming more important due to more attention being paid by legislation and the public's attitude towards chemicals. In most cases the impurities are no more harmful than the major component but it is well to know of their existence, their properties and how to detect them.

3. *General properties*. The general physical and chemical properties are summaries rather than tables of detailed information. An attempt has been made to emphasise the particular property that makes that particular surfactant different from other surfactants. Comparisons are made between the type being described and very similar types. The properties covered are:

   • Solubility. This is of primary importance to a user because most formulations are liquids in a solvent, the most common being water. The information presented is for the surfactant alone, but the solubility can be changed with mixtures of surfactants and other non-surfactants, e.g. hydrotropes.
   • Chemical stability. Surfactants will need to be stable in the

formulation in which they are used for two reasons: storage stability of the formulation, and chemical stability under the conditions of use. In aqueous solution, hydrolysis is of paramount importance particularly if the pH of the solution is far removed from 7, as is often the case for detergents. Other components such as oxidising (bleach) or reducing agents can attack the organic-based surfactant. In actual use temperatures can often be 100°C or higher for aqueous solutions, e.g. in deep oil wells, and so hydrolysis at high temperatures can be a factor determining whether or not a surfactant is used.

- Compatibility with aqueous ions. This is quite different from chemical stability, and refers principally to the effect of hard water or inorganic ions on solubility. Most formulations used in the aqueous phase are diluted with water, which may be hard or soft. Many formulations also contain added polyvalent ions, e.g. polyphosphates, and such ions can often cause considerable solubility changes in the surfactant. Such solubility changes can affect storage stability and performance, as solubility is related to surface activity.

- Compatibility with other surfactants. Where there are exceptions to the general rules (see below) mention is made of this fact, particularly where surfactants cannot be easily classified.

- Surface-active properties. Very basic surface-active properties such as the critical micelle concentration (CMC) and the surface tension in water are generally available for most major surfactant types. Such information is interesting but should not be used as the only criterion for the selection of surfactants. Some of the information on critical micelle concentrations was obtained from Makurejie and Mysels (1971). Most of the data are on purified compounds and can differ significantly from measurements on commercial samples (see Mixtures, later). It was noted, particularly on CMC, that there were very large variations in the data quoted by different authorities. Almost certainly the reason was that the products differed in composition rather than that the accuracy of the measurements differed. Wherever possible the author has given an average value given by a large manufacturer rather than the value for a pure product, as it will be the former which will be used in practically all applications rather than a well characterised purified product.

- Functional properties. The properties covered in this section are wetting, dewetting, foaming, defoaming, dispersing, and detergency. These are the most important of the general properties and are a guide to surfactant selection once the connection between the functional properties and the applications can be correlated.

SURFACTANTS COMMERCIALLY AVAILABLE

4. Applications. The most significant applications are given; these will often be detergency and cleaning, due to these applications being the major use of surfactants by volume. However, industrial uses will be mentioned where particularly relevant, as will the principal characteristic of the surfactant that determines its use, e.g. wetting property in a paint for difficult to wet surfaces.

5. Specifications. The surfactant manufacturer should give help in the analysis of his product. The user's main concern will be in the specification of the material, which will give help in checking product variability. This section contains the most useful tests found in specifications and has a twofold use. (i) To help the user in checking the surfactant for batch-to-batch variation in chemical constitution. In most cases the surfactant manufacturer will give this information and a lot more. However, the specifications given are those which the author considers most important and which the user should know in order to check on batch-to-batch consistency. Numerical data are given where it is considered appropriate. If no figure is given then the information should be recorded and used to compare with subsequent batches. In some cases there is reference to the amount of impurities; this may or may not be important, depending upon the end use. (ii) To help in identification.

6. Safety. No statements are made on safety, except in a few special cases. The main reason is that it is impossible to make a general statement on safety on a group of surfactants. Individual members of a group of surfactants can vary so much in toxicity that generalisations cannot be made. The primary source of safety data is the surfactant manufacturer or supplier.

The classification given is based on the chemical structure of the hydrophilic group:

- *Anionic* — the surface-active part of the molecule carries a negative charge, e.g. $C_{12}H_{25}CO\text{–}O^-$ Na^+, and has a long chain hydrophobe carrying the negative charge
- *Nonionic* — the surface-active part of the molecule apparently carries no charge, e.g. $C_{12}H_{25}\text{–}O\text{–}(CH_2CH_2O)_7\text{–}H$
- *Cationic* — the surface-active part of the molecule carries a positive charge, e.g. $C_{12}H_{25}N(CH_3)_3^+$ Cl^-
- *Amphoteric* — the surface-active part of the molecule can carry a positive or a negative charge or both, depending upon the conditions, e.g. $C_{12}H_{25}N^+(CH_3)_2\text{–}CH_2COO^-$

Speciality surfactants that are chemically quite dissimilar, e.g. silicones and fluorocarbons, have been described quite separately although they are strictly examples of a particular hydrophobe. There are anionic, cationic,

98 HANDBOOK OF SURFACTANTS

nonionic and amphoteric silicones, but they are all included under silicones. The classification adopted is not strictly scientific but more based on common usage in the industry. Every system of classification has merits and demerits but the index should enable the reader to find the product group required quickly.

Some generalisations are as follows, but note that exceptions can always be found

1. Anionic surfactants are generally not compatible with cationics and vice versa.
2. Nonionics and true amphoterics are compatible with each other and with anionics or cationics.
3. If the hydrophobic group is an alkyl paraffin chain, then maximum surfactant properties in aqueous solution are in the region of C10–C18 length of the hydrocarbon chain.
4. Straight-chain alkyl groups show increased viscosity, better bio-degradability and inferior solubility compared with similar branched-chain or ring-containing surfactants.
5. More than one hydrophilic group in the surfactant molecule will increase solubility and shift the optimum chain length of the hydro-phobe to higher carbon numbers.
6. If the surfactant can react with ethylene oxide then the products are more water-soluble than the starting material. The larger the amount of ethylene oxide the better the water solubility.
7. Organic sulphates are readily hydrolysed by hot acids. Organic esters are hydrolysed by acids or, more readily, by alkalis. Organic amides are more resistant to hydrolysis than esters or sulphates. Organic sulphonates are resistant to hydrolysis by hot acid or hot alkali.

In addition, commercial surfactants are invariably mixtures and are not pure. Hydrophobic groups have a distribution of chain lengths (see chapter 2). Hydrophilic groups show various forms of impurities (see chapters 6–11); in particular ethoxylated products have a distribution of chain lengths of the polyoxyethylene chain (see section 7.1.1). The chemical formula of a typical commercial surfactant very rarely describes the fact that it is a mixture. Discrepancies between published data on surfactants can often be explained by the fact that the actual surfactants being examined were different.

Reference

Makurejie, P. and Mysels, K.J., (1971) Critical micelle concentration of aqueous surfactant systems. *National Bureau of Standards (US)*, **36**, 227.

6 Anionics

Anionics are manufactured and used in greater volume than all other types of surfactants. The reason is the ease and low cost of manufacture, and they are used in practically every type of detergent, the main application of surfactants. For optimum detergency the hydrophobic group is a linear paraffin chain in the range of C12–C16, and the polar group should be at one end of the chain. Hence the majority of water-soluble anionic surfactants available on a large scale are of the type:

$$CCCCCCCCCCCCC-X$$

X is the hydrophilic group which is ionised and can be:

- Carboxylate (soap) $RCOO^-$
- Sulphonate RSO_3^-
- Sulphate $ROSO_3^-$
- Phosphate $ROPO(OH)O^-$ (monophosphate)

The main types of anionics commercially available are:

1. Carboxylates soaps; ethoxy carboxylates; ester carboxylates; amide carboxylates (sarcosinates, alky phthalamates)
2. Isethionates and taurates
3. Phosphates (ethoxylates, alcohols, amides)
4. Sulphates: alcohol; alcohol ether; alkanolamide ethoxylates; natural oils; nonyl phenol ether
5. Sulphonates: alcohol ether (ethane) or alkyl phenyl ether; paraffin; alkyl benzene; fatty acids and esters; naphthalene derivatives; olefine sulphonates; petroleum sulphonates.
6. Sulphosuccinates and sulphosuccinamates
7. Taurates

Note that soap is a generic name for the alkali metal salts of a carboxylic acid derived from animal fats or vegetable oils. Whether to classify sulpho fatty acid esters as sulphonates or as carboxylic acid derivatives can be argued either way. They have been classified as sulphonates as their properties are more similar to sulphonates than to carboxylates.

The cations most commonly used are sodium, potassium, ammonium, calcium and various amines, principally isopropylamine, monoethanol-amine, diethanolamine and triethanolamine. The triethanolamine salts give improved solubility in water over the sodium salts. Products contain-

100 HANDBOOK OF SURFACTANTS

ing ethylene oxide in the molecule have superior aqueous solubility compared with similar molecules without the ethylene oxide. The calcium salts and the amine/alkanolamine salts generally give better solubility in non-aqueous solvents. The magnesium salts, formerly hardly used at all, are now becoming more common.

Regarding solubility, carboxylic acid salts are more sensitive to low pH, polyvalent cations and inert electrolyte in the aqueous phase than the corresponding phosphates, sulphates or sulphonates.

Regarding chemical stability, sulphonates contain the C–S bond, which is more stable chemically than the C–O–S bond of the sulphates or the C–O–P bond of the phosphates. Both sulphates and phosphates are esters and are capable of being hydrolysed back to the free acid and alcohol. Thus, sulphonates and carboxylates are more stable to hydrolysis and extremes of pH than sulphates or phosphates.

The decreased use of the carboxylates (soaps) is due to the superior detergent performance of sulphonates and sulphates in hard water when properly formulated.

6.1 Soaps

Carboxylates are those surfactants where the hydrophilic (ionic) group is a carboxyl group, COOH. Soaps are the alkali metal salts of the carboxylic acids of fatty acids, i.e. $C_{17}H_{35}COONa$ is the sodium salt of stearic acid and is a soap.

The most important soap-forming fatty acids are, from a practical point of view: C12 (dodecyl) straight-chain saturated acid; C14 (myristyl) straight-chain saturated acid; C18 (stearic) straight-chain saturated acid; C18 (oleic) straight-chain, singly unsaturated acid. It is no coincidence that below C8 the products are very soluble in water, between C8 and C18 sparingly soluble and above C20 insoluble in water. Optimum surface-active properties are obtained with the sparingly soluble products. Most commercial soaps, such as the tablet of soap with which you wash your hands, will be a mixture of fatty acids obtained from tallow, coconut oil, palm oil, etc.

Soap has the following attractive features:

1. It is widely produced and used in very large volume
2. It is an excellent detergent
3. The raw materials are independent of the price and availability of petroleum
4. It biodegrades very readily
5. Its toxicology is well known.

Why then has it been replaced on such a large scale? The big disadvantage of soaps is their instability towards metal ions, particularly the calcium and magnesium salts found in hard water, and also the instability towards acids. In both cases the end result is the same; the soap comes out of aqueous solution because of the low aqueous solubility of the calcium and magnesium salt or of the free fatty acid. This shortcoming was possibly the most important single factor in spurring on the development of the newer synthetic surface-active agents.

It is well to bear in mind that in the absence of these conditions soaps can give excellent performance. Also, there has been considerable effort to use additives (known as lime soap dispersing agents) to reduce these adverse effects. Soap is of considerable interest particularly in countries with a plentiful supply of fats and oils and without petroleum.

Nomenclature

Generic:
  Soap
  Carboxylic acid salts
Examples
  Tallow soap — sodium salt of a mixture of carboxylic acids made from tallow
  Tall oil soap — sodium salt of a mixture of carboxylic acids made from tall oil

For the names of all the commonly occurring fatty acids see Appendix 1.

Description

Soaps are produced on a very large scale by a very small number of manufacturers (Procter and Gamble, Unilever, Henkel and Colgate) by the saponification of natural oils and fats. The most common carboxylates produced are:

- Tallow soaps, i.e. soaps produced from tallow (oleic 40–45%, palmitic 25–30%, stearic 15–20%)
- Coconut soaps, i.e. soaps produced from coconut oil (C12 48%, C14 17–20%, C16 8–10%, oleic 5–6%)
- Oleic soaps, i.e. soaps produced from olein
- Tall oil soaps, i.e. soaps produced from tall oil (a mixture of fatty acids and rosin acids from wood). Distilled tall oil (DTO) usually consists of 25–30% rosin acids, 70–75% fatty acids (composition: saturated 5%; oleic acid 25%; linoleic acid and other unsaturates 70%). Tall oil fatty acids are acids of similar composition but with rosin contents of 1–10%.

102   HANDBOOK OF SURFACTANTS

- Coconut acid (the starting material for many surfactants) has the following composition:

C12 (Lauric acid)	46–50%
C14 (Myristic)	17–20%
C16 (Palmitic acid)	8–10%
C8 (Caprylic acid)	small amounts
C10 (Capric acid)	small amounts
C18 (Oleic acid)	small amounts

  Stripped coconut fatty acids are often described as lauric acid.
- Naphthenates occur in crude petroleum and are complex mixtures of linear C6 and C7 acids plus alkyl and alkyl-carboxyl substituted cyclic pentanes. Products used as surfactants have mol. wt 250–350. Once widely used, but now being phased out of surfactant applications.

General properties

1. *Solubility.* C12 saturated soaps are soluble in water, C18 soaps are very slow to dissolve, and C16–C18 unsaturated soaps are soluble in water. The potassium salts of soaps are more soluble than the sodium (this is the opposite to sulphates, where the potassium salt is more insoluble). The alkanolamine salts (MEA, DEA and TEA) have better solubility. The TEA salts of lauric acid and oleic acid are milder than the sodium salts, are more soluble and have better foaming properties. Tall oil fatty acids (i.e. made with unsaturated acids) are more water soluble and give lower viscosity solutions than those from tallow. A number of liquid crystalline states exist in soap. At high soap/low water contents the crystalline phases play an important part in soap making and are also important in applications. The α, β, δ and ω phases have been identified and well described by Ferguson *et al.* (1943).
2. *Chemical properties.* A solution of a sodium salt of a C12–C18 fatty acid has pH = 9.5–10.3 (at 1% concn). Soaps are insoluble in aqueous solution below pH 7 due to the formation of the water-insoluble free fatty acid.
3. *Compatibility with aqueous ions.* All the soluble salts are readily insolubilised by electrolyte and 'salted out', e.g. by NaCl, etc. Soaps do not perform well in hard water (reduction in solubility and foaming ability). because of the insolubility of the calcium and other divalent and trivalent salts. They act as builders in conjunction with anionic/nonionics.
4. *Surface-active properties.* CMC C12 (saturated) sodium salt (mol. wt 222.3) = 2.6×10^{-2} M (0.57%); CMC C18 (saturated) sodium salt (mol. wt 306.5) = 1.8×10^{-3} M (0.055%); CMC C18 (unsaturated, oleate) sodium salt (mol. wt. 304.4) = 2×10^{-3} M (0.06%).

ANIONICS

5. *Functional properties*

- Foaming. Maximum foam production occurs with soaps made from C12–14 fatty acids. Sodium stearate gives rich creamy foam but oleate, laurates and tallates give more open foam; oleates give thick creamy foam but with less total foam, so a mixture of coconut and oleic gives copious foam with excellent stability (e.g. for shampoos).
- Defoaming. C18 soaps give defoaming of other surfactants in presence of calcium.
- Emulsifying properties. Soaps can be made *in situ* (e.g. for use as an emulsifier, by adding fatty acid to the oil phase and alkali to the aqueous phase).

Applications

1. *Personal care.* The main application is soap bars for washing; the standard bar is 80% beef tallow/20% nut oil (either coconut or palm kernel). Soap for washing in sea water and liquid soaps are made from sodium/potassium salts of coconut oil fatty acids (C12–C16).
2. *Household detergents.* Used in 1:1:1 sulphonate:soap:nonionic phosphate-free liquid heavy duty detergents, and as foam depressants in heavy duty detergents (C18–C22)
3. *Industrial laundries.* Soap can be used on its own due to most industrial laundries having water softeners. Blends with LABS and nonionics are also used with very similar formulations to those used in household heavy duty detergents.
4. *Pine oil disinfectants.* Castor oil soaps were used very widely but their use is now diminishing; a typical formula is: chlorinated phenols, 3–5%; pine oil, 5–10%; industrial alcohol, 10–20%; 25% castor oil Na salt, 20–25%; water, to 100%.
5. *Cosmetics and shampoos.* In soft water, soaps have most of the properties desirable in a shampoo, but to obtain clear soap solutions the pH must be alkaline. The alkalinity causes roughening of the scales of the hair cuticle, thereby giving a dull appearance. These disadvantages can be overcome by a mild acid rinse or by using the less alkaline alkanolamine salts. In hard water soaps also cause dullness by the deposition of calcium and magnesium soaps on the hair. This can be prevented by the addition of a lime soap dispersant or of sequestering agents for calcium and magnesium ions, e.g. EDTA or polyphosphates, but these agents have no effect on the alkalinity of the soap. Free fatty acids give conditioning in shampoos below pH 8. C18 is used as a thickener for surfactant solutions and as an opacifier.

104 HANDBOOK OF SURFACTANTS

6. *Textile industry*. Potassium oleate is used in gel foam for scouring carpet backing latexes.
7. *Paper industry*. Soap is used as a dispersing agent in paper coating mixtures.
8. *Emulsion polymerisation*. 20% Potassium oleate is used as the main emulsifier for crumb SBR rubber.
9. *Oil field chemicals*. Soap is used as an emulsifier for drilling muds.
10. *Polishes* Water-resistant film from aqueous solution — ammonia, morpholine or other volatile amine salts are used in polishes where the evaporation of the amine leaves only the free acid, which is then insoluble to water and remulsification is prevented. The same principle has been used in waterproofing textiles.

Specification

The values given in the table below give typical figures. The tests given are those used most frequently.

Test*	Saturated acid	Oleic acid
Odour	free from rancid odours	
Titre	54–67°C	flow point 5–7°C
Iodine value (unsaturation)	1–4	90
Fatty acid composition	GLC and HPLC on free acids and methyl esters	
Sap. value	197 for C18	198–205
Unsap. material	0.2–1.2	0.2–0.5

* The British Standards Institution publishes *Methods of Soap Analysis*, B.S. 1917.

6.2 Modified carboxylates

The modified carboxylates now commercially available with additional functional groups have improved solubility in hard water. The main ones available are:

1. Ethoxy carboxylates — addition of the polyoxyethylene chain (see section 6.2.1). The General structure is $RO(CH_2CH_2O)_n CH_2COO^-$
2. Ester carboxylates — addition of a hydroxyl, or multiple COOH groups (see Section 6.2.2)
3. Sarcosinates — addition of an amid group (see section 6.2.3). The products have the general structure $RCON(R')COO^-$
4. Half ester sulphosuccinates — addition of a sulphate group (see section 6.7.1)
5. Betaines — addition of an amine group (see section 9.4.1)

ANIONICS 105

6.2.1 Ethoxy carboxylates

Nomenclature

Generic
 Alkoxyalkanoic acids
 Alkyl (poly-1-oxapropene) oxaethane carboxylic acids
 Alkyl (e.g. lauryl) polyglycol ether carboxylic acids
 Alkylphenol polyglycol ether carboxylic acids
 Carboxymethylated alcohols
 Ethoxy carboxylates
 Ether carboxylates
 Polyalkoxylated ether glycollates
Example
 Lauryl alcohol + 3EO carboxylate, sodium salt: $C_{12}H_{25}O(CH_2CH_2O)_n$ $CH_2COO^- NA^+$

Description

These products are the ethers of glycollic acid $HO–CH_2–COOH$, prepared by the reaction of chloracetic acid with the corresponding ethoxylate.

$$RO(CH_2CH_2O)_nH + ClCH_2COOH \rightarrow$$
$$RO(CH_2CH_2O)_nCH_2COOH + HCl$$

A very large number of products are possible but, if prepared by the above route, they will all contain chloride ions, unless purified. The ethoxylate can have an alcohol, a nonyl phenol, a fatty amine or an acid as starting material, but the ethoxylated alcohols (as shown above) are the most common.

The reaction above does not go to completion, and therefore the starting nonionic can remain as an impurity. Salt is present, e.g NaCl in the sodium salts, unless it is deliberately removed.

General properties

1. *Solubility*. The addition of the ethoxylated grouping results in increased water-solubility in the products. HLB values vary from about 8, with excellent solubility in organic solvents, to 20, where the products are soluble only in polar solvents. Generally these are good lime soap dispersants and they have good alkali solubility.
2. *Chemical stability*. They do not hydrolyse in alkali or acid.
3. *Compatibility with aqueous ions*. They are stable to addition of electrolytes and offer an improvement of properties in hard water compared with soap.

106 HANDBOOK OF SURFACTANTS

4. *Compatibility with other surfactants*. They are compatible with nonionics, amphoterics, and some cationics. They are good lime soap dispersants.
5. *Surface-active properties*. The CMC of lauric acid + 10 EO methyl carboxylate is 0.012%. The CMC is higher for the salts. Also, the micelles formed in acid solution are smaller than for the salt. Thus, it is possible to adjust the micelle size by pH and choice of the salt, which is an important feature of emulsification. The surface tension increases with increasing pH. There is a moderate surface tension reduction — C12 acid + 2.5 EO methyl carboxylate, 33 mN/m at 1g/l.
6. *Functional properties* They retain the best properties of nonionic surfactants, i.e. wetting without the inverse solubility with temperature (no cloud point); good dispersing properties, (e.g. pigments in paint); good foam stability; good anti-corrosive action; very mild and non-aggressive action on hair and skin; they reduce the skin irritation of LES.

Applications

1. *Household products*. They are stable to hydrochloric acid and are used in toilet cleaners, as thickeners for bleach (high EO content), and as soap as additives to give liquid fatty acid soaps.
2. *Textiles*. Detergents (cloud point elevated to more than 100°C — see French Patent no. 848,529 Sandoz); kier boiling in caustic; dye bath auxiliaries; fabric softeners particularly for woollens (i.e. dialkyl adducts as base); spin finish antistatic agents, particularly if amine based.
3. *Shampoos*. Gerstein (1976) claims mild and some conditioning properties, particularly at low pH; they give creamy foam, particularly in hard water, and improve lubricity; they are used in conjunction with ether sulphates to improve foam and reduce irritation; they are compatible with cationics (including cationic polymers). The ether carboxylate can be used for the cleaning part of the formulation, with the cationics giving excellent conditioning.
4. *Paint*. Dispersing agents
5. *Aerosols*. Carpet shampoos.
6. *Metal treatments*. Boring and cutting oils, de-oiling baths.

Analysis

Solids content	90%
Active content	70–80%
Sodium chloride	1–15%
Sodium glycollate	No known values

ANIONICS   107

6.2.2 *Ester carboxylates*

Nomenclature

Generic
  Ester carboxylates
  Long chain esters of OH acids
Examples
  Sodium di(lauryl alcohol + 7EO) citrate (see Figure 6.1) or sodium
  dilaureth-7-citrate
  Sodium lauryl alcohol + 7EO tartrate (see Figure 6.1) or sodium
  laureth-7 tartrate

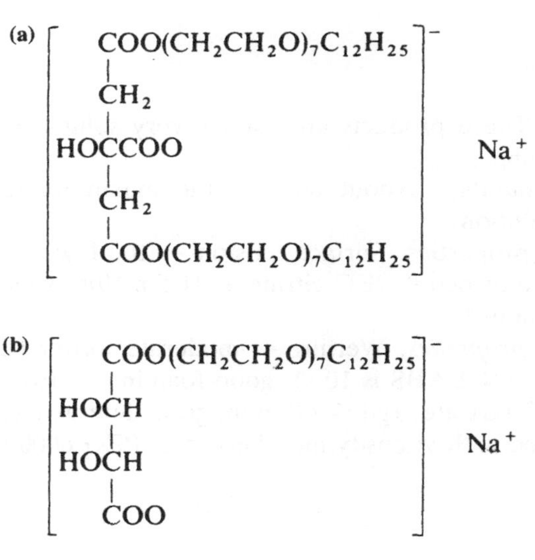

(a)
$$\left[\begin{array}{l} COO(CH_2CH_2O)_7C_{12}H_{25} \\ | \\ CH_2 \\ | \\ HOCCOO \\ | \\ CH_2 \\ | \\ COO(CH_2CH_2O)_7C_{12}H_{25} \end{array} \right]^{-} \quad Na^+$$

(b)
$$\left[\begin{array}{l} COO(CH_2CH_2O)_7C_{12}H_{25} \\ | \\ HOCH \\ | \\ HOCH \\ | \\ COO \end{array} \right]^{-} \quad Na^+$$

Figure 6.1 Examples of ester carboxylates. (a) Sodium di(lauryl alcohol + 7EO) citrate,
(b) sodium lauryl alcohol + 7EO tartrate.

Description

These surfactants are produced by the reaction of a long-chain fatty alcohol
(or ethoxylate) with a multifunctional carboxylic acid (esterification). The
multifunctional fatty acid usually contains hydroxyl groups (see Figure
6.2), These products do not contain large amounts of chloride ions as the
ethoxy carboxylates do (section 6.1.2), but they do have the disadvantage
of containing an ester group, R–CO–O–, which can hydrolyse in solution,
whereas products such as the ethoxy carboxylates have ether groups, which
are stable to hydrolysis. Typical products would be made with citric or

108 HANDBOOK OF SURFACTANTS

tartaric acid as the multifunctional carboxylic acid and lauryl alcohol + 5–7 EO.

$$R-OH + HO-X-(COOH)_n \longrightarrow HO-X \begin{array}{c} \diagup (COOH)_{n-1} \\ \diagdown COOR \end{array} + H_2O$$

Figure 6.2 Preparation of ester carboxylates.

General properties

1. *Solubility*. These products are usually very soluble in water in the C12–C16 range.
2. *Chemical stability*. About 80°C is the maximum temperature in aqueous solution.
3. Surfactant properties. Surface tension — the lowest value is found with sodium dicoco + 7EO citrate at 31.5 mN/m, which is similar to ether sulphates.
4. *Functional properties*. Wetting — moderate wetter (Draves 2g/litre 50–110 s — (NaLABS is 10 s); good foaming — Ross Miles for Na C12 + 7EO tartrate, 1g/l = 145 mm; good detergency; solutions can be thickened with viscosity modifiers, e.g. PEG 6000 distearate.

Applications

1. *Cosmetics*. Sodium lauryl alcohol + 7EO tartrate or sodium di(lauryl alcohol + 7EO) citrate can be used as a shampoo detergent, but gives a more economical formulation mixed with AES. Such mixtures show considerably reduced skin irritation with AES when only small quantities of the citrate are added.
2. *Liquid detergents*.
3. *Institutional surface cleaners*.

Specification

Activity	25° aqueous soln typical
Alkalinity value (mg KOH/g)	10–30
Sap. value (mg KOH/g)	15–18

ANIONICS 109

6.2.3 Amide carboxylates

Nomenclature

Generic
 Sarcosinates
 N-acyl derivatives of sarcosine (*N*-methylglycine CH_3NHCH_2COOH)
 N-alkylphthalamates
Examples
 Lauryl sarcosine, $C_{12}H_{25}CON(CH_3)CH_2COOH$
 Sodium *N*-octylphthalamate, see Figure 6.3

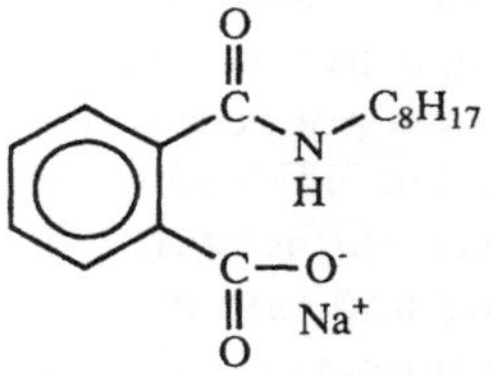

Figure 6.3 Sodium *N*-octylphthalamate.

Description

The sarcosinates are made from a fatty acid chloride and *N*-methylglycine, catalysed by alkali:

$$RCOCl + CH_3NHCH_2COONa \rightarrow RCON(CH_3)CH_2COONa + NaCl$$

The product as made contains sodium chloride. Some commercial products contain considerable amounts of sodium chloride. The free acid may be separated, and is free of sodium chloride; this can then be neutralised with various alkalis to give salts free of sodium chloride.

The *N*-alkylphthalamates are prepared from the reaction between phthalic anhydride and a primary amine, as shown in Figure 6.4.

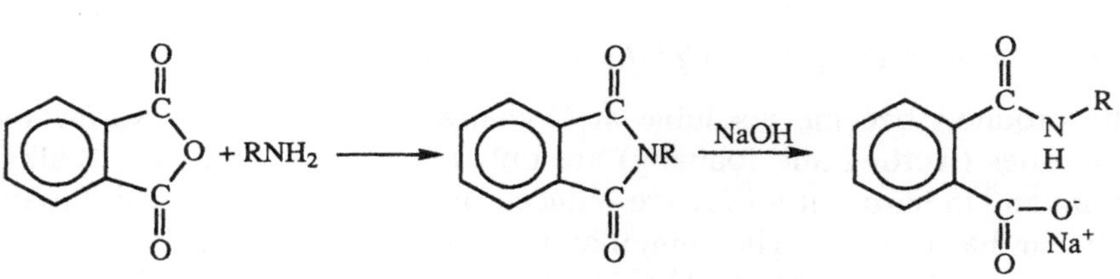

Figure 6.4 Preparation of *N*-alkylphthalamates.

110 HANDBOOK OF SURFACTANTS

From the structure, the product is a soap but with an amide group in the molecule. Unlike the sarcosinates, these products do not contain sodium chloride. The amide group in the sarcosinates is separated from the carboxyl by only one methylene group, whilst in the phthalamates, there is the aromatic ring between.

At the present time there are not many products of this type on the market. Stepan are offering the di(hydrogenated) tallow derivative.

General properties of sarcosinates

1. *General.* These compounds are more acidic than fatty acids and therefore there is less need to avoid free acid and pH control is easier. The properties are similar to those of isethionates.
2. *Solubility.* The sodium salt not very soluble at acid or neutral pH but quite soluble in alkali. The TEA salt is more soluble at neutral pH.
3. *Compatibility with other surfactants.* They give good compatibility with anionics, nonionics and cationics, where they do not adversely affect the bactericidal properties; they are compatible with quaternaries and phenolic biocides.
4. *Compatibility with aqueous ions.* Excellent, e.g. good foam with TEA salt in hard water. The surfactant properties vary with pH. Below pH5, most sarcosinates are converted to the acid form and separate from solution.
5. *Chemical stability.* They are stable in acid and moderate acid at normal temperatures, but unstable in strong acid or acid at high temperatures, when they lose foaming characteristics and thicken appreciably.
6. *Surface active properties.* Sodium lauryl sarcosinate has a CMC of 0.08% with a minimum surface tension of 24.3 mN/m. Sodium cocoyl sarcosinate has a CMC of 0.009% and a minimum surface tension of 22.7 mN/m. Above the CMC, the surface tension is dependent upon pH and decreases to a minimum between pH 6 and 7.
7. *Functional properties.* They foam well even in the presence of sebum and oils; they foam better in hard water than in soft water; they are good lime soap dispersants; some bacteriostatic activity is claimed.

General properties of N-alkylphthalamates   (Data from Bernhardt, 1992)

The products become insoluble at C chains above C12. The surfactant properties (wetting and foaming) are only apparent at around C12 alkyl chain, and in hard water they are reduced (unlike sarcosinates which foam better in hard water). The ammonium salt is unstable to heat and reverts back to the phthalimide at 100°C in 20 min, while the sodium salt is stable at 120°C for 28 h. This could be the basis for a labile surfactant.

Alkyl-group (R)	Solubility (1%)	Draves wetting (s)		Ross Miles Foam (cm of a 0.1% solution)	
	Distilled water	Distilled water	Hard water	Distilled water	Hard water
C8H17	Soluble	>120	>120	0	0
C12H25	Soluble	6.7	>120	14	6
C16H33	Insoluble	>120	>120	6.5	0
C18H37	Insoluble	>120	>120	4	0

Applications of sarcosinates

1. *Additives to anionics or nonionics.* Addition of sarcosinates to anionics depresses the Krafft point of the anionic (1% addition of sodium lauroyl sarcosinate to SDS depresses the Krafft point from 17°C to 0°C). Addition of sarcosinates to nonionics elevates the cloud point. Addition of sarcosinates to amine oxides of similar chain length depresses the surface tension.
2. *Household products.* Toothpaste ingredients; enzyme inhibiting, strong foamer, good detergents; liquid soaps.
3. *Personal care.* They are foaming agent in shampoos, they boost lather of alkyl sulphates in presence of sebum and give good foaming in the presence of soaps; detoxifying agents in shampoos; addition to anionics improves mildness, conditioning and foaming; shaving preparations; foam baths; facial cleaners.
4. *Surgical scrubs.*
5. *Corrosion inhibitors.*

Applications of N-alkylphthalamates

The only product commercially available, the di(hydrogenated tallow) *N*-phthalamate, is used as an emulsifier and suspending agent for conditioners and an anti-dandruff agent in shampoos.

Specifications of sarcosinates

Activity	30% for Na salt, 40% for TEA salt
Acid value	50–60 mg KOH/g
Sodium chloride	1–5%

6.3 Isethionates (ester sulphonates)

Nomenclature

Generic
   Acyl oxyalkane sulphonates

112 HANDBOOK OF SURFACTANTS

Esters of isethionic acid, $HOCH_2CH_2SO_3H$. Note: the taurates (see section 6.8) are amides of methyl taurine, $CH_3NHCH_2CH_2SO_3$.
Isethionates
Sulphoalkyl esters
Example
Coconut fatty acid, 2-sulphoethyl 1-ester, sodium salt, formula Coco-$COOCH_2\ CH_2SO_3^-Na^+$

Description

Preparation is by reaction of acid chloride (of the fatty acid) with sodium isethionate:

$$RCOCl + HOCH_2CH_2SO_3Na \rightarrow RCOOCH_2CH_2SO_3Na + HCl$$

The sodium isethionate is prepared from ethylene oxide and sodium sulphite. Some typical impurities are: salt (8–10%, but some products very low; <1%); soap; fatty acid; unreacted sodium isethionate.

General properties

1. *General*. The esters were originally known as Igepon A (now Hostapon — a Hoechst trade name) and have properties similar to those of alkyl sulphates with similar chain length, although the foaming properties may be slightly inferior.
2. *Solubility*. The sodium salt of C12–14 is soluble in hot water (50% soln at 70°C) but has very low solubility in cold water (0.01% at 25°C); the sodium salt of oleic acid has 11% solubility at 70°C, 2.5% at 25°C.
3. *Compatibility with aqueous ions*. They are practically unaffected by calcium ions; 50% mixtures with soap considerably reduces scum and precipitation in hard water.
4. *Chemical stability*. They are suitable for use at pH 6–8 but hydrolyse outside this range in hot water. This limits their usefulness to solids and powders.
5. *Surface-active properties*. Coconut derivative: CMC 0.06%, surface tension at 0.1% 27 mN/m. Oleic derivative: surface tension at 0.1% 28 mN/m.
6. *Functional properties*. Good foaming properties (0.05% Ross Miles initial foam coco derivative 93 mm, oleic derivative 145) but not as good as alkyl sulphates; gives excellent foaming properties when used as a mixture with soap; excellent detergents for grease and oil; good lime soap dispersants (coco derivative 15–20 index, oleic acid derivative 15–20 index).

ANIONICS

7. *Disadvantages*. Unstable in aqueous solution at high temperatures and high and low pH.

Applications

1. *Household products*. Synthetic bar soap, excellent in removing grease and oily dirt and good foam in hard water. Typical formula: 78% active powder (coco derivative), 20 parts; water, 12 parts; tallow soap/coco soap (80/20), 68 parts.
2. *Cosmetics*. Similar formulations to the synthetic bar soap but with higher quantities of isethionate are used as 'moisturising cream' in the USA but as medicated skin-care formulations in Europe.

Specification

Appearance	powder
Active	70–80
Sodium chloride (%)	1–10
Free acid (%)	1–10

6.4 Phosphate esters

Nomenclature

Abbreviations
  PE is used in this book to signify phosphate esters.
Generic
  Alkyl acid phosphates
  Alkyl ether phosphates
  Alkyl phosphates
  Dialkyl pyrophosphates
  Monoalkyl phosphates
  Phosphate esters
  Phosphated alcohols (or ethoxylated alcohols)
Example
  Lauryl alcohol + 7EO phosphate or lauryl polyethyleneglycol phosphate.

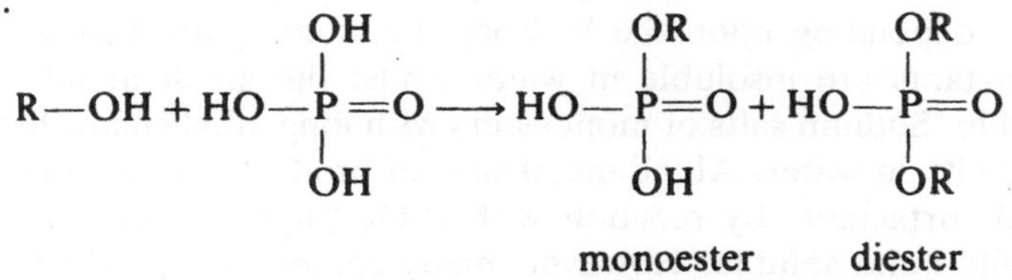

$$R-OH + HO-\underset{\underset{OH}{|}}{\overset{\overset{OH}{|}}{P}}=O \longrightarrow HO-\underset{\underset{OH}{|}}{\overset{\overset{OR}{|}}{P}}=O + HO-\underset{\underset{OR}{|}}{\overset{\overset{OR}{|}}{P}}=O$$

monoester   diester

Figure 6.5 Reaction between an alcohol and phosphoric acid.

Description

The reaction between an alcohol group and phosphoric acid is shown in Figure 6.5. This equation is an oversimplification, as a mixture of mono, di and tri esters plus polymeric esters are formed, together with some of the original phosphoric acid.

There are two phosphorylating agents in use, tetraphosphoric acid (TPA) and phosphorus pentoxide (P_2O_5). The reactions involving TPA, are now well understood and the major product is the monoester. The reactions involving P_2O_5 are more complex, but the major product is diester.

The majority of commercial products are made with P_2O_5 and are therefore predominantly diesters, but with significant amounts of mono-ester and polymeric esters present.

General properties

1. *General.* The phosphate esters have properties intermediate between the ethoxylated nonionics and the sulphated derivatives. Thus, they have good compatibility with inorganic builders, good emulsifying properties and a foaming capacity intermediate between the low-foaming nonionics and the high-foaming ether sulphates. Properties compared to the parent ethoxylate after reaction with TPA are shown in Table 6.1.

Table 6.1 Comparison of alcohol ethoxylates and their phosphated derivatives

	C12 alc.+7EO	C12 alc.+7EO+TPA
In 5% or 10% NaOH	Insoluble	Soluble
ST of 0.1% soln at 25°C (mN/m)	30.3	37
ST of 0.1% soln at 60°C (mN/m)	33.7	34.2
Draves wetting times at 25°C (s)	7.5	26
Draves wetting times at 60°C (s)	31.5	21.5
Ross Miles initial foam (mm)	104	152
Ross Miles foam (mm) after 5 min	20	133

2. *Solubility.* The free acids have good solubility in both water and organic solvents with acidity comparable to phosphoric acid. However, depending upon the hydrophobe, some phosphated nonionic surfactants are insoluble in water whilst the alkali metal salts are soluble. Sodium salts of monoesters with long alkyl chains have poor solubility in water. Alkali metal salts of products made from ethoxylated surfactants by reaction with P_2O_5 (high diester content) are usually quite soluble. However, many commercial products are the potassium salts, which have better solubility than the sodium salts.

ANIONICS 115

3. *Chemical stability*. They are resistant to hydrolysis by hot alkali, and the colour is not affected; they are stable to acid solution and stable to high temperatures.

4. *Compatibility with aqueous ions*. TPA phosphation gives products with better electrolytic compatibility than P_2O_5; the products can sequester iron and other metal ions. Polyoxyethylation gives good resistance to hard water and concentrated electrolyte.

5. *Surface-active properties*. Surface tension of the diesters decreases with increasing alkyl chain length. Branching of the alkyl chain gives lower values than straight-chain alkyl groups. Lauryl alcohol + 7EO phosphated (with TPA) gave a surface tension of 37 mN/m of a 0.1% solution at 25°C.

6. *Functional properties*. Efficient oil emulsification (better with P_2O_5 products); good detergency on fibres; diesters have good wetting properties with low molecular weight giving the best wetting properties; excellent rinsability; lower foam than the corresponding sulphate or sulphonate; corrosion inhibition; antistatic properties; good dispersing agents (better with P_2O_5 products); hydrotrope properties (particularly with short alkyl chains)

7. *Disadvantages*. Only moderate reduction in surface tension of aqueous solution; more expensive than sulphonates; sodium salts not always water soluble, so the more expensive potassium and alkanolamine salts are needed for aqueous solubility of a neutral phosphate ester. The compositions of most phosphate esters are complex and difficult to analyse and therefore the compositions of most commercial products are very ill defined and product replacement is difficult.

Applications

1. *Textiles*. Lubricant and/or emulsifier with anticorrosive properties; wetting agent in the presence of alkali, e.g. kier boiling (nonyl phenol + 13 EO +P_2O_5); dye bath additive for even dying and better penetration; short-chain alkyl esters are efficient antistatic agents in spin finishes for synthetic fibres (C8 alcohol + P_2O_5).

2. *Detergents*. Solubilisation of nonionics in high concentrations of electrolyte; dry-cleaning soaps for antistatic effect.

3. *Agriculture*. Emulsifying agents in herbicides, especially when blended with concentrated liquid fertilisers; dispersing agents for aqueous dispersions of insoluble herbicides or insecticides, to give better storage stability of the concentrated dispersion plus better flow properties.

4. *Lubricant additive and metal working*. C12 alcohol + 6 EO + P_2O_5 gives a good emulsifier for mineral oil but also good load-carrying properties.

116   HANDBOOK OF SURFACTANTS

5. *Antifoams*. C18 alkyl monoester products are good antifoams for anionic surfactants.

Specification

Specification is difficult, but very essential as the product being used is a mixture. A test of surfactant properties is often the best method of quality control. However, most of the chemical tests are adequate for showing batch to batch variation (see Table 6.2).

Table 6.2 Chemical tests for phosphate esters

% active	100% for acids; 30–45% for potassium salts
mg KOH/g to pH 5.2	85–110 an indication of high monoester concentration
mg KOH/g to pH 9.3	160–220 an indication of high diester concentration

Titration with KOH to two particular pH points is more useful than a normal acid number.

6.5 Sulphates

Sulphates were the largest and most important class of synthetic surfactants but, during 1970–90, were replaced by the sulphonates in terms of volume consumption. However, the alkyl sulphates are now being increasingly used in domestic detergents and consumption is increasing again.

Organic sulphates are the esters of sulphuric acid:

$$ROH + H_2SO_4 \rightarrow ROSO_3H$$

The sulphur atom is joined to the carbon atom of the hydrophobic chain via an oxygen atom. The acid ester is unstable and can revert back readily to the alcohol and sulphuric acid (particularly in acid conditions), whereas the neutralised salts are stable at neutral pH. In the case of a sulphonate (see section 6.6), the sulphur is joined directly to the carbon chain of the hydrophobe.

In the manufacture of sulphates the neutralisation must be carried out quickly to avoid the breakdown of the acid ester. In practice, sulphuric acid is very seldom used and chlorsulphonic acid or sulphur trioxide/air mixtures (in continuous reactors) tend to be the most common methods of sulphating alcohols. A fuller description is given in section 6.6, Sulphonation. The neutralisation is usually carried out continuously with the sulphation.

Commercial products available with the R group on the alcohol can be:

1. A saturated linear hydrocarbon from natural or synthetic (oxo or Ziegler alcohols) sources, $CH_3(CH_2)_nOH$, with n usually in the range

ANIONICS   117

8–13 and usually linear. The products are then known as the fatty alcohol sulphates, e.g. sodium dodecyl sulphate from dodecanol (see section 6.5.1).

2. Unsaturated alcohols, although sulphation/sulphonation can take place on the unsaturated bond.
3. A long-chain alcohol with ethylene oxide, $RO(CH_2CH_2O)_nH$, with R = C12–C15, and either natural or synthetic, but usually linear and with n commonly 2 or 3 but sometimes up to 10. The products are commonly known as ether sulphates, e.g. sodium dodecyl ether sulphate (see section 6.5.2).
4. An alkylated phenol ethoxylate, $RC_6H_4O(CH_2CH_2O)_nH$, with R commonly C9 based on tripropylene (nonyl) and therefore branched with n commonly 4–20. The products are commonly known as nonyl phenol ether sulphates. (see section 6.5.5).
5. A monoethanolamide ethoxylate, $RCONHCH_2CH_2O(CH_2CH_2O)_nH$, with R commonly derived from a natural oil (e.g. coconut oil) with n commonly in the region 3–6. The products are known as fatty acid alkanolamide ether sulphates (see section 6.5.3).

The properties of the sulphates depend upon the properties of the hydrocarbon chain and those of the sulphate group. The properties of the hydrocarbon chain are common to those of all other surfactants.

It is worth mentioning the secondary alkyl sulphates which were produced by reacting oleum with alpha-olefins from cracked waxes. These products had typical detergent sulphate properties but tended to have disagreeable odours and dark colours. These products are no longer available in Western Europe, but they are relatively easy to make with simple equipment and may be available in other parts of the world.

Properties of the sulphate group

1. *Solubility.* The alkali metal salts show good solubility in water.
2. *Physical properties of aqueous solutions.* Most surfactant sulphates show what is known as the 'salt effect'. The addition of inorganic salts (sodium chloride, sodium sulphate) to dilute (below about 30%) solutions increases the viscosity, while addition to concentrated solutions decreases the viscosity. Any salts formed during the sulphation and/or neutralisation can lead to viscosity changes. Sources of salts are moisture during sulphation, hydrolysis of the acid prior to and during neutralisation, and the addition of hypochlorite bleach (by the surfactant manufacturer).
3. *Chemical stability.* Sulphates are easily hydrolysed particularly by acids below pH 3.5. The acid hydrolysis releases sulphuric acid which catalyses the hydrolysis:

118 HANDBOOK OF SURFACTANTS

Acid conditions: $ROSO_3H + H_2O \leftrightarrow ROH + H_2SO_4$
Alkaline conditions: $ROSO_3Na + NaOH \leftrightarrow ROH + Na_2SO_4$

The alcohol sulphates are therefore more stable in alkali than in acid. The stability depends upon a number of factors, but for aqueous formulations to give long-term shelf stability at room temperature, the pH range 5–9.5 is preferred. At elevated temperatures ($<50°C$) sulphates are unstable. The ammonium salts on the alkaline side pH ($>$pH 7) give off ammonia, which can be irritating in household products or give odours in industrial products. The ammonium salts are also slightly superior to the sodium or triethanolamine salts with regard to resistance to acid hydrolysis.

6.5.1 Alcohol sulphates

Nomenclature

Abbreviations
AS, alcohol sulphates, used in this book
ALS, ammonium lauryl sulphate
FAS, fatty alcohol sulphate
LAS, sometimes used but it is more often used for linear alkyl benzene sulphonates or linear alkane sulphonates (paraffin sulphonates) and thus can lead to confusion
SDS, sodium dodecyl sulphate

Generic
Alcohol sulphates
Alkyl sulphates
Fatty alcohol sulphates
Example
Sodium lauryl sulphate, $C_{12}H_{25}SO_4Na$

Description

Alcohol sulphates are now generally made from primary linear alcohols, which can be natural or synthetic. Alcohol sulphates are not stable as the free acids, so only the salts are commercially available. The most common form is the sodium salt. Amine salts based on mono-, di- or triethanolamine and ammonia are also offered by most manufacturers; these are usually 30–40% active. Solutions of 60–70% can be made and are sometimes offered by manufacturers. Alcohol sulphates are now made by sulphonation of the alcohol with sulphur trioxide (see section 6.6)

General properties

1. *General*. There has been more scientific work published on AS (principally SDS) than on any other type of surfactant. The shape and

ANIONICS

size of micelles in aqueous solution affect the solution properties. A summary of the present state of knowledge of AS micelles can be found in chapter 4.

2. *Solubility*. The common sodium salts of C12–C14 (dodecyl) alcohol sulphates give gels at active concentrations above 30%. Below 30% active they will be liquid at normal temperatures but set to a soft paste when the temperature falls below 25°C (the Krafft point; see chapter 4). Amine salts (MEA, DEA or TEA) improve solubility; DEA salts are less soluble than TEA salts but give a lower-viscosity solution. The Krafft point depends upon the distribution of carbon chains in the parent alcohol and a wide distribution gives a lower Krafft point (around 10°C), giving flowable solutions at room temperature. The solutions of fatty alcohol sulphates show cloud points below room temperature, which can vary from one manufacturer to another. The main reason for the variation is the amount of electrolyte and free alcohol remaining in the product.

3. *Chemical stability*. Both primary and secondary alcohol sulphate salts are unstable to acid, as described under Sulphates (see section 6.5). The amine salts darken on storage, particularly on exposure to light.

4. *Viscosity of aqueous solutions*. The viscosity increases very rapidly around 30–40% to give a gel, but then falls at about 60–70% to give a pourable liquid and then increases again to a gel. The concentration at which the minimum occurs varies according to the alcohol sulphate used and also the presence of impurities, e.g. unsulphated alcohol. The position of the minimum can also be affected by temperature. Viscosity of aqueous solutions can be reduced by the addition of short-chain alcohols and glycols. They are easily thickened with alkanolamides (and salt). The viscosity can be increased by addition of electrolyte (e.g. increased by Cl^- and SO_4^{2-}). The MEA salt is more sensitive to added inorganic salt (particularly the Cl^-) ions than the TEA salt. The effect of salt on the viscosity is also dependent upon the concentration of unreacted fatty alcohol in the alcohol sulphate. To thicken the TEA salt, use alkanolamides or cocoamidopropyl betaine.

5. *Compatibility with aqueous ions*. Stable to hard water at low alkyl chain lengths (C10), but sensitivity to hard water increases with increasing chain length. Magnesium salts have improved stability over the sodium salts in hard water and have higher alkali tolerance.

6. *Surface-active properties*. CMC of sodium dodecyl sulphate (mol. w 288.3) = 8×10^{-3} M (0.24%). CMC of sodium octadecyl sulphate (mol. w 372.5) = 2×10^{-4} M (0.007%). Lowest surface tension obtainable with broad-cut C12 alkyl sulphate sodium salt at 1% is about 33–35 mN/m.

7. *Functional properties*. 1. Foaming properties. Foam volume and

stability increase in hard water compared to soft water; Optimum foaming is with C12–C14 mixture (especially if some free alcohol is left) in hard water for quantity and quality (forms small bubbles and a rich creamy foam). C8–C10 are foam depressants; C10 increases the flash foam; C14 gives lower volume and less stable foam than C12–C14 mixture; C16–C18 less foam. 2. *Wetting properties.* Good wetting agents, but sulphation along the chain (secondary alcohol) rather than at the end of the chain will give a smaller branched molecule (US Patents 2,422,613 and 2,423,692) with better wetting but reduced detergency. Wetting also improves as the chain length of the hydrophobe is reduced. 3. Excellent emulsifiers (particularly for sebum). 4. Excellent detergency in mixtures (see later)

8. *Disadvantages.* Poor hydrolytic stability.

Applications

1. *Household detergents.* Tallow alcohol sulphate can be substituted for LABS in heavy duty detergents to give lower foaming products but with slightly superior detergency. Tallow is:

	Beef	Mutton
Oleic acid (%)	40–50	35–40
Palmitic acid (%)	25–35	25–35
Stearic acid (%)	19	30

Heavy duty powdered detergents may contain AS (usually C16–C18 as the alkyl chain length) in conjunction with, or replacement of, LABS. They are also used in heavy-duty liquids in conjunction with nonionics.

2. *Cosmetics and shampoos.* DEA lauryl sulphate is often used with lauroaminopropionates (amphoterics, see chapter 9) to improve detergency in shampoos. SDS was once the main surfactant for shampoos, but it had poor solubility (due to inorganic content from chlorsulphonic acid manufacture) and needed alkanolamides (coco-monoethanolamide at 10–15%) or glycols to give clear solutions. It was, however, excellent for creams and pastes. Modern manufacturing methods (SO_3) give AS with low amounts of electrolyte and low alcohol, which give a low cloud point. For pastes or creams it may be necessary to add alcohol or salt to give the low solubility required. Alternatively, addition of some C16–C18 alcohol will thicken/opacify. The TEA salt gives clear products due to better solubility. The addition of sarcosinate has been recommended with TEA

ANIONICS   121

sulphate to improve lather. TEA sulphate has lost ground to the ammonium salt (which is cheaper and avoids the nitroso problem), amphoterics or monoester sulphosuccinates. At the present time, ammonium lauryl sulphate is probably used in more shampoos than any other anionic surfactant. The ammonium salt is claimed to be less irritant than the sodium salt. Optimum alcohol blend for shampoos is: 70–75% C12, 25–30% C14.

3. *Textiles*. Low temperature detergents for delicate fabrics; dye retarder when amine groups on the fibre; dyestuff dispersant in aqueous media.

Specification

Active material, usually 30–40% for aqueous solutions
Mean molecular weight, 290–310 for sodium salt of C12–14 alcohol
Carbon chain distribution, sometimes specified by user.
Free MEA, DEA or TEA, <3%
Unsulphated matter, up to 5% but most products now <1%
Sodium sulphate, up to 2%
Sodium chloride, should below 0.1% for SO_3 sulphated products
Cloud point (Krafft point), 10–25°C
Viscosity can vary over a very wide range

6.5.2   Alcohol ether sulphates

Nomenclature

Abbreviation
    AES, alcohol ether sulphates, used in this book

Generic
    Alcohol ether sulphates
    Ether sulphates
    Ethoxy sulphates
    Polyoxyethylene alcohol sulphates
        Sulphated polyoxyethylated aliphatic alcohols
Examples
    Sodium C12–C14 3 mole ether sulphate — this shorthand description means that a primary alcohol consisting predominantly of C12–14 hydrocarbon chains has been reacted with 3 moles of ethylene oxide, then sulphated and then neutralised with sodium hydroxide
    C12–C14 alcohol + 3EO sulphate sodium salt — same as the above
    Sodium laureth-3-sulphate — C12 alcohol +3EO sulphate sodium salt
    Sodium dodecane/oxyethylene/3 sulphate (nomenclature used by US

122 HANDBOOK OF SURFACTANTS

National Bureau of Standards) — this is the same as C12 alcohol + 3EO sulphate sodium salt
Sodium lauryl ether sulphate (the amount of ethylene oxide is not defined) — very often used to describe sodium C12–C14 alcohol + 3EO sulphate, particularly in dishwashing detergents, shampoo and cosmetic formulations

Description

These products are made by the sulphation of ethoxylated alcohols using sulphur trioxide/air in continuous plants (see section 6.6). It is possible to make AES in batch reactors with sulphur trioxide injection, but colours are generally very poor. The generic formula is $RO(CH_2CH_2O)_nSO_4$, where R is a hydrocarbon radical, usually linear, and n is 2 or 3, although there are a few speciality products with n in the range 4–10. Recently (1991–92), products with a narrow EO range are being produced (see section 7.1.1).

The sodium, magnesium, or ammonium salts are readily available from a large number of manufacturers where n is 2 or 3. Practically all products are based on natural or synthetic alcohols with 10–15 carbon atoms. The ranges available are: 27–29% active, aqueous, based on the sodium or ammonium salts; 60% active, aqueous/alcoholic, based on the ammonium salts; 65–70% active, aqueous, based on the sodium salts.

The hydrophobic alcohols used are similar to those described under alkyl sulphates (see section 6.5.1), but oxo alcohols with odd numbered carbon chains are used to a greater degree than with alcohol sulphates. The free acid is unstable (similar to the alcohol sulphate; see section 6.5.1) and must be neutralised as soon as it is made; 90% + active blends can be made by neutralising the ether sulphate acid with a 2:1 dialkanolamide. The highly active (65–70%) compounds have several advantages. Reduction in transport and storage costs are fairly obvious, but highly active materials exhibit increased resistance to microbiological attack even if unpreserved. Also, the level of impurities tends to be somewhat lower in the highly active materials. Minor components and impurities include the following. (i) Alcohol sulphate. This is formed from the free alcohol which remains after the ethoxylation. Ethoxylation of alcohols under normal alkaline catalysis gives a broad range of ethoxylates (see chapter 7). In the case of the 2-mole ethoxylate, there can be as much as 20–25% of free alcohol remaining, which on sulphation gives an alcohol sulphate. Narrow-range AE used for sulphation may give lower alcohol sulphate levels and, in the case of the 'reduced acetal' method (Blease et al. 1992) very low levels. (ii) Alcohol ethoxylate. Unsulphated material, when sulphation is carried out in a manner to avoid formation of 1,4-dioxane. (iii) 1,4-Dioxane is formed by the breaking of the ethoxylate chain under the conditions of sulphation. The amount is relatively small (10–100 ppm on active solids), but there has been considerable concern in Europe, mainly due to the low level required in the EEC Cosmetic Directory. The amount of 1,4-dioxane found is mainly due to the sulphation conditions, oversulphation giving higher levels. The amount of ethylene oxide in the molecule is also a critical factor, and high EO (>4 mole) can contain considerable amounts of 1,4-dioxane. High levels of 1,4-dioxane can be reduced by steam distillation.

General properties

1. *Solubility*. The presence of the polyethylene oxide (a water-soluble group) confers improved solubility on ether sulphates compared to alcohol sulphates. When mixed with alcohol sulphates, AES improve the solubility. Unlike alkyl sulphates, the ether sulphates have cloud and clear points below 0°C (see Table 6.3).

Table 6.3 Krafft points of pure $C16(EO)_n SO_4 Na$ salt

Moles of EO (n)	Krafft point (°C)
1	36
2	23
3	17

2. *Compatibility with aqueous ions*. They are stable to hard water; both the 2-mole and 3-mole ethoxy sulphates have excellent lime soap dispersing properties, due to the calcium salt being soluble; (calcium lauryl sulphate is insoluble).
3. *Chemical stability*. They hydrolyse in aqueous solution at acid pH and high temperatures (not suitable for use above 50°C), but less so than alcohol sulphates; ammonium salts are more stable than sodium at pH just less than 7 but need to be kept at below pH 7 at all times during processing. Even at neutral pH, hydrolysis may be inititiated by autocatalytic acidification. This can be prevented using phosphate or citrate buffers.
4. *Viscosity behaviour of aqueous solution*. They are very similar to alcohol sulphates in giving gels in the 30–60% concentration range, whilst being liquid above and below that concentration. Thus, diluting a concentrated solution of say 65% can first lead to a very large increase in viscosity and lead to gels. Where minimum formulation viscosity is required at maximum activity, then selection of a higher degree of ethoxylation is preferable.
5. *Thickening aqueous solutions — the 'salt effect'*. Ether sulphates show a pronounced 'salt effect', i.e. addition of sodium chloride to a

124 HANDBOOK OF SURFACTANTS

dilute solution will give an increase in viscosity. The quantitative increase in viscosity is different for each ether sulphate and can even vary from batch to batch of apparently the same product. The largest changes in viscosity on salt addition are shown by ether sulphates with low ethylene oxide content, i.e. the 2-mole ethylene oxide derivative. Ether sulphates can also be thickened with dialkanolamides, betaines, poylethylene glycol esters and amine oxides (see relevant sections). However, the foaming performance can be affected by such additives, the PEG esters tending to give lower foaming performance. If narrow-range AE (see section 7.1.2) is used for the sulphation the subsequent properties differ as there is little or no sulphate present. Blease *et al.* (1992) showed less thickening with salt in a shampoo formulation.

6. *Thinning aqueous solutions.* Alcohols (normally ethanol is used) can reduce the viscosity of ether sulphate solutions. Isopropanol is very effective but can give odours in the final products. Narrow-range AE used for sulphation require less hydrotrope in a dishwashing formulation (Blease *et al.* 1992).

7. *Surface-active properties.* CMC of sodium dodecyl alcohol + 2EO sulphate (mol. wt 376.5) = 3×10^{-3} M (0.11%); CMC of sodium dodecyl alcohol + 3EO sulphate (mol. wt 420.5) has figures quoted in the literature from 0.008% to 0.03%.
   Surface tension of C13–15 + 3EO sulphate, sodium salt: 0.01%, 40.0 mN/m (data from ICI Synperonic data sheet), 0.1%, 36 mN/m (data from ICI Synperonic data sheet).

8. *Functional properties.*
   Excellent detergents, do not show loss of detergency in brine or hard water (unlike AS or LABS) and therefore can be used 'unbuilt' (no polyphosphates); detergency is enhanced by the addition of magnesium salts; the good solubility characteristics make them especially useful for use in alkaline and phosphate-built liquid formulations, where they can improve the aqueous solubility of other less polar surfactants. Excellent emulsifiers (particularly for sebum). Excellent foaming agents, particularly in the presence of electrolytes; compared with alkyl sulphates, the foam is lighter and more open, and collapses more readily in the presence of grease and oils. In general ether sulphates do not give stable foams on their own in the presence of sebum (unlike alkyl sulphates, which do), but the addition of sarcosinates, betaines or alkanolamides can stabilise the foam in the presence of sebum. The C12 hydrophobe is claimed to give better foaming properties than the C14 (see Adam and Neumann, 1980). Products from different manufacturers with apparently very similar composition can often vary significantly in foaming performance. This is mainly due to differences in the ethoxylated alcohol feed-

ANIONICS 125

stocks. The presence of C9, C10 and C11 as alcohol sulphates is likely to depress foaming performance.

9. *Disadvantages*. Hydrolytically unstable.

Applications

1. *General*. The sulphated ethoxylated alcohols began to replace alcohol sulphates in hand dish-washing and as shampoos in the 1950s on the basis of improved solubility, foaming, hard-water tolerance of foam, better build of viscosity with salt and decreased irritation to eyes and skin.

2. *Household products*. 3-mole ether sulphate is used as: foam stabiliser/ detergent in dish-washing liquids; major component of low-temperature or low phosphate-content heavy-duty fabric laundry liquids and powders; carpet shampoos; hard surface cleaners; chlorphenolic disinfectant concentrates.

3. *Shampoos and cosmetics*. Shampoos, bubble baths, liquid soaps and shower gels are the main applications; the 2-mole ether sulphate is usually used in shampoos to give a stable voluminous foam (without alkanolamide) when the level of sebum is not high; 3-mole ether sulphates produce excellent 'flash' foam, less stable and more open than the 2-mole ether sulphates, and more suitable for foam baths and washing-up liquids; cleansing properties and viscosity building abilities fall off as the amount of ethylene oxide increases. On the other hand, increasing ethoxylation gives reduced eye and skin irritation; products with EO up to 12 are used in shampoos as mild foaming and cleansing agents and also as detoxifying agents; they also have good solubilising properties.

4. *Oil field chemicals*. 3-mole ether sulphates used as foaming agents in foam drilling where tolerance to salt and hard water is important, but they have limited temperature stability (see Section 9.2, dialkyl betaines).

5. *Gypsum board production*. Foaming agent with tolerance to high concentrations of calcium ions.

6. *Emulsion polymerisation*. Emulsifying agent for rubber latexes.

Specification

Active content, 27–28% or 60–70%
Carbon chain distribution
Amount of ethylene oxide
Unsulphated material, typical 3–8% (on 100% active)
Colour, high unsulphated material gives good colours whilst low unsulphated material gives poor colours.

126 HANDBOOK OF SURFACTANTS

pH, can show pH drift, i.e. pH changes with time
Sodium sulphate, 0.5–3%
Sodium chloride, up to 0.05%
Flash point; some products contain alcohol so flash point can be as low as
30°C (Abel closed cup)
Ethyl alcohol, 10–15% (on 60% active ammonium salt); see flash point,
above.
1,4-Dioxane, 500 ppm (on 100% active), would be high at time of writing.
Viscosity; viscosity figures, particularly on high actives, should be treated
with caution as the viscosity is extremely dependent upon the rate of shear
and thus the method of measurement.

6.5.3 Sulphated alkanolamide ethoxylates

Nomenclature

Generic
    Sulphated alkanolamide ethoxylates
    Sulphated ethoxylated alkanolamides
    Sulphated polyoxyethylene amides
Example
    Sodium salt of dodecyl monoethanolamide+3EO sulphate, $C_{12}H_{25}$
    $CONH(OCH_2CH_2)_3OSO_4^-Na^+$

Description

These products are prepared by sulphation of the ethoxylated monoalkanolamides. Complete sulphation gives very high-viscosity products, so
sulphation is only carried out to approx. 50–70% of the theoretical value.
Products described as sulphated alkanolamide sulphates are therefore
mixtures of sulphated ethoxylated monoalkanolamides and ethoxylated
monoalkanolamides. Sulphation is normally carried out as if the starting
materials were ethoxylated alcohols.

General properties

1. Solubility. Very soluble in water.
2. *Chemical properties*. Similar chemical stability to the ether sulphates
   (see section 6.5.2).
3. Functional/surface active properties. The surface tension of sulphated
   dodecylmonoamide + 8.6EO is 33.8 mN/m compared with 35.1 mN/
   m of the unsulphated derivative. The products are excellent wetting
   and foaming agents and show good detergency and good emulsifying
   properties (see Kassem, 1984).

ANIONICS

Applications

1. *Detergents*. The products are excellent detergents particularly for dish washing but too costly and therefore not used.
2. *Shampoos*. Were used in shampoos and are probably now being replaced wholly or partly by betaines.
3. *Alkyd emulsions*. The polyunsaturated linseed (or soybean) amide ethoxylates can be used as polymerisable surfactants e.g. in alkyd emulsions. The unsaturated grouping reacts with the alkyd binder in the presence of catalysts (see Brink *et al.*, 1992).

Specification

Anionic content, typically 15–20%
Unsulphated content, typically 10–15%
Viscosity, generally high and can vary from batch to batch.

6.5.4 Sulphated oils and glycerides

Nomenclature

Generic
 Fatty monoglyceride sulphates
 Mono-, di- or tri-glyceride sulphates
 Sulphated mono-, di- or tri-glycerides
 Sulphated oils
Examples
 Salt of lauryl monoglyceryl sulphate, $C_{11}H_{23}COOCH_2CH(OH)$ $CH_2OSO_3^-M^+$ Turkey red oil (sulphated castor oil)

Description

These are produced by sulphation (usually with oleum) of the glyceride of a fatty acid, i.e. an animal fat or vegetable oil. There are three possible reactions:

1. Reaction with the OH group in the hydrocarbon chain, if it contains a hydroxyl group (e.g. castor oil).
2. Reaction with a double bond in the hydrocarbon chain of the fatty acid (e.g. oleic acid)
3. Hydrolysis of the ester group to give a free hydroxyl group on the glyceride portion and subsequent sulphation of the hydroxyl group, or sulphation of a previously prepared fatty monoglyceride. The formation of the monoglyceride and the sulphate may be combined

128 HANDBOOK OF SURFACTANTS

by reaction with oleum in the presence of the appropriate amount of glycerol. There are numerous patents on these types of preparations.

Sulphonation of the saturated hydrocarbon chain is possible when sulphur trioxide is used (see section 6.6.4)

Even with one specific oil a large number of products with different properties can be made. The degree of sulphation can be altered, thus resulting in varying mixtures of sulphated products, soap and free oil. It is probably this varying consistency of the products that has been a factor in their declining use.

The most common sulphated oils are: Turkey red oil (ricinoleic acid triglyceride) made from castor oil; sulphated methyl and ethyl ricinoleate; sulphated methyl esters, e.g. butyl oleate; fish oil, lard oil, tallow, palm kernel, tall oil and rape seed oil sulphates.

General properties

Although the products are complex mixtures, they are all sulphate esters, and subject to acid and alkaline hydrolysis (see section 6.5). In addition, any ester groups from the glyceride portion of the molecule will also be susceptible to hydrolysis.

The sulphated monoglycerides are excellent detergents. Partial glycerides containing unsaturated acids or hydroxy acids have more than one site available for sulphation and possess specific properties such as wetting and emulsification. Taking sulphated castor oils (Turkey red oil) as an example, the main properties can be grouped under two headings:

1. Good wetting, penetrating and emulsifying obtained by a high degree of sulphation.
2. Plasticising or softening properties of the oil obtained by a low degree of sulphation.

Group 1, above, are the products used for surfactant uses, while products of Group 2 are unlikely to be very useful as surfactants.

Sulphated methyl esters have good wetting properties but usually with low foam, which is rare in anionic surfactants.

Applications

1. *Household products.* Detergents: the sulphated monoglycerides of coconut fatty acids were once made on a very large scale for use in household detergents, but they have been replaced by petrochemical-based surfactants. Disinfectants: sulphated castor oil was used as emulsifier for pine oil and creosote, but has been superseded by synthetics, mainly due to the price volatility of castor oil.

ANIONICS

2. *Shampoos and cosmetics.* The ammonium salt of cocoacid monoglyceride sulphate was used as the basis for a popular USA shampoo. The product is very similar to the equivalent lauryl sulphate, but due to the additional hydroxyl groups it has slightly better water solubility. Sulphated castor oil is also used in deodorants.
3. *Textiles.* Emulsifying and wetting agents in dyeing and printing (Turkey red oil); detergent, kier boiling.
4. *Metal working.* Emulsifying agent for kerosene in hand gel cleaner (sulphated castor oil); emulsifying agents in cutting oils (sulphated butyl oleate)
5. *Leather manufacture.* Fat liquoring
6. *Printing inks.* Pigment dispersing agents and wetting agents.
7. *Mineral processing.* Until recently the sulphated monoglycerides were used in ore flotation.

Specification

Active anionic content, free fatty acid, total alkali, combined sulphate, inorganic sulphate, solubility in water and solubility in white spirit are useful tests. It is difficult to give representative figures as products vary so widely in composition.

6.5.5 Nonylphenol ether sulphates

Nomenclature

Genetic
    Alkylphenol ether sulphates
    Nonyl phenol ether sulphates
    Polyoxyethylene nonyl phenol sulphates
    Sulphated nonyl ethoxylates
Example
    Nonyl phenol + 7EO sulphate ammonium salt, $C_9H_{19}C_6H_4(OCH_2CH_2)_7 SO_4^- NH_4^+$

Description

These products are very similar to the alcohol ether sulphates but have been produced by two different routes: (i) reaction with sulphamic acid gives the ammonium salt, and the product contains equimolar quantities of ammonium sulphate; (ii) reaction with sulphur trioxide and neutralisation with sodium hydroxide gives the sodium salt, and the products are relatively free from inorganics. The products based on nonyl phenol + 4–9 EO, and using sulphamic acid were at one time commonly used in washing-

130   HANDBOOK OF SURFACTANTS

up liquids but are now too expensive. However, products made from nonyl phenol + 4–15 EO, sulphated with sulphur trioxide, do find speciality uses. The products with high amounts of EO contain high quantities of 1,4-dioxane after sulphation, which is normally stripped off before sale.

Typical impurities are ammonium chloride (from sulphamic acid-made material); 1,4-dioxane (from high ethylene oxide raw materials and sulphur trioxide sulphation).

General properties

Properties are similar to those of ether sulphates, but there are special properties, possibly due to ring sulphonation as well as sulphation of the hydroxyl group.

1. *Disadvantages*. Doubt exists on biodegradability of the alkyl phenol residue. The ammonium salt (made via sulphamic acid) gives off ammonia in alkaline solutions and is therefore restricted in detergent use.

Applications

1. *Household products*. The largest use of sulphated nonyl phenol ethoxylates (4EO) made with sulphamic acid was in blends with LABS for use in hand dish-washing liquids, where they gave high foam, low skin irritation and low cost; they have now been super-seded by more cost/effective LABS/AES formulations.
2. *Emulsion polymerisation*.
3. *Agricultural emulsifiers*.

Specification

Active matter, 30–40%
Unsulphated, 1–5% (can be high if manufacturer wishes to avoid ring sulphonation)
Inorganic sulphate, <1% for sulphur trioxide production, 6–8% for ammonium salt made from sulphamic acid.
1,4-Dioxane, should be below 500 ppm for stripped material.

6.6 Sulphonates — general

As explained in section 6.5 the sulphate has the sulphur atom indirectly linked to the carbon of the hydrophobe via an oxygen atom. In the case of sulphonates, the sulphur atom is linked directly to the carbon atom, which is usually but not always, in an aromatic ring (See Figure 6.6).

This difference in structure gives significant differences in properties between the sulphonate and sulphate groups. The most practical difference

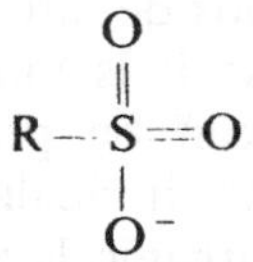

ANIONICS   131

$$R - \overset{\overset{\displaystyle O}{\|}}{\underset{\underset{\displaystyle O^-}{|}}{S}} = O$$

Figure 6.6 The sulphonate group.

is that the ester link in sulphates is very readily hydrolysed under acidic conditions, which means that the free acids are not stable (see section 6.5). On the other hand, the sulphonic acids are quite stable, and therefore the surfactant sulphonic acids are available commercially, which means that the user can carry out neutralisation and make a variety of salts with only one surfactant raw material.

The preparation of sulphates and sulphonates is carried out using the same sulphating/sulphonating agents, the difference being the chemical structure of the material to be reacted. If the reaction occurs with hydrogen attached to oxygen (an alcohol) then a sulphate is produced, whilst if the reaction occurs with a hydrogen attached to carbon then a sulphonate is produced (see Figure 6.7).

$$R - OH + SO_3 \longrightarrow R - O - SO_3^-$$

a   sulphate

$$R - H + SO_3 \longrightarrow R \quad SO_3^-$$

a   sulphonate

Figure 6.7 Sulphation and sulphonation.

The main sulphation/sulphonation agents are:

1. Chlorsulphonic acid. Chlorsulphonic acid will react with long-chain alcohols to produce sulphates, but inorganic chloride is produced as a by-product. Chlorsulphonic acid is not suitable for sulphonating aromatic rings because chlorination of the aromatic nucleus is likely to occur.
2. Oleum (fuming sulphuric acid). Liquid sulphuric acid or oleum is used to sulphonate alkyl aryl hydrocarbons, e.g. xylene, toluene, dodecyl benzene, naphthalene, and alkyl naphthalene.
3. Gaseous sulphur trioxide. Diluted with air or otherwise, gaseous sulphur trioxide can be used for most sulphonations and sulphations, but needs special reactors due to the very fast and exothermic reaction. Reaction with alcohols gives sulphates whilst reaction with hydrocarbons (aliphatic or aromatic) gives sulphonates. Methods

132 HANDBOOK OF SURFACTANTS

using gaseous sulphur trioxide are now dominant for most raw material, not only because it is more economic but also because better quality materials can be produced, compared with using oleum. The major difference in quality from oleum sulphonation (or sulphation) is the low inorganic levels and paler colours obtained when using sulphur trioxide/air on a continuous plant.

However, sulphur trioxide is a powerful, dehydrating and oxidising agent as well as a sulphonating agent, and side reactions can occur, e.g. formation of 1,4-dioxane in sulphating ethoxylates (see Section 6.5.2). Processes have been developed using sulphur trioxide in liquid sulphur dioxide, but most processes use sulphur dioxide diluted with air in a continuous plant. The major sulphonation processes are known by the name of the sulphonation plant manufacturers; some of the best known are Chemithon, Ballestra, Mazzoni, Stepan, and Meccaniche Moderne. Some of the sulphonators have produced their own plant designs, e.g. Albright and Wilson, Stepan, Berol, and Unilever. These plants are invariably continuous and basically consist of a sulphur trioxide generator, the sulphonation reactor and a continuous neutraliser. A diagrammatic view of a typical sulphonation/sulphation plant is shown in Figure 6.8.

The products will differ more as a result of raw material variation than from different plants, with the exception of colour. Dark colours occur when the product is over sulphonated/sulphated and/or the sulphonation/ sulphation rate is not controlled.

The aim of the sulphonator is to react equimolar quantities of the raw material and sulphur trioxide and to obtain 100% conversion. This is not possible in practice, so a slight excess of sulphur trioxide is used (4–10%), the excess gaseous sulphur trioxide being taken off with the air leaving the sulphonic acid. Excess sulphur trioxide gives oxidation, which gives dark coloured material. If poor colour is to be avoided then the degree of sulphonation/sulphation must be kept below 100%. Consequently there is a conflict between low sulphonation/sulphation giving good colour but high unsulphonated/unsulphated material and high sulphonation/sulphation giving poor colour but low unsulphonated/unsulphated material. In modern plants extremely good colours can now be obtained with up to 99% conversion (based on the material being sulphated/sulphonated).

These plants have been designed to give optimum conditions with the major surfactants commercially produced, i.e. alkylbenzene sulphonates with the alkyl group in the region of C10–14. Such continuous plants have limitations on the type of raw material which they can handle:

1. There is a lower limit of molecular weight where the exotherm volatalises the material being sulphonated. Thus benzene, toluene, xylene, cumene and naphthalene are not usually sulphonated on the large-scale continuous plants used for LABS.

ANIONICS

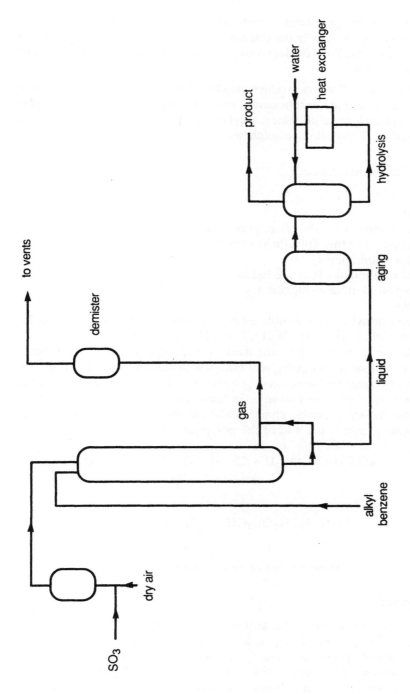

Figure 6.8 Continuous sulphonation/sulphation plant.

134   HANDBOOK OF SURFACTANTS

2. There is an upper limit of molecular weight where the raw material is too viscous to flow in the sulphonation area of the plant and oxidation occurs. This limit is plant variable, but is in the region of molecular weight 500.

Thus there are still many products produced with oleum, such as: benzene sulphonates; toluene sulphonates; cumene sulphonates; xylene sulphonates; naphthalene sulphonates; alkyl naphthalene sulphonates; petroleum sulphonates; heavy alkylate sulphonates.

6.6.1 Ethane sulphonates

Nomenclature

Generic
  Alcohol ether (or ethoxy) sulphonates
  Alkyl phenol ether (or ethoxy) sulphonates
  Ethane sulphonates
  Ether sulphonates (but see below)
  Ethoxylated ethane sulphonates
Example
  Sodium nonyl phenol 2-mole ethoxylate ethane sulphonate, which has the formula $C_9H_{17}C_6H_4OCH_2CH_2OCH_2CH_2SO_3^-Na^+$

Note that some authors give the name 'ether sulphonates' to the reaction product of unsaturated polyglycol ethers with sulphur trioxide. These are mixtures of hydroxyalkane alkyl polyglycol ether sulphonates and alkenyl alkyl polyglycol ether sulphonates. In such products, the sulphonate groups are along the hydrophobic chain, whilst the ethane sulphonates have the sulphonate group at the end of the polyglycol chain (see Figure 6.9).

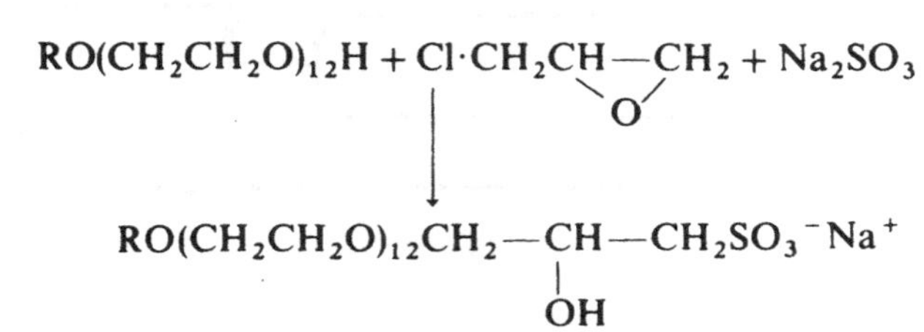

$$RO(CH_2CH_2O)_{12}H + Cl\cdot CH_2CH-CH_2 + Na_2SO_3$$

$$RO(CH_2CH_2O)_{12}CH_2-CH-CH_2SO_3^-Na^+$$
$$OH$$

Figure 6.9 Preparation of ethane sulphonates.

Description

At the time of writing, the author does not know of any large-scale producer of ethane sulphonates, although several companies are offering on the market. There has been a small specialist use for many years, but the products have recently (1985–1990) aroused considerable interest due to their potential use in enhanced oil recovery. The products cannot be

ANIONICS 135

made by conventional methods of sulphonation (oleum, chlorsulphonic acid or sulphur trioxide).

Published methods of preparation are:

1. The Strecker reaction:
   $RCl + Na_2SO_3 \rightarrow RSO_3^-Na^+$
   The usual method of preparing the chloride is by reaction of phosphorous pentachloride or thionyl chloride with an alcohol, ethoxylated alcohol or ethoxylated nonyl phenol. The products from the Strecker reaction contain a number of impurities: fatty alcohol; fatty chloride; sodium sulphate and sodium sulphite. The chlorination step can involve a reduction in the polyethylene oxide chain.
2. Using epichlorhydrin. This is the reaction between fatty alcohol, epichlorhydrin and sodium sulphite (see Figure 6.9), described in a number of patents (Rohm and Haas, 1937, 1938a, b).

General properties

The sulphonate group has better chemical stability than the corresponding sulphate group on alcohols or ethoxylated alcohols. Sulphonating ethoxylated alcohols gives products with excellent water solubility, excellent thermal and chemical stability, high salinity tolerance and high surface activities. By choice of the appropriate hydrophobic group they can give very low interfacial tension with a wide range of hydrocarbons.

1. *Disadvantages*. High price, at the present time, compared to linear alkylbenzene sulphonates or ether sulphates. The high cost of manufacture by the Strecker route is due to the low yields obtained.

Applications

1. *Industrial detergents*. Used for many years in surgical scrubs.
2. *Oilfield chemicals*. Enhanced oil recovery (see Shupe, 1977).

6.6.2 Paraffin sulphonates

Nomenclature

Abbreviations
SAS, secondary alkane sulphonates, used in this book
LAS, occasionally used, more often used for LABS.
PAS
PS
Generic
Alkane sulphonates
Paraffin sulphonate — should not be confused with petroleum sulphonates (see section 6.6.7)
Secondary n-alkane sulphonates

136 HANDBOOK OF SURFACTANTS

Example
There are very few manufacturers. Hostapur SAS60 (Hoechst) is a 60% active surfactant with the active component shown in Figure 6.10.

$$CH_3 - (CH_2)_n - CH - (CH_2)_m - CH_3$$
$$| $$
$$SO_3Na$$
$$n+m=10\text{--}14$$

Figure 6.10 Active ingredient of Hostapur SAS 60 (Hoechst).

Description

These sulphonates are produced by sulphoxidation of normal linear paraffins with sulphur dioxide and oxygen, catalysed with ultraviolet light or gamma-radiation (see Figure 6.11).

$$\text{UV light}$$
$$R\text{--}H + 2SO_2 + O_2 + H_2O \longrightarrow R\text{--}SO_3H + H_2SO_4$$

Figure 6.11 Sulphonation of paraffins.

There are very few plants world-wide, as the plants are specific to SAS and capital intensive. The conversion after sulphoxidation is poor, so unreacted paraffin must be removed and recycled. The alkane sulphonic acid is neutralised with sodium hydroxide solution and then the excess water and any remaining paraffin are removed by distillation, leaving an alkane sulphonate melt. This can be processed to make flake, or diluted with water to obtain a paste of approximately 60% concentration.

The paraffin used is usually a mixture with a carbon chain distribution C14–C17, which is the optimum chain length for heavy duty detergents in SAS. The products are available as 30% aqueous solution, 60–65% paste or 90%+ prills. Impurities include disulphonate and sodium sulphate.

General properties

1. *General.* Excellent aqueous solubility and excellent biodegradability.
2. *Solubility.* The C14–C17 mixture has 31% solubility at 20°C, which is superior to LABS or lauryl ether sulphate (28%); synergistic solubility with lauryl ether sulphate is 40% (80 SAS/20AES) at 20°C. This

ANIONICS 137

means that the mixture exhibits better solubility than either of the components. Solubility is not affected by addition of amine oxides (unlike LABS, where the solubility decreases with addition of amine oxides).

3. *Compatibility with aqueous ions.* Excellent with calcium and magnesium; if high concentrations of alkali are added, surface activity and solubility can be improved by adding nonionic surfactants (alcohol ethoxylates).

4. *Surfactant properties.* CMC = 0.035% for Hostapur SAS60; surface tension of 0.1% aqueous soln = 30 mN/m.

5. *Functional properties.* Wetting optimum at C15, with C12 and C14 inferior (see Trautman and Jurges, 1984). This is quite unusual, as wetting is usually at an optimum chain length well below the detergency chain length optimum; wetting time 5–10 s at 0.1% at 20°C (DIN 53 901); foaming 600 ml foam for 0.1% in demineralised water, 150 ml of foam for 0.1% in 220 ppm $CaCO_3$ (foaming measured by DIN 53 902/1).

6. *Disadvantages.* Difficult to thicken by addition of electrolyte unless ether sulphates are present.

Applications

1. *Household products.* Powdered heavy-duty detergents; liquid heavy-duty detergents; washing-up liquids. The SAS products are used in highly concentrated formulations where the overall formulation is cheaper using SAS rather than LABS, which needs the addition of hydrotropes and alcohol. Hard surface cleaners (general purpose) with builders but without the need for hydrotropes, due to the good solubility.

2. *Industrial cleaning.* High electrolyte-containing formulations, e.g with caustic soda, phosphoric acid, or sulphuric acid.

3. *Emulsion polymerisation.* Emulsifier in the polymerisation of vinyl chloride and as a post-stabiliser after polymerisation; in the production of carboxylated butadiene–styrene copolymers

4. *Leather processing.* Used in fat liquoring to replace the older sulphonated/sulphated natural oils.

5. *Mineral separation.* Ore flotation agent.

6. *Petroleum industry.* Enhanced oil recovery.

Specification

Active content	60%
Appearance	Pale yellow paste
Di- and poly-sulphonates	6–7%
Sodium sulphate	2–4%
Unsulphonated material	Less than 1%

138   HANDBOOK OF SURFACTANTS

6.6.3   *Alkyl benzene sulphonates*

Nomenclature

Abbreviations
  LABS, Linear alkyl benzene sulphonates, used in this book
  ABS, alkylbenzene sulphonates, often, but not always, used for branched acid
  DDBS, dodecyl benzene sulphonic acid or sulphonate
  DOBS, dodecyl benzene sulphonic acid or sulphonate derived using Dobanes (alkyl benzene made by Shell Chemicals)
  LAS, often used, not to be confused with paraffin sulphonates or alcohol sulphates.

Generic
  Acid, alkyl (usually linear C12–14) benzene sulphonic acid
  Alkylbenzene sulphonates
  Broad cut acid, alkyl (usually linear, wide distribution of alkyl chain length) benzene sulphonic acid
  Dodecylbenzene sulphonates
  Hard acid, alkyl (branched chain) benzene sulphonic acid, non-biodegradable
  Narrow cut acid, alkyl (usually linear C12–14, narrow distribution of alkyl chain length) benzene sulphonic acid.
  Soft acid, alkyl (linear) benzene sulphonic acid, biodegradable.
Examples
  Sodium dodecylbenzene sulphonate
  Dodecyl diphenyl oxide disulphonate

Description

The empirical formula of a sulphonate can be expressed as $RC_6 H_4 SO_3^-$ M^+, where R is a linear hydrocarbon in the range C9–C15 and M is an alkali metal ion or an amine derivative. The two major products produced world-wide are where R = C9–C14, known as broad cut alkyl benzene sulphonic acid, and R = C10–C13, known as narrow cut alkyl benzene sulphonic acid. The free acid is quite stable and also freely available; it has the advantage of reduced transport costs (as it is 100% active) and reduced storage costs, as the salts can easily be made *in situ*. The most common salt is the sodium salt. It is usually sold as a 30% aqueous solution. The 95% flake form is also available.

There are the following variations in alkyl groups and isomers:

1. Variation in the length of the chain — C8 up to C15
2. Substitution of the benzene ring in different positions on the chain: 2-phenyl, 3-phenyl and 4-phenyl (see Figure 6.12). This shows three isomers of the same alkyl chain length, C_{12}, but differing in the point

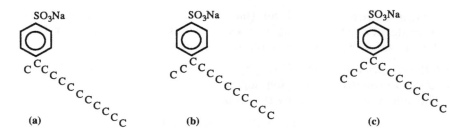

Figure 6.12 Substitution of the benzene ring at different positions on the chain: (a) 2-phenyl; (b) 3-phenyl; (c) 4-phenyl.

of attachment to the benzene ring. There will be a 5-phenyl isomer and also a 6-phenyl isomer, but note that the latter is exactly the same as the 7-phenyl isomer. Thus **each** alkyl chain length will have a number of isomers which depend upon the position of attachment between the benzene nucleus and the alkyl chain length. A typical LABS could have alkyl chains of the following distribution:

<C10	0.2%
C10	8.6%
C11	31.2%
C12	30.9%
C13	23.8%
C14	1.8%

Each chain has 5–8 isomers. Therefore, there will be 40–60 isomers in a typical LABS. However, a small number will predominate and a LABS descibed as C12–C14 with high 2-phenyl content will have most of the isomers of chain length 12, 13 and 14 and with the majority of the attachment at the 2 position.

The sulphonate group is shown at the para-position to the alkyl side chain, but it can be found also on the ortho-position, but only to a much smaller degree.

25–30% of the 2-phenyl alkane is produced by using aluminium chloride as catalyst with the amounts of 3-, 4-, 5-, and 6-phenyl isomer decreasing steadily from about 22% to 15% respectively. 15–20% of 2-phenyl alkane is produced by using hydrogen fluoride as catalyst with the rest of the isomers distributed fairly evenly. On sulphonation these isomers are retained and the two different alkylation catalysts give what are now known as 'high 2-phenyl content' and 'low 2-phenyl content' sulphonates.

3. Variation in chain branching. Branched hydrocarbons are very rarely used now in Europe, the USA and Japan because of poor biodegradability.

140 HANDBOOK OF SURFACTANTS

4. Linear narrow or broad cut (based on linear narrow or broad cut alkylate derived from linear cracked olefins (wax cracking) or from linear olefins by polymerisation of ethylene).

All these variations can affect physical properties, e.g. solubility, and functional properties to varying degrees.

The sulphonation is usually shown as:

$$RC_6H_5 + SO_3 \rightarrow RC_6H_4SO_3H$$

However it is more complex and side reactions occur. These side reactions give products that can affect the properties of the finished product. The literature gives various different mechanisms. The following description gives the overall effect and is not intended to describe the mechanism.

Excess SO_3 is invariably used in practice to reduce the unreacted alkylate to a minimum. This excess SO_3 forms a disulphonic acid and a anhydride:

$$RC_6H_5 + \text{excess } SO_3 \rightarrow RC_6H_4SO_3SO_3H$$
$$\text{disulphonic acid or pyrosulphonic acid}$$
$$\rightarrow (RC_6H_4SO_2)_2 \text{ O (anhydride) or } RC_6H_4(SO_2)C_6H_4R \text{ (a sulphone)}$$

These products can be removed by:

1. Ageing — the disulphonic acid reacts with unreacted alkylate:

$$RC_6H_4SO_3SO_3H + RC_6H_5 \rightarrow 2RC_6H_4SO_3H$$
$$\text{LABS}$$

2. Hydrolysis. The presence of water will hydrolyse the disulphonic acid, the anhydride and any oleum:

$$RC_6H_4SO_3SO_3H + H_2O \rightarrow RC_6H_4SO_3H$$
Disulphonic acid $\qquad$ LABS
$$(RC_6H_4SO_2)_2O + H_2O \rightarrow RC_6H_4SO_3H$$
Anhydride $\qquad$ LABS
$$H_2SO_4nSO_3 + nH_2O \rightarrow (1+n)H_2SO_4$$

The sulphone is not easily removed but its formation depends upon the SO_3:alkylate ratio (see Berna and Moreno, 1987).

Thus, sulphonation plants will have an ageing stage, and also the addition of about 0.5% water will eliminate the anhydride. As the great majority of sulphonic acids are used as the aqueous salts these side reactions should give no problems, but it is best to be aware of them, as the by-products can affect analyses of active contents and some physical properties such as sodium salt slurry viscosity (see Moreno *et al.*, 1988). These side reactions are not only dependent upon the raw materials but also on the running of the sulphonation plant, e.g. variation in the SO_3 excess.

ANIONICS 141

There can also be small quantities of tetralin and indanes which will be found in the unsulphated material (or 'free oil'). The amount of tetralin in the alkylate is determined by the catalyst used for its manufacture. Aluminium chloride gives considerably higher tetralin concentrations than does hydrogen fluoride.

General properties

1. *General.* The physical properties and surfactant properties are mainly influenced by the average molecular weight and the spread of carbon number of the alkyl side chain. However the isomers (e.g. the 2-phenyl content) can also play a part.
2. *Solubility.* The acids are soluble in water and soluble/dispersible in organic solvents. However, on dilution with water the acid will form a high-viscosity liquid or a gel at between 30–80% acid in water. For salts, see Table 6.4.

Table 6.4 Solubility of salts of LABS

	sodium	isopropylamine	calcium	TEA
10% in water	S	D	D	S
10% in ethanol	S	S	S	S
10% in mineral oil	I	S	S	I
10% in white spirit	I	S	S	I
10% in aromatic solvent	I	S	S	I
10% in perchlorethylene	I	S	S	I

LABS broad cut acids give better sodium salt solubility, higher active concentrations can be made (typically 32%) and with lower viscosities than with narrow cut acid (typically 25%). TEA salts are more soluble in water (50–60% active) than sodium (30% maximum). The larger the ionic radius of the counterion, the lower the solubility of the LABS, regardless of the starting alkylate; thus NH_4 <K<Na<Li salt solubility.

Calcium and isopropylamine salts have better solubility in hydrocarbon solvents than the sodium salts. Increased solubility results in lower cloud points and less response to salt. The cloud point can be lowered by addition of urea, sodium xylene sulphonate or alcohol. Solubility and cloud points of the aqueous salts depends upon the size and shape of the hydrophobic group, i.e. length of chain — solubility decreases with increasing chain length; 2-phenyl content — the higher the 2-phenyl content the more soluble the LABS.

The Krafft point of different alkyl chains in pure LABS (sodium salt) is:

$$C10 = 30°C \quad C12 = 42°C \quad C14 = 54°C$$

142 HANDBOOK OF SURFACTANTS

3. *Chemical stability*. Resistant to hydrolysis in hot acid or alkali. The acids evolve heat on dilution with water and generate hydrogen on contact with some metals (e.g. zinc).
4. *Compatibility of aqueous solution to ions*. Completely ionised, and the free sulphuric acid is water soluble, so the solubility is not affected at low pH; calcium salts are precipitated from solution, but in the presence of nonionics the active content is not reduced even at high hardness levels; sodium salts are reasonably soluble in the presence of electrolyte (salt and/or sodium sulphate).
5. *Surfactant properties*. CMC of sodium dodecylbenzene sulphonate (mol. wt 348.4) = 5 × 10^{-3} M = 0.18%; surface tension of a 1% solution of sodium dodecylbenzene sulphonate = 36 mN/m.
6. *Functional properties*. Wetting: C11–C13 is optimum; good wetting is dependent upon concentration; 0.5% solution of sodium salt is necessary to obtain less than 10s on Draves test. Foaming: in blends with AES optimum foam stability (not Ross Miles, which measures 'flash foam') is at C11–C12. Detergency: C13–C14 is optimum.
7. *Disadvantages of LABS*. LABS used on its own for personal care gives severe defatting action on skin, but reduced skin irritation can be achieved by addition of sulphosuccinates, sarcosinates or amphoterics; sodium salts need builders or sequestrants in hard water for stability.

Applications

1. *General*. Used in practically all detergents, domestic and industrial, where heavy duty performance is required.
2. *Heavy duty clothes washing detergents*. Powdered detergents, both high and low foam; C13–C14 used in free flowing powders. C9–C11 give best solubility in liquid detergents, also high 2-phenyl content gives best solubility.
3. *Light duty detergents (washing up liquids, hard surface cleaners)*. Broad cut acids give better solubility and so give higher actives and/ or lower viscosities than narrow cut. Low active detergents use narrow cut to give a higher viscosity. High 2-phenyl content gives medium to high active detergents (due to better solubility), whereas low 2-phenyl content are used in low active detergents (poorer solubility than high 2-phenyl).
4. *Industrial detergents*. Commercial laundries; vehicle washing where considerable foam is required.
5. *Plaster board manufacture*. Foaming agent for plaster board due to stability to calcium ions.
6. *Emulsifying organic non-water-soluble materials*. Isopropylamine salt is used as oil-soluble emulsifier for solvent degreasers, hand gels, emulsion cleaners and dry-cleaning charge soaps.

ANIONICS

143

7. *Agriculture*. In conjunction with nonionics, the calcium salt is used for agricultural herbicide formulations with self-emulsifying properties.
8. *Cellulose and paper industry*. Pulp washing agents; dispersing agent for dyes.
9. *Textiles*. Washing of fabrics, dye dispersing agents.
10. *Solubiliser*. Isopropylamine derivative added to fuel oil to solubilise traces of water.
11. *Emulsion polymerisation*. Emulsifying agent for vinyl acetate/acrylate copolymerisation.

Specification

Acid (typical narrow cut)
Appearance, brown viscous liquid with acidic odour
Active material, 97–98%
Free oil (neutral material), 1–3%
Free sulphuric acid, 1–3%
Molecular weight, 315–340
2-phenyl isomer, 15% (low) to 30% (high)

Sodium salt (from narrow cut acid)
Active material, 25%
Unsulphated organic material, 0.5–2%
Inorganic sulphate, 0.3–1%
Cloud point, 19–40°C
Molecular weight, 337–362
2-phenyl isomer, 15% (low) to 30% (high)

Sulphones are found in the free oil of the acid. They are supposed to be hydrolysed when salts are formed but Moreno *et al*. (1988) reported them as being found in the free oil. The above figures are for products made by sulphonation with sulphur trioxide, which represents the majority of LABS on the market. If oleum is used then the inorganic content will be very much higher.

6.6.4 *Fatty acid and ester sulphonates*

Nomenclature

Type 1. Sulphonated unsaturated acids where the sulphonate group is **not** near the end of the chain.

Generic
   Sulphonated acids
Example
   Sulphonated oleic acid (see Figure 6.13)

144 HANDBOOK OF SURFACTANTS

$$CH_3(CH_2)_7CH(CH_2)_8COOH$$
$$|$$
$$SO_3H$$

Figure 6.13 Sulphonated oleic acid.

Type 2. Salts of alpha-sulphonated fatty esters (or acids).
Abbreviations
  FES, Fatty ester sulphonates, used in this book,
  FAS, Salts of alpha-sulphonated fatty acids
  ES, ester sulphonates

Generic
  Alpha sulphonated fatty acids (or esters)
  Ester sulphonates.
  Fatty acid sulphonates
Example
  Sodium tallow methyl ester alpha-sulphonate (this is a mixture, see
  Figure 6.15)

Type 3. Omega-sulphonated fatty acids (or esters). Not commercially
available but may be in the future.

Description

There are two quite distinct types of sulphonated fatty acids/esters on the
market. These are:
  Type 1. *Sulphonated unsaturated acids.* These are produced by the
sulphonation of unsaturated fatty acids (or esters). There are not many
products of this type available, but an example is sulphonated oleic acid
(see Figure 6.13). These should not be confused with the sulphated oils and
esters (see section 6.5.4), which have very different properties. The
sulphonic acid group is in the middle of the chain. The acid is quite stable
but the potassium salt is the normal salt offered for sale due to its higher
solubility than the sodium salt.
  Type 2. *Alpha sulphonated acids or esters.* These were first produced
commercially in Japan and then in Europe in the early 1980s. At the
present time (1993) there are very few large-scale producers in the world
but the potential of this type of sulphonate is likely to be very large. The
products are available as high active pastes (50–60% active) but such
products often contain added ether sulphates or the salts of short-chain
carboxylic acids to facilitate handling. The chemistry and properties of the
products were fully described by the US Department of Agriculture in the
1950s, but practical large-scale production on continuous sulphonation
plants has only been realised in the 1980s. Sulphonation at the alpha

ANIONICS 145

position of the fatty acids does not occur readily due to the weak activation of the carbon atom. Strong sulphonating agents (sulphur trioxide) must be used, leading to problems of poor colour, particularly if the starting esters have unsaturated groups. The sulphonation reaction is complex (see Figure 6.14).

Fast Step 1

$$R_1CH_2COOR_2 + 2SO_3 \longrightarrow R_1CHCOOSO_2OR_2$$
$$|$$
$$SO_3H$$

disulphonate (or anhydride or sulphoanhydride)

Slow step 2

$$R_1CHCOOSO_2OR_2 + R_1CH_2COOR_2 \longrightarrow 2R_1CHCOOR_2$$
$$| \qquad\qquad\qquad\qquad\qquad\qquad\qquad\qquad |$$
$$SO_3H \qquad\qquad\qquad\qquad\qquad\qquad\qquad\qquad SO_3H$$

alpha sulphonated methyl ester ($R_2 = CH_3$)

Figure 6.14 Sulphonation of fatty esters.

The above reactions do not fully describe the published work but indicate the complexity of the reaction and the formation of the disulphonates. At the end of the reaction there is a mixture of sulphoester acid and disulphonate. Neutralisation with sodium hydroxide gives the results shown in Figure 6.15.

Thus the sodium salt will be a mixture of:

1. The monosodium sulphonate salt of the ester, about 80% of the active
2. The disodium salt of the alpha-sulphonated carboxylate, about 20% of the active.
3. The sodium salt of the fatty acid produced by hydrolysis of the starting ester which was not sulphonated, i.e. soap, about 3% of the active.

These three products are different in solubility, physical properties and surfactant behaviour, and therefore users should expect differences from one manufacturer to another. However, consistent manufacturing conditions should give a consistent material, but it would be advisable to check for reproducibility from any manufacturer until confidence in batch-to-batch variation can be established.

The majority of data have been obtained using distilled methyl esters of fatty acids, e.g. tallow fatty acid. However, the presence of some unsaturated triglycerides in the methyl ester seems to have a surprisingly

146 HANDBOOK OF SURFACTANTS

1. With the alpha sulphonated ester

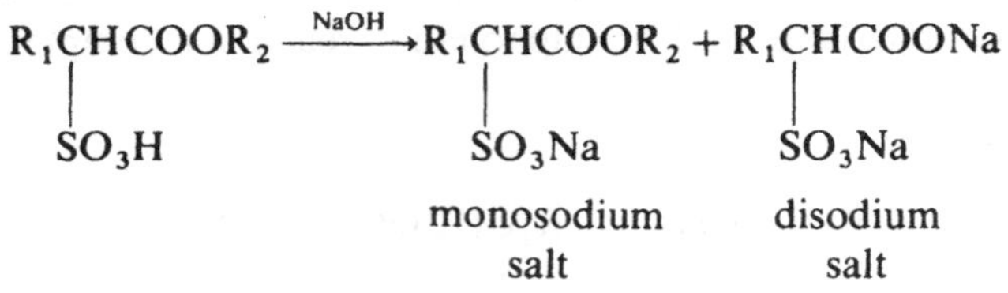

$$R_1CHCOOR_2 \xrightarrow{\text{NaOH}} R_1CHCOOR_2 + R_1CHCOONa$$

$$\underset{SO_3H}{|} \qquad \underset{SO_3Na}{|} \qquad \underset{SO_3Na}{|}$$

monosodium   disodium
salt   salt

2. With the disulphonate

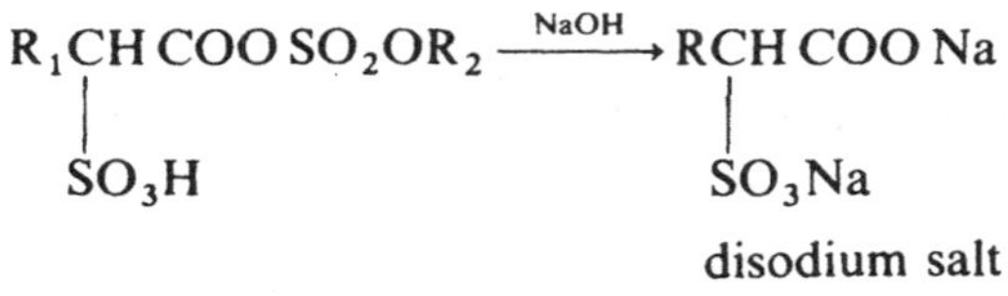

$$R_1CH\,COO\,SO_2OR_2 \xrightarrow{\text{NaOH}} RCH\,COO\,Na$$

$$\underset{SO_3H}{|} \qquad\qquad \underset{SO_3Na}{|}$$

disodium salt

Figure 6.15 Neutralisation of sulphonated esters.

low effect on colour and this could well permit the use of undistilled saturated methyl ester for sulphonation in order to reduce costs. Even under optimum sulphonation, dark-coloured products are reported in the literature and special bleaching of the acid and also of the salts are often needed.

Impurities/by-products include disulphonate, soaps, bleaching by-products.

Type 3. *Omega ester sulphonates.* These products are different in chemical structure from Type 1 and Type 2 described above. They are products made by sulphonation of saturated fatty esters using UV radiation as catalyst. The sulphonic acid group is statistically distributed at random along the chain. Such products are not made on a large scale, but they are included here as they may well be commercialised in the future. They will not be discussed any further.

General properties

1. *General.* Type 1: excellent wetting; good stability and solubility in high concentrations of electrolytes. Type 2: excellent cold and hot water detergency in hard water; FES behave in a similar manner to fatty alcohol sulphates with a similar chain length but with better hydrolytic stability.
2. *Solubility.* Type 1 (sulphonated oleic acid): poor aqueous solubility of acid and sodium salt; potassium salt more soluble. Type 2: solubility

ANIONICS 147

of the salts of C12–C14 FES is excellent but the solubility of the salts of C16–C18 FES is poor compared with other detergent sulphonates.

Surfactant	Krafft point (°C)
LABS (C10–C13)	<0
AOS (C15–C18)	<0
SAS (C14–C17)	<0
FES (C14)	10
FES (C16–C18)	39

Salts of the disulphonate will tend to crystallise out of solution. The commercial products can be solubilised with hydrotropes.

3. *Chemical stability.* Type 2: the principal active product is the sulphonate of a methyl ester of a fatty acid, which would be expected to be hydrolytically unstable. However, it is claimed that the methyl ester is stable between pH 3 to 9.5 at 80°C. The product is stable in a neutral detergent, and minimal hydrolysis occurs in spray drying of an FES-containing slurry and storing at 60 °C for some months (see Knaggs *et al.* 1965).

4. *Compatibility with aqueous ions.* Type 2: show excellent sequestration of hard water in presence of soap, i.e. lime soap dispersion. Thus soap as an impurity could be beneficial.

5. *Compatibility with other surfactants: viscosity behaviour.* Type 1: the disodium salt of sulphonated oleic acid has been shown to be lower in viscosity than other anionics and mixed anionic/nonionic systems (see Henkel, 1992). Type 2: the addition of the disodium salts of FES (acid), particularly C10–C12, can lower the viscosity of LABS, soap, alkyl sulphates, ether sulphates and FES. This can be utilised in reducing the viscosity of slurries for spray drying.

6. *Surfactant properties.* The CMC of FES is: for the C14, 0.018%; for the C16, 0.0016%; and for the C18, 0.0007%. The lowest surface tensions obtainable are: for the C14, 34 mN/m; for the C16, 33 mN/m; and for the C18, 33 mN/m.

7. *Functional properties.* Type 2: foaming properties are better with C12–C14 than with C16–C18 fatty acids; thus alpha-sulphomethyl tallowate gives low foam. C16–C18-based detergents showed superior detergency to LABS in the absence of polyphosphates at low concentration and at high water hardnesses. Soil suspending power was good, but foam was too high in horizontal washing machines, although foam problems were not encountered with C12–C14, which is unexpected. Some doubt exists, on fabric incrustation in the complete absence of polyphosphates.

148   HANDBOOK OF SURFACTANTS

8. *Solubilisation.* Excellent for polar compounds compared with LABS or AS.

Applications

Type 1. Wetting agents, particularly in high pH solutions and high concentrations of electrolyte; give very low foam compared to most other anionics of similar wetting properties. Applications in: textile processing, metal cleaning, industrial detergents, emulsion polymerisation.
Type 2. Excellent detergency with good lime soap dispensibility; particularly, phosphate free detergents can be manufactured to give excellent performance if soap is included; optimum detergency found at C16–C18 fatty acids (with the methyl ester); synthetic soap bars can be made with cocomonoethanolamide and inorganic builders; and as an emulsifier for emulsion polymerisation (pvc).

Specification (Type 2 only)

Active, pastes at 40% or 60–65% concentration
Disodium salt, 20% (of the active content)
Soap, 2–5% of the active content
The viscosity/active curve for the C16/C18 FES shows two pronounced minima at 40% and 60–65% active.

6.6.5 Alkyl naphthalene sulphonates

Nomenclature

Generic
Alkyl naphthalene sulphonates
Phenol formaldehyde sulphonic acid condensates — see polymeric surfactants (chapter 11).
Naphthalene sulphonic acid–formaldehyde condensates — see polymeric surfactants (chapter 11)
Dinonylnaphthalene sulphonates — see petroleum sulphonates (section 6.6.7)
Examples
Sodium isopropylnaphthalene sulphonate
Sodium dibutylnaphthalene sulphonate

Description

Alkylation of naphthalene followed by sulphonation gives a wide variety of products depending upon the size of the alkyl group and the amount of

ANIONICS 149

sulphonation. If the alkyl group contains fewer than five carbon atoms, the products are water-soluble and show surfactant properties depending on the size and number of alkyl groups. If the products have alkyl groups of nine carbon atoms and above, the products are not water-soluble (with one sulphonic acid group) but begin to show solubility in mineral oil. These latter products therefore show properties more similar to the petroleum sulphonates, and are included in section 6.6.7. The remainder of this section is therefore devoted to the water-soluble alkyl naphthalene sulphonates.

The water soluble alkyl naphthalene sulphonates are made on simple batch reactors (glass lined) by reaction of a short-chain alcohol with naphthalene in the presence of oleum. The oleum acts as a catalyst for the alkylation of the naphthalene by the alcohol as well as sulphonating the naphthalene (see Figure 6.16).

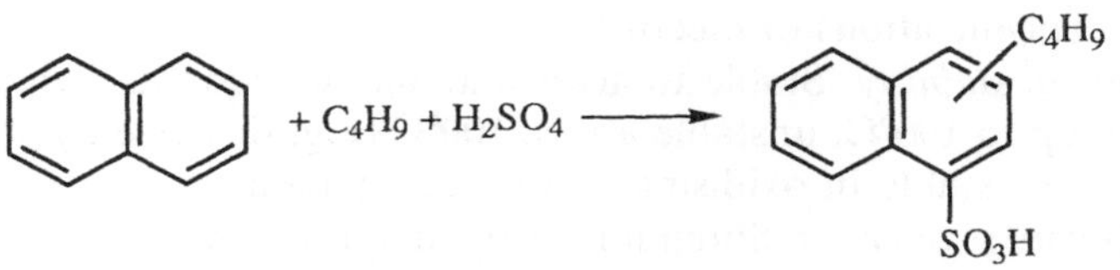

Figure 6.16 Alkylation of naphthalene.

The degree of alkylation and the degree of sulphonation can be controlled by the proportions of starting ingredients. The position of sulphonation on the naphthalene ring will depend upon the temperature of sulphonation, the sulphonate group being in both the 1- and 2-positions with low temperatures favouring the 1-position and high temperatures ($>100°C$) the 2-position. There will also be some disulphonation, and the alkyl groups can be anywhere on the naphthalene ring. Thus the reaction products are complex mixtures. One way of characterisation is by molecular weight:

Alkyl derivative of monosulphonate	Molecular weight of sodium salt
Methyl	244
Ethyl	258
Isopropyl	272
Butyl	286
Di-isopropyl	313
Dibutyl	351

Note that all kinds of mixtures are possible, so a molecular weight of 272 could mean alkylation with isopropyl alcohol or with a mixture of methanol

150 HANDBOOK OF SURFACTANTS

and butanol. The two main commercial products are butyl naphthalene sulphonate and isopropyl naphthalene sulphonate. Both are available in powder form as the sodium salt, but most products contain large quantities (5–15%) of sodium sulphate. However, some manufacturers offer products with low sodium sulphate content.

General properties

1. *Solubility.* Excellent solubility in water, depending upon the type and amount of alkyl substitution; most products can be made at 40% active aqueous solutions; the dibutyl derivative is less soluble than the monobutyl or di-isopropyl, but has better wetting and emulsifying properties.
2. *Compatibility with aqueous ions.* Tolerant to hard water; good lime soap dispersibility; compatible with strong alkalis, strong acids and high concentrations of electrolytes.
3. *Chemical stability.* Stable to acids and alkalis; aqueous solutions are stable up to 100°C; unstable above 100°C (e.g. disperse dyeing under pressure); stable in oxidising and reducing agents.
4. *Surfactant properties.* Sodium isopropylnaphthalene sulphonate has a CMC of 0.7% and a surface tension of 37 mN/m at 1% concentration.
5. *Functional properties.* Excellent wetting properties with the higher molecular weight (di-isopropyl and dibutyl) derivatives; best detergent properties with the dibutyl derivative; low foam with low molecular weight products (250–260 — methyl, ethyl), but high foam with molecular weight 280–300; hydrotrope properties claimed for low–medium molecular weight products.
6. *Disadvantages.* Doubts on biodegradability, but there is a lack of information.

Applications

1. *General.* The alkyl naphthalene sulphonates were developed as soap substitutes before and during the 1914–1918 war and were mainly used in textile processing. They were extensively used during the 1930s and up to 1950 as general-purpose surfactants and detergents but were replaced by petrochemical-based anionics (LABS). The present uses are restricted to their excellent wetting and dispersing properties rather than the detergent properties.
2. *Household products.* Products with molecular weight 280–290 have been recommended for use with lauryl sulphate in rug shampoos in order to depress the cloud point and improve the friability of the residue on drying.

ANIONICS 151

3. *Textiles*. Wetting agent and dispersing agent for disperse dyes, but limited to dyeing temperatures of 100°C maximum.
4. *Industrial detergents*. Used as a wetting agent in cheap acid cleaners, e.g. brick cleaning, metal cleaning and metal pickling; stable to 5% hydrochloric acid.
5. *Agricultural*. Act as combined wetting and dispersing agent for wettable powder pesticides; use in fertilisers where an alkyl naphthalene sulphonate is incorporated into a urea melt to give protection against caking.
6. *Paints*. Wetting agent in water-based paints.
7. *Electroplating*. Wetting agent for acid electroplating baths.

Specification

Active content, 80–95% powder
Sodium sulphate, 3–20%
Unsulphonated material, 1–3%
Molecular weight, 240–290
Occasionally sold as a liquid with active content 20–40%.
Compare with LABS of molecular weight 340–360

6.6.6 Olefin sulphonates

Nomenclature

Abbreviation
    AOS, alpha olefin sulphonates, used in this book

Generic
    Alkene sulphonates/hydroxy alkane sulphonates
    Alpha-olefin sulphonates
Examples
    Sodium olefin (C14–16) sulphonate
    Sodium C14–16 olefin sulphonate

Description

The products are manufactured by the sulphonation of alpha-olefins using sulphur trioxide on continuous plants similar to those producing LABS. There are usually two ranges offered, the more common being based on C14–C16 olefins and the other based on C16–C18 olefins. The olefins made from ethylene are preferred raw materials, giving better colour and fewer impurities than olefins from paraffins by wax cracking. Depending upon the manufacturing process and the olefin carbon number, alpha-olefins produced from chain-growth processes often contain a considerable

152 HANDBOOK OF SURFACTANTS

amount of 'linear branched' olefins, known as vinylidene olefins. Typical branching for vinylidene olefins are ethyl, butyl and hexyl groups. Such branches will occur in the sulphonates.

The sodium salts are available as 40% liquids, high active slurries (60–70%) and >90% spray dried powders or flake.

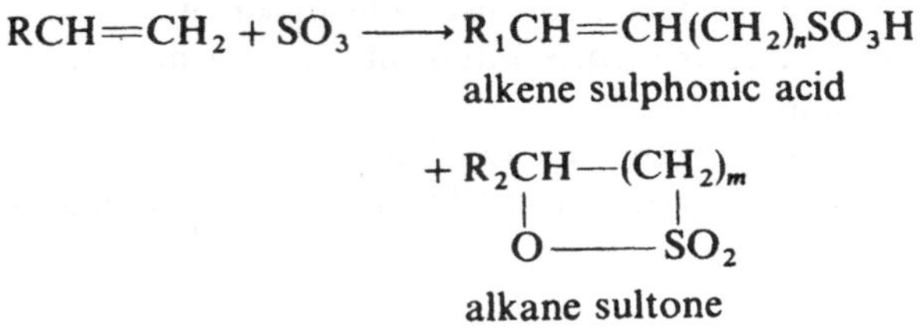

$$RCH{=}CH_2 + SO_3 \longrightarrow R_1CH{=}CH(CH_2)_nSO_3H$$
alkene sulphonic acid

$$+ R_2CH{-}(CH_2)_m$$
$$O{-}{-}SO_2$$
alkane sultone

Figure 6.17 Sulphonation of alpha-olefins.

1. Alkene sulphonic acid

$$R_1CH{=}CH(CH_2)_nSO_3H \xrightarrow{NaOH} R_2CH_2CH{=}CHCH_2SO_3Na$$
2-alkenyl sulphonate
+
$$R_2CH{=}CHCH_2CH_2SO_3Na$$
3-alkenyl sulphonate

2. Alkane sultone

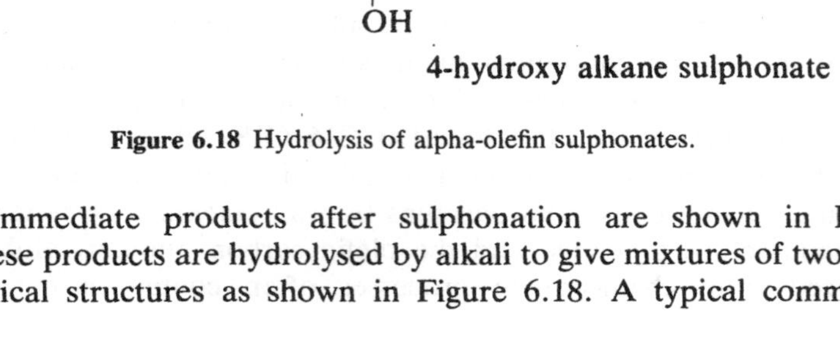

$$RCH{-}(CH_2)_n \xrightarrow{NaOH} R_2CH_2CHCH_2CH_2SO_3Na$$
$$O{-}{-}SO_2 \qquad OH$$
3-hydroxy alkane sulphonate
+
$$R_2CHCH_2CH_2CH_2SO_3Na$$
$$OH$$
4-hydroxy alkane sulphonate

Figure 6.18 Hydrolysis of alpha-olefin sulphonates.

The immediate products after sulphonation are shown in Figure 6.17. These products are hydrolysed by alkali to give mixtures of two types of chemical structures as shown in Figure 6.18. A typical commercial

ANIONICS 153

product could be 70% alkene sulphonates and 30% hydroxy alkane sulphonates. If a C14–16 olefin was used as the starting material, then R = C 10–C 12. If a C16–C18 olefin was used as the starting material, then R = C12–C14.

The properties of the hydroxy alkane sulphonate and the alkene sulphonates differ. (see Table 6.5)

Table 6.5 Properties of the hydroxy alkane sulphonate and the alkene sulphonates

Property	100% alkene sulphonates	100% hydroxy compound
Viscosity of 35% solution (cS)	11	22 000
Solubility	More soluble	—
On cooling the solution	Crystallises	Thickens
Foaming ability	Better	—
Detergency	Better	—

Sulphonation is carried out on the conventional continuous plant using sulphur trioxide/air mixtures. The reaction is extremely exothermic, and more care has to be taken to prevent over sulphonation than in the case of alkyl benzene sulphonation. Colour degradation occurs very easily and bleaching is often necessary. This, together with the fact that the products are mixtures, could well lead to considerable differences between manufacturers and also between deliveries. Analysis is complex, and it is recommended that some simple functional tests. (e.g. foaming, solubility, wetting) be carried out on any new materials and also on every batch from new suppliers until confidence in reproducible properties can be obtained.

Impurities/by-products include disulphonates (these can be up 1–5% on actives), sultones, breakdown products of bleach. (hydrogen peroxide), and olefins.

General properties

1. *General.* The C16–C18 products are similar in properties to the C14–C16, but they foam less and have lower solubility, better detergency and better emulsifying properties.
2. *Solubility.* AOS are more soluble than LABS or AS due to the hydroxyl group in the hydroxy derivative and the unsaturated links in the alkenyl derivatives.
3. *Compatibility with aqueous ions.* Calcium salts have low solubility; small degrees of hardness (50 ppm $CaCO_3$) improve dishwashing performance, but higher degrees of hardness reduce performance.
4. *Chemical stability.* Good chemical stability, similar to other sulphonates (see section 6.6); good hydrolytic stability, can be used in strong acid or alkaline formulations.

154 HANDBOOK OF SURFACTANTS

5. *Viscosity behaviour*. The viscosity of the hydroxyalkanes and alkenes differ markedly (see above). It is difficult to obtain a high viscosity with salt at low active (expected due to the excellent solubility). Addition of ammonium chloride will give higher viscosity than sodium chloride and oleamide diethanolamide will give higher viscosity than cocodiethanolamide. Partial replacement of cocodiethanolamide with cocoamidopropylbetaine will increase viscosity significantly. Viscosity can also be increased by the addition of hydroxyethyl cellulose but only low levels (<1.5%) of hydroxyethyl cellulose can be used as high concentrations are incompatible.

6. *Surfactant properties*. CMC for C15–C18 is at 0.03%. Surface tension decreases with increasing chain length up to C14 but is constant thereafter. Lowest surface tension for C14–C16 = 33 mN/m in water but drops to 28 mN/m in hard water (125 ppm $CaCO_3$).

7. *Functional properties*. The C14–C16 give excellent cleaning and excellent foaming in hard water. They foam better in hard water than in soft water (see Hart and DeGeorge 1982); foam well in the presence of sebum; flash foaming increases rapidly with length of alkyl chain up to C15 then more slowly. The type of foam can also depend upon the degree of branching in the olefin, with foams with smaller cell sizes being generated by branched sulphonates. Wetting optimum is at C16 alkyl chain.

Applications

1. *General*. AOS based on C16-C18 olefins were first commercialised by Lion Fat and Oil in heavy-duty powders in Japan in 1968, and were claimed to be superior in detergent ability in hard water. The main use in the USA has been as a partial replacement for ether sulphates in household products, particularly in bath additives and liquid soaps, due to the foam stability in the presence of soap. AOS has found a market in synthetic soaps, both toilet and household, in South East Asia due to the cold-water conditions of use in that geographical area. AOS have not found significant applications (at the time of writing) in Europe.

2. *Household products*. Dishwashing detergents (good solubility for high active).

3. *Personal care*. Bubble baths (foam stability good in presence of soap); shampoos (problems in low actives, see viscosity behaviour, above); claims that sodium C14-C16 alpha-olefin sulphonate will not 'strip' the hair of natural oils; when combined with alkanolamides, betaines or amine oxides, performance equivalent to an alcohol sulphate can readily be achieved; used in liquid soaps (foam stability

good in presence of soap) for cold water washing; used in synthetic detergent bar soaps for cold water washing.

Specification

Active matter, 35–40%
Free oil, 1–2%
Sodium sulphate, 0.5–2.0%
Sodium chloride, up to 1%

It should be stated that these tests will not give any information relating to changes in the composition, such as the relative amounts of alkene sulphonates, hydroxyalkane sulphonates and disulphonates. There is no suggestion that a full analysis for these components should be carried out on a routine basis, and simple functional tests (e.g. foaming) might be more relevant.

6.6.7 Petroleum sulphonates

Nomenclature

Generic
  Alkyl benzene bottoms sulphonates
  Dialkyl benzene sulphonates
  Heavy alkylate sulphonates
  Overbased sulphonates or overbasified sulphonates
  Petroleum sulphonates
  Sulphonates
  Synthetic petroleum sulphonates
  Synthetic long chain alkyl benzene sulphonates
Examples
  Calcium petroleum sulphonate mol. wt 470 — no defined chemical structure
  Sodium salt of dialkyl (C10–14) benzene sulphonate (see Figure 6.19)
  Barium dinonylnaphthalene sulphonate (see Figure 6.20)

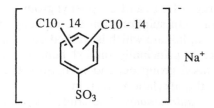

Figure 6.19 Sodium salt of dialkyl (C10–C14) benzene sulphonate.

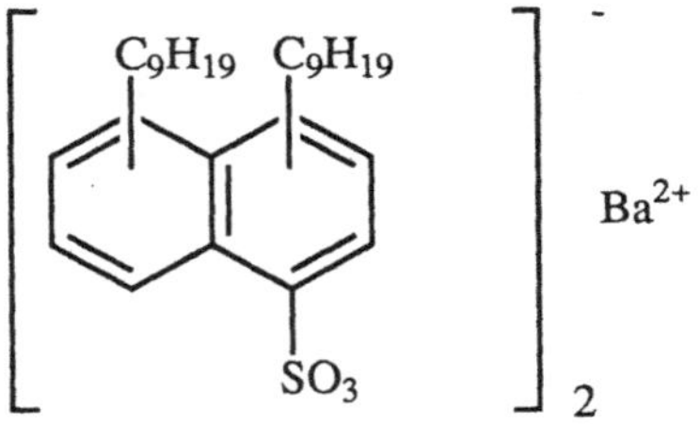

HANDBOOK OF SURFACTANTS

156

Figure 6.20 Barium dinonylnaphthalene sulphonate.

Description

This is a group of sulphonates whose common characteristic is their molecular weight and oil solubility. The molecular weight is higher than of the normal 'detergent' sulphates and sulphonates, making the acids and salts generally insoluble in water but soluble in mineral oil. The various types of sulphonates are characterised by different molecular weight, the particular salt and the alkalinity. These products originated as a by-product in the refining of petroleum, and were thus referred to as 'petroleum sulphonates'. Petroleum sulphonates are sulphonated extracts from lubricating oil fractions. The products are sulphonates of fused ring compounds and there are usually two fractions, the 'green acids' which are water-soluble and the 'mahogany acids' which are water-insoluble but soluble in mineral and natural oils. However, due to shortages of the natural sulphonates, plus certain disadvantages, chemical manufacturers have attempted to find synthetic replacements of these natural petroleum sulphonates. Such products have become known as 'synthetic' petroleum sulphonates. Synthetic petroleum sulphonates are sulphonated alkyl and dialkyl derivatives of benzene, xylene or naphthalene, and are offered by suppliers as a substitute for the petroleum sulphonate. The main types available are:

1. *Alkyl benzene bottoms sulphonates*. These are the sulphonates of the residues from the distillation of the alkylation of benzene to make detergent alkylate. The composition of such residues is a mixture of dialkyl (averaging C12) benzene and diphenyl alkane (the alkane having an average of C12 and the phenyl groups being at both ends of the alkyl chain). On sulphonation the diphenyl alkane gives a very water-soluble sulphonate which is washed out in the processing. Thus the sulphonate is principally dialkyl benzene sulphonate with the length of the alkyl group dependent upon the distribution of chain length used in the alkylation reaction.
2. *Dialkyl benzene sulphonates*. The crude alkyl benzene bottoms (see above) can be fractionated and the dialkyl benzene separated and

ANIONICS 157

then sulphonated. The products are similar to the sulphonates in (1) above.

3. *Synthetic long chain alkyl benzene.* The long alkyl chain is synthetically produced at C20–C26 and then reacted with benzene.
4. *Synthetic long chain alkyl xylene.* The alkyl chain is similar to that in (3), but is generally shorter to obtain oil solubility.
5. *Synthetic dialkyl (normally C9) naphthalene.* Oil solubility can be achieved at lower alkyl chain length when naphthalene is the nucleus for sulphonation.

The products are all made by conventional sulphonation reagents, using either oleum or sulphur trioxide. Oleum has been the major reagent used in batch reactors but there are difficulties in removing the unreacted acid or sodium sulphate. After neutralisation, the product becomes water-insoluble and washing can give considerable problems in emulsification. In recent years, considerable effort has been made to sulphonate the various alkylates by using continuous sulphur trioxide reactors which have been designed to manufacture water-soluble detergent sulphonates. This method has the advantage of reducing problems of unreacted acid, but the high viscosity of the starting alkylates (much higher than conventional dodecyl benzene) can give severe oxidation, poor conversions and considerable sludge formation.

All products are normally produced and sold as 40–60% solutions in a mineral oil. The sodium, calcium and barium salts are the most common, although zinc, magnesium and salts are available for special applications. The divalent salts, e.g. calcium, can also be found with high levels of alkalinity from finely dispersed calcium hydroxide or calcium carbonate; such products are know as overbased or overbasified salts.

General properties

1. *Solubility.* Soluble in mineral oil, particularly where predominantly paraffinic with low aromatic content.
2. *Chemical properties.* In general these products have the properties of sulphonates, namely good stability to acid, alkali and heat; over-basified salts will combat acidic conditions, e.g. automotive crank case oils.
3. *Functional properties.* Emulsifying mineral oil in water; also the ability to form microemulsions in the presence of low molecular weight alcohols and/or soap and alkanolamides; wetting–ability to replace water on a metallic surface and adhere to that surface to give a degree of protection against rust; excellent dispersing agents for insoluble dispersions in mineral oils.

158 HANDBOOK OF SURFACTANTS

4. *Disadvantages*. The natural petroleum sulphonates are chemically very ill defined and subject to variability in some functional uses. This is not easily detected by using chemical tests. The 'synthetic' petroleum sulphonates (groups 1–5 in Description, above) do not exactly replace the natural products and each group has its specific properties; many of the technically best products for use in lubricating oils and as rust preventatives are the barium salts; however, at the time of writing there is growing concern over the use of barium products in industry in Europe.

Applications

1. *Lubricating oil additives*. Ammonium and sodium salts are detergents, dispersants and rust inhibitors; overbasified salts of barium, calcium and magnesium are used in conditions of high acidity, e.g. crank case oils for large marine diesels.
2. *Metal working*. Oil-in-water emulsifying agents (sodium salts) for mineral oil in producing soluble cutting oils; they also give some degree of rust protection; they are usually used in admixture with other surfactants with emulsifying properties such as fatty acid soaps, fatty alkanolamides and ethoxylated alcohols or alkyl phenols; degreasing of metal surfaces is possible using petroleum sulphonates modified with fatty acid soaps and coupling agents to make emulsion or emulsifiable cleaners.
3. *Rust inhibitors*. Calcium or barium salts for temporary metal protection; the higher equivalent weight sulphonates (>500) are the most efficient.
4. *Fuel oils and petrol*. Rust inhibitors and sludge dispersant (ammonium, calcium, barium and magnesium salts).
5. *Pigment dispersants*. Dispersing aids for pigments in organic solvents.
6. *Petroleum production*. Oilfield drilling emulsifiers; crude oil demulsifiers; emulsifier for microemulsions in enhanced oil recovery (tertiary recovery).
7. *Mining*. Petroleum sulphonates function as promoters or collectors and lower molecular weight products act as frothing agents; the main ores processed with sulphonates are iron and silica sand.
8. *Textiles*. The type of formulations used in metal working are used in textile processing oils to give fibre-metal and fibre-fibre lubrication.

Specifications

Appearance, dark red–brown viscous liquids
Active material, 50–60%
Equivalent weight of acid, 400–600
Molecular weight distribution, critical for some uses

Water, 0.5–5%
Acid number, 0.1–1% (as sulphuric acid)
Inorganic salts (sodium sulphate), 0.1–1%
Total base number, 0–400 mg KOH/g
Metal content (calcium and barium), 2–3% normal salts, 8–13% overbased.
Inorganic chloride content, should be very low (<100 ppm) for rust inhibitors

Equivalent weight equals molecular weight for monovalent salts, e.g. sodium, but is equal to half the molecular weight for the divalent salts, e.g. calcium. The equivalent weight of the sulphonic acid before neutralisation is nearly always in the range 400–600.

High molecular weight material gives good rust preventative properties, whilst lower molecular weight material gives good emulsifying properties. If both emulsifying and rust preventative properties are required, the molecular weight distribution must be carefully controlled.

The normal salts usually have total base numbers 0–20 mg KOH/g, but overbased products have typically barium 70, calcium 300 and magnesium 400. However, 500–600 total base numbers are now probably available.

6.7 Sulphosuccinates and sulphosuccinamates

6.7.1 Sulphosuccinates

Nomenclature

$$\begin{array}{c} CH_2COOH \\ | \\ HSO_3CHCOOH \end{array}$$

Figure 6.21 Sulphosuccinic acid.

These products are esters of sulphosuccinic acid (see Figure 6.21). The acid groups can be either the monoester (Figure 6.22) or the diester (Figure 6.23). The monoesters are often called half esters.

$$\begin{array}{c} CH_2COOR \\ | \\ NaSO_3CHCOONa \end{array}$$

Figure 6.22 Monoesters.

Generic
  Dialkylsulphosuccinate — a diester

160 HANDBOOK OF SURFACTANTS

$$CH_2COOR$$
$$|$$
$$NaSO_3CHCOOR$$

Figure 6.23 Diesters.

Fatty alcohol ether sulphosuccinate — could be mono- or di-ester, but is almost always mono if the alcohol is ethoxylated
Half ester sulphosuccinate — a monoester
Sulphosuccinates — could be mono- or di-ester
Examples
Sodium di(2-ethylhexyl)sulphosuccinate — a diester
Disodium coco alcohol ethoxylate(3) monosulphosuccinate — a monoester formed from coconut alcohol (average C12 alkyl chain) ethoxylated with 3 moles of ethylene oxide (see Figure 6.24)

$$CH_2COO(CH_2CH_2O)_3C_{12}H_{25}$$
$$|$$
$$NaSO_3CHCOONa$$

Figure 6.24 Disodium coco alcohol ethoxylate(3) monosulphosuccinate.

A monoester formed from coconut monoethanolamide ethoxylated with 5 moles of ethylene oxide (see Figure 6.25)

$$CH_2COO(CH_2CH_2O)_nCH_2CH_2NHCOC_{11}H_{23}$$
$$|$$
$$NaSO_3CHCOONa$$

Figure 6.25 Disodium coco monoethanolamide ethoxylate(5) monosulphosuccinate.

Description

These products are made by esterification of maleic anhydride and then reacting with sodium bisulphite. The diester is made using 2 moles alcohol/ mole of maleic anhydride and the monoester using 1 mole alcohol/mole of maleic anhydride (see Figure 6.26). The most common diesters are made with C8 and C9 alcohols (which are readily available in bulk as they are used on a large scale for the production of phthalate plasticisers for pvc). These alcohols can be isomeric mixtures, e.g. iso-octanol, or fairly pure products, e.g. 2-ethylhexanol.

A bigger variety of alcohols is used in the preparation of monoesters, as the normal chain length of the alcohol is C12–C18, and thus detergent

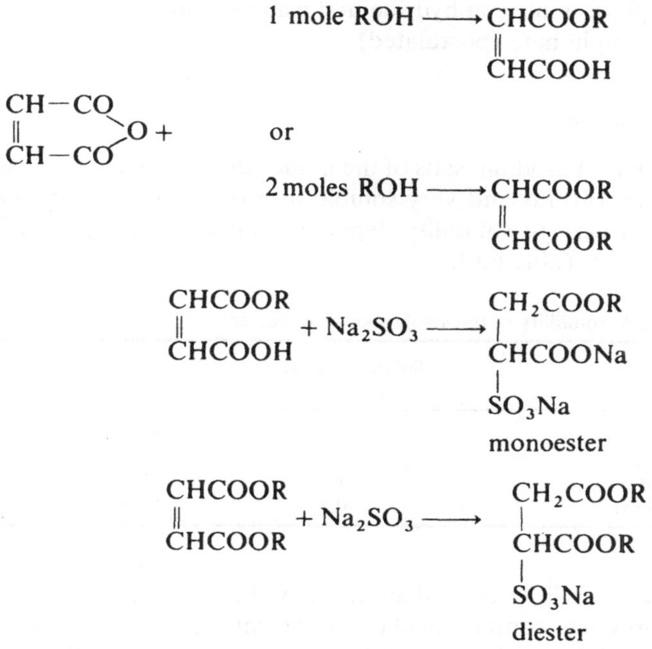

ANIONICS   161

$$1 \text{ mole ROH} \longrightarrow \begin{array}{l} \text{CHCOOR} \\ \| \\ \text{CHCOOH} \end{array}$$

$$\begin{array}{l} \text{CH—CO} \\ \| \qquad \rangle \text{O} + \\ \text{CH—CO} \end{array} \qquad \text{or}$$

$$2 \text{ moles ROH} \longrightarrow \begin{array}{l} \text{CHCOOR} \\ \| \\ \text{CHCOOR} \end{array}$$

$$\begin{array}{l} \text{CHCOOR} \\ \| \qquad + \text{Na}_2\text{SO}_3 \longrightarrow \\ \text{CHCOOH} \end{array} \begin{array}{l} \text{CH}_2\text{COOR} \\ | \\ \text{CHCOONa} \\ | \\ \text{SO}_3\text{Na} \end{array}$$

monoester

$$\begin{array}{l} \text{CHCOOR} \\ \| \qquad + \text{Na}_2\text{SO}_3 \longrightarrow \\ \text{CHCOOR} \end{array} \begin{array}{l} \text{CH}_2\text{COOR} \\ | \\ \text{CHCOOR} \\ | \\ \text{SO}_3\text{Na} \end{array}$$

diester

Figure 6.26 Production of sulphosuccinates.

alcohols, ethoxylated alcohols or ethoxylated monoalkanolamides have all been used to make the half ester. Practically any ethoxylate or propoxylate with a primary hydroxyl end group can be used in the esterification. The most common monoesters are based on ethoxylated coconut fatty alcohols or ethoxylated alkanolamides.

The sulphation is carried out after esterification, using sodium bisulphite (usually a solution of sodium metabisulphite) in aqueous solution in the presence of alcohol. Sulphation is very easy with the monoester because the starting materials (due to the free carboxyl group) have some degree of water solubility. Sulphation of the diester is not easy because the starting esters are generally insoluble in water and the reaction is difficult to start. Once product is formed the reaction proceeds rapidly due to the sulphosuccinate acting as solubiliser for the ester. The exothermic reaction can be very violent, and care must be taken in large-scale production. Excess sulphite (present in metabisulphite) or bisulphite can be removed by oxidation with hydrogen peroxide, but the bisulphite oxidation product is sodium bisulphate which is a strong acid. This strong acid or excess alkali can give hydrolysis, resulting in hazy products. Impurities include unreacted hydrophobe, trisodium sulphosuccinate (from the reaction of

162 HANDBOOK OF SURFACTANTS

sodium sulphite, maleic anhydride and water), unreacted sodium sulphite and maleic sulphonate (postulated)

General properties

1. *Solubility.* Disodium salts of the monoesters are generally insoluble in organic solvents and very soluble in water. The sodium salts of the diesters vary in solubility depending upon the chain length of the ester, (see Table 6.6).

Table 6.6 Solubility of the sulphosuccinate diesters

Ester group	Solubility in water at 2°C (%)	Solubility in paraffinic hydrocarbons
Dicyclohexyl	20	Only soluble when warm
Dihexyl	30	Soluble
Dioctyl	1	Very soluble
Ditridecyl	0.1	Very soluble

The solubilities quoted are those of the diesters, but most commercial products contain alcohol as solvent, and therefore commercial dialkylsulphosuccinates will be more soluble in small quantities of water than the above figures would suggest.

2. *Chemical stability.* Monoesters are hydrolytically unstable at acid pH and high temperatures, and are also unstable to alkali, but to a more limited extent; on storage they drift to acid pH with alkanolamides; this tends to increase viscosity but not to affect the foaming properties, whereas hydrolysis of alkyl sulphates gives reduced foam and opacity (due to the alcohol released); keep pH between 5 and 9 preferably 6–8, in formulated products.

For diesters, the di-isooctyl sulphosuccinate is stable between pH 1 to pH 10 at room temperature, i.e. it is more unstable to alkali than acid; if long-term stability is required in dilute solution, then pH 8 is probably a safe maximum.

3. *Compatibility with aqueous ions.* Monoesters have excellent mineral acid and inorganic salt tolerance; calcium tolerance of 0.5% solution >2000 ppm $CaCO_3$; good tolerance to Fe, Al and Mg ions. Diesters have moderate tolerance to hard water; calcium tolerance of 0.25% solution = 300 ppm $CaCO_3$

4. *Compatibility with other surfactants.* As anionics they are generally incompatible with cationics, but the C12 dialkanolamide monoester will tolerate small amounts of cationics without interference.

5. *Surfactant properties.* The C6–C8 diester sodium salts have a CMC of 0.06% for relatively pure products but commercial products show

ANIONICS   163

CMCs of 0.1–1.0%. Monoesters containing EO have CMCs of 0.02–0.1%, depending upon the number of moles of EO. Minimum surface tension of 26 mN/m is obtainable with a 1% C8 diester sodium salt. The commercial C13 diester sodium salts have a CMC of 0.0005–0.0015%, with a surface tension of 27 mN/m at 0.1% concentration. Minimum surface tension of monoesters is higher at approximately 30 mN/m, but it varies from 28 to 35 mN/m depending on the ester group.

6. *Functional properties.* Wetting: diesters are outstanding in having excellent and penetrating properties; Draves wetting time for 0.025% solution <25 s; monoesters are average to poor wetters. Solubilising and emulsifying: both mono- and di-esters are excellent solubilising agents, with the diesters being outstanding in their capacity to form microemulsions (see chapter 4).

7. *Skin irritation.* The monoesters are claimed to give lower irritation to eyes and skin than other anionics, and equivalent to the imidazolines. This is why the ethoxylated alcohols or ethoxylated alkanolamides are used, as these starting materials give the lowest skin irritation, but they do give somewhat poorer foaming performance than the non-ethoxylated alcohols or alkanolamides. In practice, the monoesters are often used as part replacement of ether sulphates in shampoos, as they have a detoxifying action on the ether sulphate.

8. *Disadvantages.* Poor detergency (although there are patents such as USP 4,434,087 on the use of diesters of mixed C6 and C8 chain length, which are claimed to give excellent dishwashing detergents when used with ether sulphates); hydrolysis in alkaline conditions; poor solubilisation of perfumery oils by the monoester.

Applications

Monoesters are used as follows:

1. *Shampoos and bath additives.* They behave as extremely mild foaming agents but do not have the detergent power to be the sole ingredient; synergistic effect with other surfactants, e.g. fatty alcohol sulphates and/or alpha-olefin sulphonates; low irritation is claimed for these mixtures (see Goldemberg, 1979). The longer the alcohol chain and the more EO the better the irritation reducing properties (see Sass, 1974). Alkanolamide-based monoesters have lower irritation than the fatty alcohol monoesters. Products based on *undecylenic* acid monoethanolamide as foaming agents also have fungicidal activity (see Hunting, 1981).

2. *Rug shampoos.* Good detergent, copious foam; when used with lauryl sulphate gives a dry residue easy to vacuum-clean.

164 HANDBOOK OF SURFACTANTS

3. *Polymer applications*. Emulsifier in emulsion polymerisation to give intermediate particle size in vinyl acetate/acrylate copolymers; emulsifier for acrylate emulsion polymerisation for small particle size latexes; foaming agents for latexes.

Diesters are used as follows:

1. *Textiles*. Wetting agent to improve dyestuff dispersibility and speed up the wetting and penetration of resin treatments.
2. *Dry cleaning*. Emulsifier/wetter for charge systems.
3. *Paints and printing inks*. Dispersing and flushing agents for pigments and colours into organic media.
4. *Metal treatments*. Dewatering agent with mineral oil (WD-40 type products)
5. *Agriculture*. Dispersing and wetting agent in wettable powders.
6. *Polymer applications*. Emulsifier and particle size control in both suspension and emulsion polymerisation; dispersing pigments, colours and dyes in plastics

Specification

	Monoester (Note 1)	Diester (Note 2)
% Solids	40–45	65–75
Solvent	Water	Water/isopropanol
Acid number (mg KOH/g)	—	1–3
Flash point	—	30–50°C
Solubility in water	V. soluble	Low (1% at RT)

Note 1 — disodium sulphosuccinate fatty alcohol ethoxylate(3) ester.
Note — monosodium sulphosuccinate bis(2-ethylhexyl) ester.

6.7.2 Sulphosuccinamates

Nomenclature

Generic
  Sulphosuccinamates
Example
  Sodium *N*-octadecyl sulphosuccinamate (see Figure 6.27)

$$CH_2COONa$$
$$|$$
$$NaSO_3CHCONHC_{18}H_{37}$$

Figure 6.27 Sodium *N*-octadecyl sulphosuccinamate.

ANIONICS 165

Description

The most common commercial products are: *N*-octadecyl (stearyl), a paste at 35% active; *N*-oleyl, a liquid at 35% active. The products are manufactured by reacting a long-chain fatty amine with maleic anhydride and then sulphating with bisulphite (see Figure 6.28).

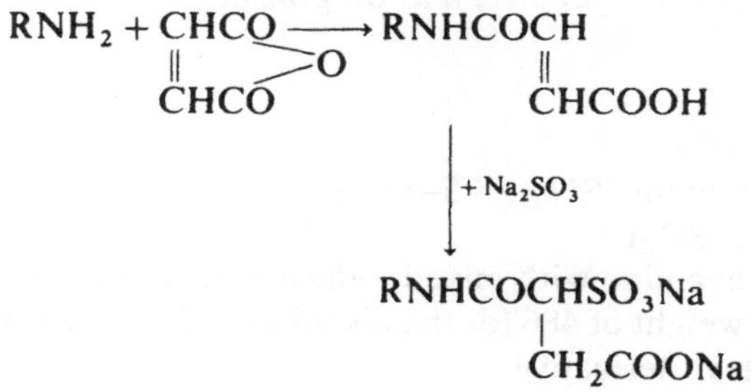

$$RNH_2 + \begin{matrix} CHCO \\ \| \\ CHCO \end{matrix} \!\!\! > \!\! O \longrightarrow \begin{matrix} RNHCOCH \\ \| \\ CHCOOH \end{matrix}$$

$$\Big\downarrow + Na_2SO_3$$

$$\begin{matrix} RNHCOCHSO_3Na \\ | \\ CH_2COONa \end{matrix}$$

Figure 6.28 Production of sulphosuccinamates.

Impurities include unreacted hydrophobe, unreacted bisulphite, and long-chain amino acids formed by addition of the amine group across the double bond of the maleic anhydride.

General properties

1. *Solubility.* Similar to the equivalent monoester sulphosuccinates, but the usual commercial sulphosuccinamates are the *N*-stearyl and *N*-oleyl, whilst the most common monoesters are made from ethoxylated alcohols or amides. Thus the sulphosuccinamates available are not very soluble in water but do disperse. They are also insoluble in most organic solvents.
2. *Compatibility with aqueous ions.* Excellent electrolyte compatibility.
3. *Chemical stability.* More stable to alkali (due to the amide bond) than the esters and hence their use in high pH latex compounds.
4. *Surface-active properties.* CMC of tetrasodium *N*-(1,2-dicarboxyethyl)-*N*-octadecyl sulphosuccinamate is 0.035–0.045%, and the lowest surface tension in water is 40 mN/m.
5. *Functional properties.* Sulfosuccinamates show good dispersing and solubilising properties with moderate wetting power. The solubilising action improves the tolerance of soaps and sulphonated oils to the effect of electrolytes.

166 HANDBOOK OF SURFACTANTS

Applications

1. *Textiles*. Carpet backing — these products were developed on a large scale for one specific end, as the foaming agent for carboxylated styrene–butadiene latex which is applied to the back of tufted carpets. These aqueous latexes were heavily loaded with chalk and the sulphosuccinamates showed outstanding ability to give stable foam in alkaline conditions.
2. *Emulsifying agent for wax and oil polishes*.

Specification

Solids content, 35%
Surface-active content (SO_3), 3.5–5.0%
Free sulphite (as SO_2)
Titration using hyamine with mixed indicator (ISO Method 2271), assuming a molecular weight of 486 for the disodium tallow derivative and 493 for the disodium oleyl derivative.
Conversion rates can be low and therefore surface-active component must be measured rather than assuming that solids represent active material.

6.8 Taurates (amide sulphonates)

Nomenclature

Generic
    Derivatives of methyltaurine, $CH_3-NH-CH_2-CH_2-SO_3$
    Sulphoalkyl amides
    n-Acyl-*N*-alkyl-taurate has the structure $RCON(R)CH_2CH_2SO_3^-M^+$
Example
    n-Cocoyl-*N*-methyltaurine has the structure $CocoCON(CH_3)CH_2CH_2$
    $SO_3^-Na^+$

Description

These compounds are prepared by reaction of fatty acid chloride on methyltaurine (see Figure 6.29). The methyl taurine is made from sodium isethionate and methylamine (CH_3NH_2). Taurine, the similar product but without the methyl group, can be made in the same way but has inferior detergent properties to methyltaurine.

Impurities include soap and sodium chloride. The sodium chloride can be removed by reverse osmosis (see Kubo *et. al.*, 1992)

General properties

1. *General*. These products are similar to the corresponding fatty acid soaps in soft water but are more effective in hard water, and are not

ANIONICS

$$RCOCl + CH_3NHCH_2CH_2SO_3 \xrightarrow{\text{NaOH}} RCONCH_2CH_2SO_3Na$$
$$| \atop CH_3$$

Figure 6.29 Preparation of taurates.

sensitive to low pH (unlike the isethionates) and better wetting agents; they are similar in many properties to the corresponding isethionates but the taurines are more hydrolytically stable than the isethionates; the taurates can be looked upon as amide derivatives of sulphates just as sarcosinates are amide derivatives of soap; both are milder than their counterparts because of the amide group; taurates perform better in hard water than sarcosinates.

2. *Solubility.* The sodium salt of C12-C14 is soluble in hot water and cold water ($>50\%$ solution at 70°C, $>50\%$ at 25°C); oleic acid derivative shows $>50\%$ solubility at 70°C, 40% at 25°C.

3. *Compatibility with aqueous ions.* Calcium and magnesium salts are soluble in high concentrations of eletrolyte.

4. *Chemical stability.* Very stable to hydrolysis under acid and alkaline conditions hot or cold.

5. *Surface active properties.* Coconut derivative surface tension at 0.1% = 32 mN/m, oleic derivative 32 mN/m.

6. *Functional properties.* Good foaming properties (0.05% concentration Ross Miles: initial foam coco derivative 115 mm, oleic derivative 98) but not as good as alkyl sulphates; does not reduce foam of soap solutions. Excellent detergents for grease and oil; also effective detergents for solid soil (better than isethionates). Excellent wetting agents even under extreme conditions of temperature and pH. Good suspending power for dirt particles; reduce redeposition when used in formulated detergents. Good lime soap dispersants. Low foam anionic (there are very few such products) when the alkyl group is cyclohexyl and the palmitoyl derivative is used, i.e.:

$$C_{15}H_{31}\text{--}CO\text{--}N(C_6H_{11})\text{--}CH_2\text{--}CH_2\text{--}SO_3^- \ Na^+$$

7. *Disadvantages.* Most commercial products contain appreciable amounts of salt.

Applications

1. *General.* Mild cleaner, foaming agent, foam boost stabiliser, conditioner; formerly used widely in shampoos but replaced by lauryl sulphates and ether sulphates.

2. *Household products.* Sea water laundering.

3. *Shampoos.* The cocoyl methyl taurate has been suggested as an ideal

168 HANDBOOK OF SURFACTANTS

ingredient to blend with AOS in shampoos, and it is claimed as being as effective as alkanolamides in boosting and stabilising the foam height of AOS; bubble baths and toilet bars, both with soap mixtures; taurates can be used with soap without affecting the foam, unlike ether sulphates which are defoamed with soap.

4. *Agriculture*. Pesticide powders — give good wetting and dispersibility.

5. *Textiles*. Disperses pigments and facilitates removal of loose colour from dyestuffs and printed goods; oleic derivative used in textile scouring (kier boiling in alkali) and in dyeing to remove loose colour and give a levelling effect on acid dyestuffs applied to wool and nylon.

Specification

Appearance, white paste or liquid
Active, 20–25%
Sodium chloride, 1–10%
Free fatty material, 1–5%

References

Adam, W.E. and Neumann, K. (1980), *Fette, Seifen, Anstrichm.* **82**(9), 367–70.
Berna, J.L. and Moreno, A. (1987) *Communicaciones a las XVIII Journadas del Comité Espanol de la Detergencia* (CED-AID) Barcelona, pp. 79–99.
Bernhardt, R.J. (1992) *Proc* 3rd. *CESIO International Surfactants Congress*, London, Vol. 3, p. 310.
Blease, T.G. *et al.* (1992) *Proc. 3rd International Surfactants Congress*, London, Vol. 3, pp. 275–280.
Brink, C. *et al.* (1992) *Proc. 3rd CESIO International Surfactants Congress*, London, Vol. 3, pp. 211–216.
Ferguson, R.H. *et al.* (1943) *Ind. Eng. Chem.* **35**, 1005–1012.
Gerstein, T. (1976) USP 3,990,991.
Goldemberg, R.L. (1979) *J. Soc. Cosmet. Chem.* **30**, 415–427.
Hart, J.R. and DeGeorge, M.T. (1982) 'The effect of conditioning ingredients on the lathering potential of anionic surfactants. Presented at the Society of Cosmetic Chemists Annual Scientific Seminar, Memphis, Tenn., USA, May 13–14, 1982.
Henkel (1992) Ger. Offen. DE 4,107,414.
Hunting A.L.L. (1981) *Cosmetics and Toiletries* **96**, 29–34.
Kassem, T.M. (1984) *Tenside, Surfactants, Detergents*, **21** (3), 144.
Knaggs, E.A, Yeager, J.A. Varenyl, L. and Fischer E. (1965) *J. Am. Oil Chem. Soc.* **42**, 805.
Kubo, M. *et al.* (1992) JP 04,149,169; *Chem. Abs.* 117:253938q.
Moreno, A., Cohen, L., and Berna J.L. (1988) *Tenside, Surfactants, Detergents*, **25**, 216–221.
Rohm and Hass, (1937) USP 2,098,203.
Rohm and Haas (1938a) USP 2,106,716.
Rohm and Haas (1938b) USP 2,115,192.
Sass, C. (1974). Sulphosuccinates and the cosmetic types available. Presented at the Society of Cosmetic Chemists Annual Scientific Seminar, Chicago, Ill., USA, May 9, 1974.
Shupe R.D. (1977). USP 4,018,278.
Trautman, T. and Jurges, P. (1984) *Tenside, Surfactants, Detergents*, **21**(2), 57–61.

7 Nonionics

7.1 General introduction

Nonionic surfactants are surfactants that do not have a charged group. The hydrophilic group is provided by a water soluble group which does not ionise to any great degree. Those groups used in practice are shown in Table 7.1.

Table 7.1 Water-soluble groups

Hydroxyl	C–OH	Poor hydrophilic properties
Ether	C–O–C	Poor hydrophilic properties
Amine oxide	N→O	Excellent hydrophilic properties
Phosphine oxide	P→O	Excellent hydrophilic properties
Sulphoxide	S→O	Excellent hydrophilic properties
Triple unsaturation	C≡C	Very poor hydrophilic properties
Ester group	COO–	Very poor hydrophilic properties
Amide group	CONH–	Very poor hydrophilic properties

One may argue that some of these groups have some ionic character, e.g. the amine oxides, which will act as cationics in acid solution. However in neutral and alkaline solution, which is where they are normally used, they behave similar to ethoxylates and polyhydroxy compounds and therefore they are included in the nonionic section.

The most common nonionic groups are the hydroxyl group (R–OH) and the ether group (R–O–R'). The water solubilising properties of a hydroxyl group or an ether group are low compared with the sulphate or sulphonate groups. If only one hydroxyl or one ether group is present the chain length of the hydrocarbon R will be only 6–8 before the product becomes insoluble and has poor surfactant properties. Thus dodecyl alcohol is practically insoluble, and aqueous solutions show poor foaming, poor detergency, poor wetting, etc. Surfactants showing desirable properties are obtained by using multiple-hydroxyl groups or multiple-ether groups to increase water solubility. In practice, the most versatile method of using ether groups is by the reaction of ethylene oxide with the hydrophobe. Ethylene oxide will react with a hydrogen atom attached to a hydrophobic group (see Figure 7.1). The amount of ethylene oxide added can be controlled by varying the amount of ethylene oxide added in proportion to that of the hydrophobe. The reaction is almost quantitative. Any free ethylene oxide can easily be removed at the end of the reaction (it is a gas

170   HANDBOOK OF SURFACTANTS

$$R-H + nCH_2-CH_2 \underset{O}{\longrightarrow} R-(CH_2-CH_2O)_nH$$

Figure 7.1 Ethoxylation.

at room temperature). The larger the amount of ethylene oxide the more water-soluble the product. The properties of the ethoxylates will depend mainly upon the amount of ethylene oxide in the hydrophilic group. Table 7.2 shows the effect of varying the ethylene oxide content.

Table 7.2 Ethylene oxide content and water solubility

Moles EO added to $C_{12}H_{25}OH$	Water solubility of product
0	Insoluble
1	Insoluble
2	Insoluble
3	Insoluble
4	Partly miscible
5	Partly miscible
6	Soluble with difficulty
7	Quite soluble

The alternative method of using multiple hydroxyl groups has so far not been utilised to the same degree in practice because there is no easy cheap method of attaching multiple hydroxyl groups on to a hydrocarbon. Nevertheless, many surfactants are based on this principle because of the widespread occurrence of multiple hydroxyl products in natural products, such as the saccharides, polysaccharides and other carbohydrates. The chemistry is complex and the intermediates are often high-melting solids which can degrade on heating. A very large amount of research has been carried out and many surfactants based on multi-hydroxyl groups are on the market. However, the polyhydroxy derivatives do not offer the formulator the same variety of liquid products with variable properties as are obtained from the ethoxylated derivatives. Nevertheless, there are significant problems surrounding ethylene oxide and its derivatives, and more efforts will probably be made in the future to find alternatives to ethylene oxide. Some space is therefore devoted to various alternatives to ethylene oxide. One relatively new product group, the alkyl polyglycosides, is being made in commercial quantities and will probably be produced in significant quantities in the future.

Examples of ethoxylates (from ethylene oxide reacting with a hydrophobe) include:

- Alcohol ethoxylates
- Mono alkanolamide ethoxylates
- Fatty amine ethoxylates

NONIONICS 171

- Fatty acid ethoxylates
- Ethylene oxide/propylene oxide copolymers
- Alkyl phenol ethoxylates

Examples of multiple hydroxyl products (from reaction of a hydrophobe with a multiple hydroxyl product by esterification) include:

- Glucosides
- Glycerides
- Glycol esters
- Glycerol esters
- Polyglycerol esters and polyglycerides
- Polyglycosides
- Sorbitan esters and sorbitan ester ethoxylates
- Sucrose esters

The description of nonionic surfactants will follow closely that of anionics, i.e. by describing groups of surfactants which have a similar hydrophilic group. However, ethoxylated products share many common characteristics which are independent of the hydrophobe and it is simpler and avoids repetition if these common characteristics are described in this section.

7.1.1 The chemistry of ethoxylation

An examination of the chemistry of the reaction between ethylene oxide and the various hydrophobes will give considerable insight into the properties of the various ethoxylates. Figure 7.2 shows the reaction mechanism in the presence of an alkaline catalyst.

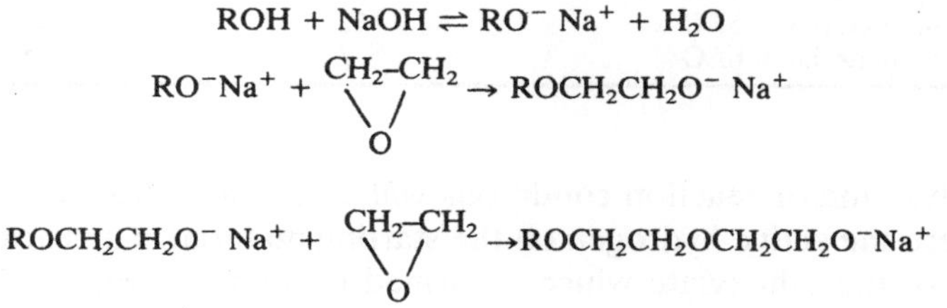

$$ROH + NaOH \rightleftharpoons RO^- Na^+ + H_2O$$

$$RO^-Na^+ + \underset{\underset{O}{\diagdown \diagup}}{CH_2\text{--}CH_2} \rightarrow ROCH_2CH_2O^- Na^+$$

$$ROCH_2CH_2O^- Na^+ + \underset{\underset{O}{\diagdown \diagup}}{CH_2\text{--}CH_2} \rightarrow ROCH_2CH_2OCH_2CH_2O^-Na^+$$

Figure 7.2 Mechanism of ethoxylation with an alkaline catalyst.

The mechanism of ethoxylation depends upon the catalyst used, but most common ethoxylates are made using an alkaline catalyst. Using alkaline catalysts, the rate of ethoxylation is dependent upon the ionisation of the active hydrogen:

172 HANDBOOK OF SURFACTANTS

Acid ionization constant of alcohol group in water is 10^{-15}
Acid ionization constant of phenol in water is 10^{-9}
Acid ionization constant of carboxylic acid in water is 10^{-5}

There are three different situations:

1. Where the active hydrogen on the starting material (ROH) is equal in reactivity to that of the ethoxylate which is formed in Step 2 in Figure 7.2, e.g. the starting material is an alcohol, an amide or water. As more ethylene oxide is added to the alcohol ethoxylate the ethylene oxide adds on in a random manner to **ANY** hydroxyl group, and therefore free alcohol remains until a very large amount of ethylene oxide has been added.
2. Where the active hydrogen on the starting material is more acidic than that of the ethoxylate which is formed in Step 2 in Figure 7.2, e.g. the starting material is a phenol, mercaptan or carboxylic acid. As ethylene oxide is added to an alkyl phenol it is **preferentially** added to the hydroxyl group attached to the benzene ring, i.e. the starting material. Thus, all the phenol group reacts before any EO goes on to the hydroxyl group on the product. The difference is shown on comparing the ethoxylation of an alcohol and an alkyl phenol with 3 moles of ethylene oxide (see Table 7.3). A change in

Table 7.3 Ethoxylation of alcohols and phenols with 3 moles of EO

EO content	Dodecyl alcohol (%)	Nonyl phenol (%)
Free starting material	22	0
Starting material + 1EO	10	10
Starting material + 2EO	14	24
Starting material + 3EO	16	27
Starting material + 4EO	15	20
Starting material + 5EO	12	11
Starting material + 6EO	7	5

catalyst and/or reaction conditions will affect these figures.

3. Where the active hydrogen on the starting material is less acidic than that of the ethoxylate which is formed in Step 2 in Figure 7.2, e.g. starting material is an amine. Using basic catalysts, very little reaction with ethylene oxide would take place because ionisation of the hydrogen on the amine in the presence of a base does not take place to any practical extent. However, amines will react readily with ethylene oxide if either water or an acid is present to give an ethanolamine (see Figure 7.3). The ethanolamine formed can then be reacted with further ethylene oxide using a basic catalyst.

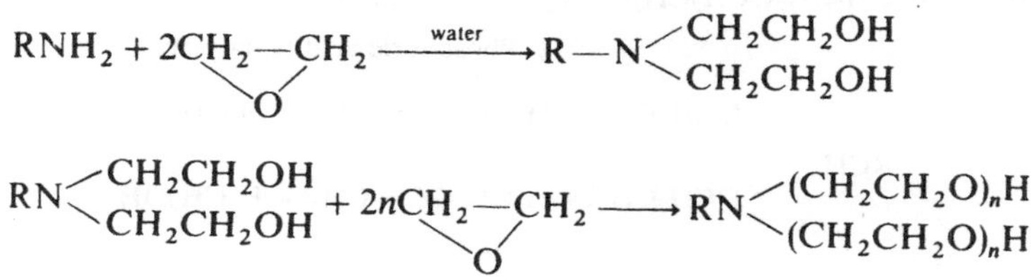

$$RNH_2 + 2CH_2\!-\!CH_2 \xrightarrow{\text{water}} R\!-\!N\!\!<^{CH_2CH_2OH}_{CH_2CH_2OH}$$

$$RN\!\!<^{CH_2CH_2OH}_{CH_2CH_2OH} + 2nCH_2\!-\!CH_2 \longrightarrow RN\!\!<^{(CH_2CH_2O)_nH}_{(CH_2CH_2O)_nH}$$

Figure 7.3 Reaction of ethylene oxide with amines.

Distribution of the polyoxyethylene chains

The ethoxylation of an alcohol or alkyl phenol gives a distribution of chain lengths. This distribution is dependent upon the catalyst used and the conditions of ethoxylation. In many applications the exact distribution is not critical, but there are specific applications where the distribution can affect the surfactant properties (see Dillan et al. 1985). The easiest way to broaden the distribution is by using a blend of two ethoxylates, and blends are used in many applications, e.g. emulsifying systems (see chapter 4).

Narrowing the distribution can be done by fractionation and separation of the different chain lengths, by synthetic routes or by changing the conditions of the ethoxylation reaction. The fractionation and synthetic routes are normally employed to make 'pure' ethoxylates for academic research, but the products so obtained are not commercially viable for practical applications.

The most practical commercial method to narrow the distribution is by varying the method of ethoxylation. A considerable number of patents have been published in the last ten years on the effect of different catalysts to control the distribution. The most common type of catalysts is the alkaline earth catalysts but these tend to be insoluble in the reaction medium. Typical catalysts are barium phosphate (Kemp, 1989), calcium acetate (Wu and Cheng, 1992; Gao et al., 1992), and barium soap (Sanracesaria et al., 1992). Other catalysts that have been proposed are antimony pentachloride (Wimmer, 1991), lanthanide nitrates (Kemp, 1992), Lewis acids (Crass, 1992), and calcined talc (Raths 1992).

Blease et al. (1992) also proposed a method for narrowing the distribution of alcohol ethoxylates by first reaction of a fatty aldehye with a glycol, then hydrogenating the acetal intermediate to give an intermediate alcohol ethoxylate and, then reacting this ethoxylate with ethylene oxide in the normal manner (see Figure 7.4). The products obtained contain very small quantities of the starting alcohol, but the distribution of the higher ethoxylates is close to a Poisson distribution and similar to that of the nonyl phenol ethoxylates. Such products are being produced in commercial quantities.

174 HANDBOOK OF SURFACTANTS

$$RCHO + HOCH_2CH_2OCH_2CH_2OCH_2CH_2OCH_2CH_2OH$$

↓ Phosphoric acid (catalyst)

$$RCH \Bigg\langle {OCH_2CH_2OCH_2CH_2OCH_2CH_2OCH_2CH_2OH \atop OCH_2CH_2OCH_2CH_2OCH_2CH_2OCH_2CH_2OH}$$

↓ H_2

$$ROCH_2CH_2OCH_2CH_2OCH_2CH_2OCH_2CH_2OH$$

↓ nEO + catalyst

$$ROCH_2CH_2OCH_2CH_2OCH_2CH_2OCH_2CH_2O(CH_2CH_2O)_nH$$

Figure 7.4 Synthesis of narrow range ethoxylates.

Ethoxylation gives high yields of ethoxylates; there are, however, four by-products which can cause problems.

1. *Polyglycols*. The hydroxyl group on water can take part in all the reactions, as ethylene oxide will react with water, particularly under basic catalyst conditions, to form polyglycols. Thus, it is necessary to remove water from the starting materials, (except for amines, where the level must be controlled), or the formation of polyglycols is inevitable. Polyglycols can be formed, however, in the absence of water. Polyglycols are often insoluble in the nonionic surfactant, and show as a haze or even separate out. Addition of water to the finished nonionic will often clear the haze, so a clear ethoxylate is no guarantee that it does not contain polyglycols. For the majority of detergents applications, small quantities of polyglycols are not detrimental, although there are a few applications where they can affect surfactant performance. They can interfere with cloud-point determinations giving an indistinct end point. In the normal ethoxylation with an alkaline catalyst, the water present and the amount of ethylene oxide will determine the amount of polyethyene glycols, but with an acid catalyst, (e.g. a Lewis acid) then much higher amounts of glycols are formed. Table 7.4 illustrates this effect.

Table 7.4 Polyglycols formed using acid and alkaline catalysts

	Products formed from C16–18 alcohol + EO (with 0.05% water)					
	C16–18(EO)$_x$H		Polyglycols		Others (1,4-dioxane, etc.)	
Moles EO	NaOH	BF$_3$	NaOH	BF$_3$	NaOH	BF$_3$
5	96	68	4	19	0	19
20	88	41	12	32	0	27

NONIONICS 175

2. *Catalyst residues*. Ethoxylates used as surfactants have the catalyst neutralised and left in the final product. The most common acids for neutralisation are acetic acid or phosphoric acid, which leave sodium acetate or sodium phosphate in the product if sodium hydroxide or sodium methoxide is used as the catalyst.

The catalyst can be removed entirely by neutralisation with an acidic clay, but this is rarely done as it is more expensive than *in situ* neutralisation. The catalyst residue itself will rarely interfere with normal applications in the presence of other electrolytes and large volumes of water. However, the catalyst residues can sometimes give hazes or even precipitates in concentrated products.

3. *Ethylene oxide*. Unreacted ethylene oxide is left at the end of the reaction and is readily removed by vacuum and heating. However, very small traces (1–25 ppm) of ethylene oxide remain in the ethoxylate, and this quantity slowly reduces with time. Very low levels of ethylene oxide are demanded for cosmetic use.

4. *1,4-Dioxane*. This is formed in small amounts (<50 ppm) in most ethoxylates, and can be removed by steam distillation. Very low levels are required for cosmetic use. As shown in Table 7.4, the type of catalyst used can affect the level of 1,4-dioxane very significantly.

7.1.2 General properties of nonionics

7.1.2.1 Solubility in water and cloud point. The solubility of EO derivatives is due to the hydrogen bond between water and the EO group. The energy of a hydrogen bond is approximately 7 kcal/mole, and heating can impart enough energy to destroy the bond. On heating an ethylene oxide derivative, dehydration takes place and the product comes out of solution; the temperature at which this takes place is known as the **cloud point**. On cooling the product dissolves. A 1% solution is usually used for determination of the cloud point, as it is dependent upon the concentration (Table 7.5).

Table 7.5 Cloud points of ethoxylates

Octyl phenol + 8.5 EO (% Concentration)	Cloud point (°C)
0.01	>100
0.015	>100
0.02	38
0.03	48
0.05	48
0.10	49
0.50	50
5.0	50

176 HANDBOOK OF SURFACTANTS

The water solubility increases as the amount of ethylene oxide increases. There is a simple rule of thumb relating the amount of ethylene oxide with the number of carbon atoms (N) in the hydrophobe to achieve water solubility; water solubility just achieved at $N/3$ moles of EO; fairly good water solubility at $N/2$ moles of EO; very good water solubility at $3N/2$ moles of EO.

The cloud point of a range of nonionics will increase as the percentage weight of ethylene oxide increases so long as the hydrophobic group is constant. This is illustrated in Table 7.6.

Table 7.6 Cloud point of C12–14 fatty alcohol ethoxylates

Moles of EO	Concn/solvent	Cloud point (°C)
2	10%25% BDEG*	50
3	10%/25% BDEG*	59
4	10%/25% BDEG*	67
5	10%/25% BDEG*	73
6	10%/25% BDEG*	77
7	2%/water	54
8	2%/water	67
9	2%/water	82
10	2%/10% NaCl	56
15	2%/10% NaCl	73
20	2%/10% NaCl	76
30	2%/10% NaCl	77

* Butyldiethylene glycol

The cloud point is very sensitive to EO content at low EO levels, and therefore glycol/water mixtures are used at very low EO levels whilst sodium chloride solutions can be used at high EO levels.

Nonionics tend to have maximum surface activity near to the cloud point. Addition of alkalis and/or inorganic salts generally lowers the cloud point (as shown for NaCl in Table 7.6), but there is no consistent pattern on the addition of inorganic acids. Addition of large quantities of inorganic salts can cause precipitation at room temperature ('salting out'). The order of salting out of inorganic anions and cations has been summarised by Meguro *et al.*, (1987) as:

$$SO_4^{2-} > Cl^- > Br^- > NO_3^- \text{ and } Na^+ > K^+ > Li^+$$

Addition of mineral acids does not usually cause reduction in cloud point and solubility. Addition of non-polar liquids generally raises the cloud point but there are exceptions (see Table 7.7). Addition of aromatic and polar aliphatic compounds generally reduces the cloud point, particularly aliphatic alcohols, fatty acids, phenols and glycols. (see Table 7.6).

Double cloud points are sometimes observed with mixtures of nonionics and some EO/PO copolymers. On increasing the temperature the solution first becomes turbid, then less turbid, then hazy and then turbid again.

NONIONICS

Table 7.7 Effect of added non-polar liquids on cloud points

Compound added to saturation	Cloud point of 1% nonyl phenol + 9EO (°C)
None	56
n-Heptane (C_7H_{16})	71
n-Decane ($C_{10}H_{22}$)	79
n-Dodecane ($C_{12}H_{26}$)	79
n-Hexadecane ($C_{16}H_{34}$)	80
Cyclohexane	54
Ethyl benzene	31
Ethylene tetrachloride	31

Nonionics with hydrophobic groups with two or three side branches have decreased cloud points compared with near linear products and do not form spherical micelles. Nonionics with the end group on the polyethylene oxide which has been 'end-blocked' have reduced cloud points. End blocking is where the end hydrogen atom on the hydroxyl group has been replaced by an alkyl or aryl group (see Figure 7.5). The replacement of the hydrogen atom by an organic group will reduce the acqueous solubility and hence reduce the cloud point. Methyl, ethyl and benzyl groups are the most common used in practice.

$$R(OCH_2CH_2)_n\, OH + R'Cl \longrightarrow R(OCH_2CH_2)_n\, OR' + HCl$$

Figure 7.5 End blocking of a nonionic and effect on cloud point.

The cloud point of most nonionics markedly increases by the addition of small quantities of ionic surfactants, whatever the charge. Table 7.8 gives some examples.

Table 7.8 Effect of other surfactants on cloud point of a 1% solution of octylphenol + 9.5 EO

Addition of other surfactant	Cloud point (°C)
Nil	66
0.02% SDS	84
0.04% SDS	89
0.02% CTAB*	78
0.04% CTAB*	88

* CTAB = cetyltrimethyammonium bromide

7.1.2.2 Phase diagrams and liquid crystals. Ethoxylated products are prone to form liquid crystal structures (see chapter 4) in aqueous solution. The very viscous solutions or even gels at concentrations of 40–70% in water (see Figure 7.6) are due to these liquid crystal structures. In order to prepare dilute solutions, ethoxylates must be added to well stirred water. To prepare concentrated solutions, water must be added to well stirred ethoxylates.

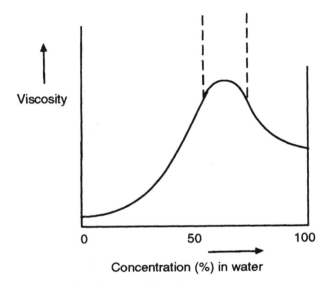

Figure 7.6 Viscosity of ethoxylates.

7.1.2.3 *Compatibility with other surfactants.* Ethoxylates are compatible with all other surfactants. This does not mean that they are inert to other surfactants. In fact, synergy is very strong with anionics, and mixed micelles with other surfactants are well known.

7.1.2.4 *Chemical stability of the polyoxyethelene chain.* Nonionics show excellent chemical stability in a very large number of aqueous formulations, particularly household products. Nevertheless, the polyethoxy chain shows behaviour similar to that of the simple ethers and undergoes oxidation very readily. Oxidation can cleave the polyethoxy chain, which will then change the surfactant properties as the degree of hydrophilicity has been reduced. The chemical stability of products where the polyethoxy chain is attached to the hydrophobe by an ester or amide group is described in the relevant section dealing with ethoxylated esters or amides. In the majority of cases oxidation will attack the polyethoxy chain before the hydrophobic chain unless unsaturation is present in the hydrophobic chain.

When free radicals are introduced into a system containing polyethoxy groups they initiate oxidation so long as oxygen is present. A chain

NONIONICS 179

reaction is set up, which is propagated by the regeneration of new free radicals. Hydroperoxy groups are intermediates which accumulate as they are more stable than the free radicals. Catalysts can have a profound influence on the formation and the decomposition of the hydroperoxides. Transition metal ions (e.g. copper, cobalt, manganese) even at very low concentrations (a few parts per million) can induce decomposition of the peroxide. Reducing or oxidising agents (e.g. ferrous ions, bleach) can accelerate peroxide decomposition. Acid catalysts can accelerate peroxide decomposition to form aldehyde groups, which can give rise to coloured compounds.

On oxidation the peroxide concentration first increases to a maximum and then decreases as the peroxy compounds degrade into the final product. A variety of final product groups can be found which depend upon the conditions of oxidation, the temperature, the catalyst, etc. The principal organic groups to be found are: carboxylic acids; aldehydes; alcohols; lactones; and esters. Some of these groups may have a better water solubility than the ether group but scission of the polyoxyethylene chain commences almost from the start of the oxidation and whatever the product groups formed. The overall effect will affect the surface-active properties, but the interpretation of the chemical effect by the observed physical effect is extremely difficult if not impossible. An example would be viscosity, which should fall if chain scission took place and nothing else changed except the molecular size. However, a change in the water solubility could change the size of the micelle, which would influence the viscosity. As the cloud point is a measure of water solubility of the hydrophilic chain, oxidation and scission of the chain would be expected to give decreased cloud points, which has been shown to be true in practice. Thus the cloud point can be used as a measure of oxidation, but again care should be taken in interpretation, as there can be many reasons for the reduction in cloud point.

The most common methods for determining hydroperoxide are iodine or arsenite titrations. However, in the presence of nonionic surfactants, as much as 40% of the iodine liberated from potassium iodide by the hydroperoxides was found to be unavailable for titration, (see Henderson and Newton 1966, 1969; Hugo and Newton, 1963). Hydroperoxide formation is increased by:

- Decreasing the surfactant concentration
- Increasing exposure to light
- Increasing temperature (e.g. sterilisation by autoclaving)
- Bleaching with hydrogen peroxide
- Decrease in pH below 6
- Presence of transition metal ions

Stabilisation of polyoxyalkylene derivatives is achieved by:

- Storage in the dark
- Minimal air access, storage under nitrogen if possible; temperature as low as possible
- Buffered aqueous solutions to neutrality
- Low concentrations avoided wherever possible
- Addition of antioxidant, but check whether the nonionic already contains an antioxidant; get the nonionic manufacturer's recommendation on the antioxidant to use.

The great majority of aqueous surfactant formulations containing polyoxyethylated surfactants will not need any further stabiliser added than that added by the nonionic manufacturer. However, the formulator should be aware of the tendency of such products to oxidise as many formulations are used in oxidising conditions.

7.1.3 Surface-active properties of nonionics

7.1.3.1 *Surface tension.* The surface tension decreases with increasing concentration in a similar manner to all surfactants, but the minimum surface tension obtained increases as the degree of ethoxylation increases, i.e the lowest water solubility gives the lowest surface tension (see Figure 7.7).

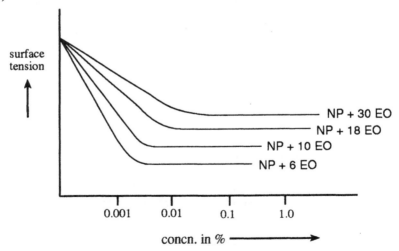

Figure 7.7 Effect of EO content on surface tension.

Measuring of surface tension has already been discussed in chapter 4, where it was noted that the value of the surface tension of surfactant solutions can depend upon the time scale of the measurement. The effect of homogeneity of the EO chain might well affect the time dependence.

NONIONICS   181

There is some evidence, as Tabata *et al.* (1984) showed, that the presence of a small amount of diether $(RO(EO)_{12}R$, where $R = C10$, C12 or C14) in homogeneous $RO(EO)_{12}H$ affected the time taken for the surface tension to reach equilibrium. In the absence of the diether, equilibrium surface tension was achieved in seconds, whilst with the diether present times were of the order of 50–100 min. The EO distribution can also affect surface tension (see section 7.1.3.3)

7.1.3.2   Micelles.   The size of the micelle for nonionics is very much larger than that for anionics or cationics (see Chapter 4) and older references gave measurements of the micelle to show that it markedly increased in size when nearing the cloud point. However, more recent work has shown that micellar size and shape is very much dependent upon the particular ethoxylate being investigated.

7.1.3.3   *Critical micelle concentration (CMC).*   In general the CMCs for ethoxylated products are much lower than for ionic materials of similar activity, being of the order of 10^{-4} mol/l. Factors affecting the CMC are:

1. *The effect of varying ethylene oxide content.* From Figure 7.4 the CMC obviously increases with increasing degree of ethoxylation but when the molar concentration is calculated the effect is by no means so pronounced (see Table 7.9).

Table 7.9 CMC and degree of ethoxylation

Product	CMC in (mol/l)	Molecular weight	CMC (%)
NP + 4EO	29.5×10^{-6}	380	0.0011
NP + 5EO	61×10^{-6}	424	0.0026
NP + 6EO	61×10^{-6}	468	0.0028
NP + 7EO	70×10^{-6}	512	0.0035
NP + 8EO	44×10^{-6}	556	0.0025
NP + 9EO	67×10^{-6}	600	0.0046
NP + 10EO	68×10^{-6}	644	0.0044
NP + 20EO	79×10^{-6}	1113	0.0088

The nonyl phenol ethoxylates were commercial unfractionated materials

2. *The effect of EO distribution.* The CMC will vary depending upon the homogeneity of the EO chain. This is illustrated in Table 7.10. The relative amounts of the isomers of the C12 alcohol mixtures were adjusted so that the average molecular weight was equal to $C_{12}E_6$. Note that the mixtures most widely separated gave the lowest surface tension. Commercial ethoxylates which contain wide EO distributions may then have advantageous properties over narrow EO distribution in certain applications.

182   HANDBOOK OF SURFACTANTS

Table 7.10 CMC and EO distribution

Product	CMC (mol/l)	Surface tension at CMC (mN/m)
$C_{12}E_6$	7.1×10^{-5}	31.5
$C_{12}E_5 + C_{12}E_8$	6.5×10^{-5}	31.0
$C_{12}E_4 + C_{12}E_8$	6.7×10^{-5}	29.8
$C_{12}E_2 + C_{12}E_8$	8.2×10^{-5}	28.58
$C_{12}E_1 + C_{12}E_8$	12.0×10^{-5}	27.3

3. *The effect of temperature.* The CMC decreases with increasing temperature; the molecule becomes more insoluble (dehydration of the polyoxyethylene chain) and thus becomes more surface-active. For anionics the CMC increases with an increase in temperature because the molecule becomes more soluble and hence less surface-active.

   What is more important is that the aggregation number or size of the micelle increases rapidly with temperature near the cloud point. Thus the micelle is considerably larger to solubilise other products. At or near the cloud point the size of the micelle becomes so large that it is visible, i.e. the nonionic comes out of solution. Thus nonionics are most efficient at solubilisation near the cloud point.

4. *The effect of added inorganic salts.* In general, added inorganic salts will result in a decrease in CMC, the effect being more dependent on the anion than the cation.

5. *The effect of added polar solvents.* If a polar solvent (e.g. ethyl alcohol) is added to lauryl alcohol ethoxylates, micelle formation no longer occurs above 25% alcohol (sodium lauryl sulphate micelles vanish at about 40% ethanol in water).

6. *The effect of side groups in the hydrophobic chain.* These increase the CMC compared to a linear chain, due to inhibition of micelle formation by the bulky side groups.

7.1.3.4 Functional properties

1. General. Excellent detergents: moderate foaming properties (this does not mean that they cannot create and/or stabilise foam in the presence of other surfactants); moderate wetting; good emulsification; do not adsorb on charged surfaces.

2. *Wetting.* For aqueous solution, HLB (see chapter 4) value approximately 7–9 is the optimum but the smaller molecular weights give better wetting (see Table 7.11). The wetting times of the best were in the order shown in Table 7.11 i.e. C8+2EO is the best and

NONIONICS

183

C18+10EO the worst. Addition of electrolytes generally reduces the wetting power.

Table 7.11 Wetting and EO content

Product	EO content for optimum wetting[a]
C8 alcohol	2EO
C10 alcohol	6EO
C16 alcohol	9EO
C18 alcohol	10EO

[a] These conclusions were made using a sinking method with a canvas disk. Different test methods give slightly different results but there is no doubt that the smaller hydrophobes give the better wetting properties.

3. *Foaming.* As a class, nonionics are moderate to low foamers, but at the optimum EO content and at their optimum temperature for foaming they are nearly as good as LABS (see Table 7.12). The data in Table 7.11 are the result of looking at various published work and experience. The exact composition of some samples is not known, and the temperature of test is not always given. Therefore these results should be treated with caution, but they do illustrate the very different results obtained with different tests for foaming.

Table 7.12 Foaming and EO content

	EO content for optimum (at 25°C)	
Product	Foam height[a]	Foam stability[b]
C8 saturated alcohol	15	4
C12 saturated alcohol	8–12	4–10
C16 saturated alcohol	20	4–8
C18 saturated alcohol	15–20	4–10
Nonyl phenol	25	5–6

[a] Ross–Miles method gives flash foam.
[b] Aeration method to measure dynamic foam stability.

Nonionic surfactants can be defoamers when they are practically insoluble (i.e. at or above their cloud point) in the system, but are more often employed as foam stabilisers for anionics, e.g. alkanolamides (see section 8.4). Foam stabilisers are soluble in the system, so a foam stabiliser for LABS at room temperature would be a C12 alcohol + 8–10 EO. LABS could be defoamed with a EO/PO

184 HANDBOOK OF SURFACTANTS

copolymer containing sufficient PO that the product is just insoluble at the temperature of use.

4. *Solubilisation*. Solubilisation is the dissolving of a water-insoluble substance in an aqueous solution of a surfactant to form a clear homogeneous solution, although microemulsions apparently show very similar characteristics (see chapter 4). Solubilisation takes place in the micelles, and as nonionics form micelles at concentrations appreciably lower than anionics or cations, it would be expected that nonionics would therefore solubilise organic compounds at lower concentrations than the charged species. In addition, the larger size of the nonionic micelles together with the solubilising property of the polyoxyethylene glycol chain as well as the hydrophobe would suggest that nonionics should be very useful for solubilising organic compounds in aqueous solution.

The solubilisation end-point for ionic surfactants can be detected by the formation of turbidity or an emulsion. This is not the case for nonionics, because it is difficult to differentiate between the limit of solubilisation and the haze formed by depression of cloud point. Thus many published data should be regarded as suspect in terms of the interpretation. From the practical point of view, whether the limit of solubility is reached by cloud point or by solubilisation would not seem to matter. However, the reason for the incompatibility needs to be understood. In general, maximum solubilisation is of the order of 1–3 moles of solubilisate per mole of surfactant. Thus, do not expect a 0.1% solution of a nonionic to solubilise 25% of a hydrocarbon. A microemulsion, however, could 'solubilise' a much higher amount of insoluble material.

Some generalisations as a guide for dilute solutions (0.1–1% surfactant solution) are as follows:

- Hydrocarbons show least solubilisation decreasing with increasing length of the hydrocarbon chain.
- Aromatic hydrocarbons show moderate solubilisation.
- Polar compounds, e.g. alcohols, amines, ketones, and fatty acids, show the highest solubilisation, decreasing with increase in molecular weight.
- Mixed surfactant systems can show synergism.
- Solubilisation is most effective with high HLB values (>15) (see chapter 4).

For solubilisation with concentrated solutions, the concept of a globular micelle surrounded by an aqueous phase is no longer valid, and thus solubilisation depends upon the structure of the surfactant in solution (see chapter 4).

For solubilisation in nonaqueous solutions, surfactants form

NONIONICS 185

micelles in nonaqueous solution. Thus nonionics which will dissolve in mineral oil can solubilise water or aqueous solutions in the mineral oil.

Data on solubilisation are not often published, but a useful list of early literature references is given by Nakagawa (1967). References are given for nonionics that will solubilise saturated hydrocarbons, unsaturated hydrocarbons, halogenated hydrocarbons, alcohol, phenol, acids, esters, ethers, amines, vitamins, steroids, dyes, iodine and water.

5. *Emulsification*. Emulsification, in this section, is where the emulsion is an opaque white liquid with two distinct separate phases. For microemulsions, see chapter 4. Nonionics have found wide use as emulsifying agents. In practice the most stable emulsions are often made by two or more surfactants of differing hydrophobic/ hydrophilic properties (HLB, see chapter 4). The addition of ethylene oxide in various proportions can provide such variations and the use of emulsification systems based on products with varying ethylene oxide content is the best practical way of choosing surfactants for any emulsification problem. The major systems in use are:

- Sorbitan esters and their ethoxylated derivatives (section 7.11)
- Alkyl phenol ethoxylates (section 7.15)
- Alcohol ethoxylates (section 7.3)
- Fatty acid ethoxylates (section 7.10)

In each of these systems there are products of varying solubility from soluble in mineral oil to soluble in water, enabling both water in oil (W/O) and oil in water (O/W) emulsions to be made (see chapter 4). Nonionics are also less affected by the presence of electrolytes and pH changes in the water phase. For choice of emulsifier, see Table 7.13).

Table 7.13 Emulsification and HLB number

HLB number	Application
2–7	Water in oil
7–18	Oil in water

There are many published lists of HLB values for nonionics, and the HLB value is an excellent way of indicating the degree of water or oil solubility of a product. The author considers that knowing the ethylene oxide content and hydrophobic structure and chain length is

186 HANDBOOK OF SURFACTANTS

probably of greater value in assessing the emulsifying properties of a nonionic surfactant. Lacking that information, the HLB value is useful. For HLB values required to emulsify various oil phases, see Schick, (1967)

6. *Dispersing properties.* Nonionics help to disperse organic or inorganic particles in aqueous or non-aqueous systems. They help in wetting out, reducing the work needed for dispersion, and prevent aggregation of the particles and reduce flocculation and settling. Polymeric water-soluble compounds (e.g. starch, polyvinyl alcohol) are used for this purpose and the polyoxyethylene chain in the nonionics behaves in a similar manner.

7. *Detergency.* There is a very large volume of information on the optimum carbon chain length of hydrophobe and the ethylene oxide content of nonionics for detergency, much of it contradictory. The main problems are the test method used and the other components in the detergent. Work on distilled fractions of ethoxylated dodecyl alcohol showed that whiteness retention and soil removal were optimised at different ethylene oxide levels, i.e. soil removal was best at C12 alcohol + 7–8 EO but whiteness retention was best at C12 alcohol + 4–5 EO. Thus the wide distribution of the ethoxylated nonionics may be an advantage for practical detergency. If one examines the published data on nonyl phenol ethoxylates, the optimum detergency for cotton would seem to be in the area of 8–10 moles of ethylene oxide. HLB (see chapter 4) values in the range 13–15 have been quoted as the optimum. Comparison of ethoxylated fatty acids, fatty acid amides, fatty alcohols and mercaptans (see Kassem, 1984) showed that the C12 chain was the optimum hydrophobe chain length for all the products examined. However, other people (Shell Technical Data Sheets) have shown that optimum detergency is obtained with 50/50 mixtures of C14 and C15 with EO content at 11 moles.

The type of soil also has a big effect on the efficiency of the detergent, and nonionics, in general, are more efficient in removing oily and organic dirt than inorganic or polar dirt. The effect of concentration is important, with optimum concentration at 0.1% or greater particularly if the nonionic is well below the cloud point. Note that this is several orders of magnitude above the CMC. The temperature is important, with the optimum detergency at or near the cloud point.

In practice, nonionics are usually used in blends with anionics for heavy duty performance, with the anionics in the larger proportion. More recently, the amount of nonionics has increased relative to the anionics in lower temperature powder detergents. In liquid detergents the nonionics are used at even higher concentrations.

NONIONICS 187

7.2 Acetylenic surfactants

Nomenclature

Generic
  Acetylenic diols
  Acetylenic glycols
  Acetylenic surfactants
  Surfynols — a trade mark of Air Products Inc.
Examples
  Dimethyl octynediol or 3,6-dimethyl-4-octyne-3,6 diol (see Figure 7.8)
  2,4,7,9-tetramethyl-5-decyn-4,7-diol(TMDD) (see Figure 7.8)

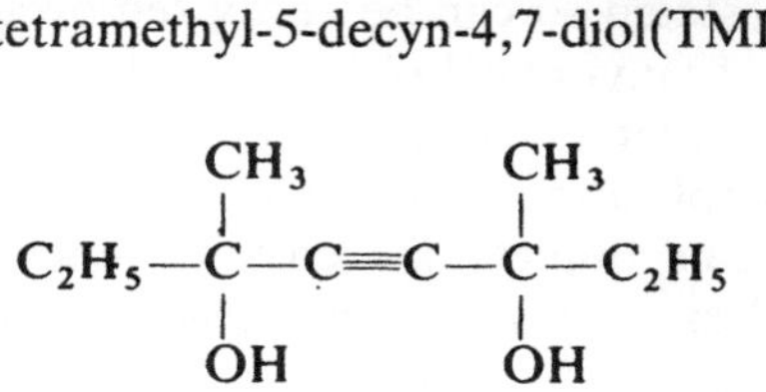

$$C_2H_5-\underset{\underset{OH}{|}}{\overset{\overset{CH_3}{|}}{C}}-C\equiv C-\underset{\underset{OH}{|}}{\overset{\overset{CH_3}{|}}{C}}-C_2H_5$$

3,6-dimethyl-4-octyne-3,6 diol

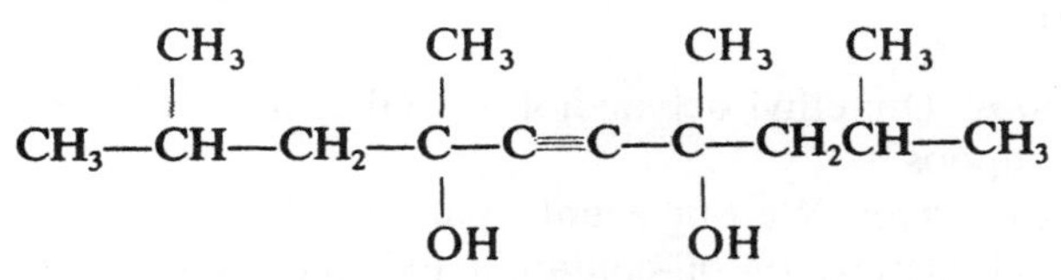

$$CH_3-CH-CH_2-\underset{\underset{OH}{|}}{\overset{\overset{CH_3}{|}}{C}}-C\equiv C-\underset{\underset{OH}{|}}{\overset{\overset{CH_3}{|}}{C}}-CH_2\overset{\overset{CH_3}{|}}{CH}-CH_3$$

2,4,7,9 tetramethyl-5 decyn-4,7 diol

Figure 7.8 Acetylemic surfactants.

Description

The products are difficult and expensive to make from acetylene. They are usually in the form of a solution in a hydroxylic solvent, but are also available in the form of free flowing powders as adsorbed active surfactant on finely divided silica.

The hydroxyl groups may also be reacted with ethylene oxide. Ethoxylation of hydroxyl groups increases water solubility without loss in surface-active properties, but the products become liquid and lose steam volatility.

General properties

1. *Appearance*. Crystalline solids (rare with nonionics). Dimethyl octynediol is a white odourless crystalline solid, m.p. 50°C.
2. *Solubility*. Low solubility in water unless ethoxylated; soluble in

188   HANDBOOK OF SURFACTANTS

alcohols and glycols; soluble in polar solvents, e.g. ketones, butyl cellusolve.

3. *Surface properties*. Equilibrium surface tenstion of 0.1% TMDD = 33.1 mN/m. Dynamic surface tension (see chapter 4) of 0.1% TMDD at 6 bubbles/sec = 35–36 mN/m (compared to octyl phenol+7EO of 42 mN/m).

4. *Functional properties*. Low foam; reduces foam of anionic and nonionics whilst increasing wetting; excellent wetting agents at low concentrations; in combination with other surfactants, there is a synergistic effect giving even better wetting; low molecular weight products are volatile with steam so can be easily removed from the system; reduces viscosity of vinyl plastisols and starch solutions.

5. *Disadvantages*. Expensive on an active basis compared to commodity surfactants, but in some applications very small quantities of active are needed, and therefore can compete on a cost/efficiency basis; very low solubility in water unless ethoxylated; unstable to acids.

Applications

1. *Shampoos*. Dimethyl octynediol is used as a solubiliser and clarifier for shampoos.

2. *Surface coatings*. Wetting agents, particularly for aqueous systems of industrial coatings on oil-contaminated steel to give better coverage and good recoatability; defoamer and pigment dispersing agent.

3. *Oil field chemicals*. Corrosion inhibitors.

4. *Agriculture chemicals*. Additive to wettable powders to give low foam, good wetting and increased redispersability.

Specification

Appearance: White waxy solid with sharp melting point (usually 100% active), liquid (50–75% active in alcohols or glycols) or free-flowing powder (adsorbed on silica at about 0.5% active).

Cloud point: <100°C — likely to be an ethylene oxide adduct.

7.3 Alcohol ethoxylates

Nomenclature

Abbreviations
    AE, used in this book
    FAE

NONIONICS

Generic
Alcohol ethoxylates
Alkoxyalkanols
Alkyl polyoxyethylene glycols
Ethoxylated fatty alcohols
Monoalkylpolyethylene glycol ethers.
Polyoxyethylene alcohols
Polyoxyethylated fatty alcohols
Polyoxyethylenated straight-chain alcohols
Example
Coco alcohol + 5EO — this represents the alcohol derived from coconut
fatty acid reacted with 5 molecules of ethylene oxide.

Description

The main commercial products are based on the following alcohols:

- Mixed coconut oil fatty acid fractions, hydrogenated C12–C14 chain length
- Synthetic straight chain C12–C18, various fractions based on Ziegler alcohols, i.e. even numbered chain lengths
- Synthetic straight chain C12–C15, various fractions based on oxo alcohols, i.e. odd and even numbered chain lengths
- Natural alcohols, e.g. castor oil
- Tallow derived C16–C18 products with high oleyl content, which are more fluid than the saturated derivatives
- Hardened tallow C16–C18, hydrogenated products with low unsaturation

The ethoxylation of a primary alcohol gives a distribution different from that of a nonyl phenol or carboxylate (see section 7.1.1). The result is that free alcohol remains in most ethoxylates when the degree of ethoxylation is not high; thus for the reaction between dodecanol and 3 moles of EO the resulting ethoxylate has the composition shown in Table 7.14. However, there are now AE appearing on the market with very low amounts of free alcohol, particularly via the acetal route (Blease *et al.*, 1992), and a much narrower EO distribution (see section 7.1.2).

In end-blocked nonionics and low foam alcohol ethoxylates, the terminal OH may be reacted with propylene oxide to give reduced foam but the degree of biodegradability decreases as the propylene oxide content increases. In addition, the terminal H on the EO or PO chain can be replaced with an alkyl group or the benzyl group. These are used to chemically stabilise the products. If the end group is bulky, e.g benzyl, then foam reduction is also obtained. A number of newer low-foam alcohol ethoxylates are now available:

HANDBOOK OF SURFACTANTS

$$R-O-(EO)_n - (PO)_m - R'$$
$$R-O-(EO)_n - R''$$
$$R-O-(EO)_n - benzyl$$
$$R-O-(EO)_n - (BuO)_m - H$$

where R = aliphatic alcohol C10–C16, R' = alkyl group C1–C3, R'' = alkyl group C3–C5, BuO = butylene oxide

Table 7.14 Dodecanol + 3EO

Component	%
Free dodecanol	22
Dodecanol + 1EO	10
Dodecanol + 2EO	14
Dodecanol + 3EO	16
Dodecanol + 4EO	15
Dodecanol + 5EO	12
Dodecanol + 8EO	3
Dodecanol + 12EO	0.5

Minor components and impurities are free alcohol, polyglycols, free ethylene oxide (very small quantities) and 1,4-dioxane (very small quantities).

General properties

1. *Solubility.* When the EO content increases beyond 5 moles EO/mole of alcohol, there will be very little difference in solubility between a C12, a C12–14 or a C12–15 alcohol ethoxylate. For solubility of a C12–14 alcohol ethoxylate, see Table 7.15. For solubility of oleyl and tallow ethoxylates, see Table 7.16.

Table 7.15 Solubility of a C12–C14 alcohol ethoxylate (S = soluble, D = dispersible, I = insoluble)

	+1EO	+3EO	+5EO	+7EO	+15EO
HLB	3.5	8	10	12	15
10% in water	I	D	D	S	S
10% in mineral oil	S	S	D	D	I
10% in white spirit	S	D	D	D	D
10% in aromatic solvent	S	S	S	S	S
10% in perchlorethylene	S	S	S	S	S

2. *Cloud point.* Highly di- or tri-methyl branched alcohols are significantly more hydrophobic than their more linear counterparts, and more EO has to be added to achieve the equivalent cloud point.

NONIONICS 191

Table 7.16 Solubility of oleyl and tallow ethoxylates

	Oleyl+2EO	Oleyl+10EO	Tallow+10EO	Tallow+30EO
HLB	5	12	12	17
10% in water	I	S	D	S
10% in mineral oil	S	D	I	I
10% in white spirit	D	D	D	I
10% in aromatic solvent	S	S	D	I
10% in perchlorethylene	S	D	D	D

S = soluble; D = dispersible; I = insoluble.

3. *Chemical stability*. Excellent to acids, but the free hydroxyl group is sensitive to concentrated alkali and turns brown in powdered products. Unstable to high pH and oxidising agents (e.g. hypochlorite bleach) (see section 7.1.2).
4. *Compatibility with aqueous ions*. Excellent to hard water.
5. *Volatility*. AEs contain free fatty alcohol and a high proportion of low EO chains. When such products are heated they volatalise, e.g. in textile lubricants they can give off vapours that can affect the eyes, and 'pluming' occurs during spray-drying manufacture of detergent powders. The narrow-range ethoxylates (section 7.1.2) reduce or even eliminate these problems.
6. *Surface-active properties*. CMC increases as the EO content increases. For a C12–C14 alcohol mixture the minimum CMC is at about 0.001–0.003% at low EO levels, rising to 0.02–0.04% at high (15–20) EO; For a C12–14 alcohol mixture the surface tension at 0.01% concentration is at a minimum level of 29 mN/m at 3EO increasing to 50 mN/m at 30 EO. C16 to C18 linear alcohols show similar behaviour in that the minimum surface tension obtainable decreases with decreasing EO content. Lowest surface tension is 29 mN/m at 5–7EO, i.e. where AE is nearly insoluble, rising to 40 mN/m at 15 EO. The branched-chain alcohols C10–C12, however, have a constant surface tension (29–33 mN/m) whilst the EO content varies from 5 EO to 30 EO.

Table 7.17 Functional properties

Alcohol	Moles EO	Applications
Lauryl	3	Viscosity control; emulsifier
Lauryl	12	Foaming agent; emulsifier; solubiliser; detergent
Lauryl	23	Solubiliser
Ceto/stearyl	5	Thickener; opacifier; emulsifier
Oleyl	20	Solubiliser for perfumes

192 HANDBOOK OF SURFACTANTS

7. *Functional properties*. See Table 7.17. By choosing the appropriate alcohol and the appropriate degree of ethoxylation the alcohol ethoxylates can give the following properties.

Excellent wetting — linear alcohol C8–C14 with 7–12 EO. For each alcohol there is a sharp optimum of EO content with the wetting power decreasing rapidly with increasing EO content above the optimum; branched-chain alcohol ethoxylates show better wetting than straight-chain alcohols of the same molecular weight.
Good flash foam but poor foam stability; linear alcohols generally give better foam than branched-chain alcohols, most stable foams with C10–14 alcohols obtained with 8–12 EO content; hard water has only a minor effect on flash foam but decreases foam stability.
Excellent dispersing power — optimum alkyl chain and EO content depends upon the material being dispersed, but high molecular weight products are better than low molecular weight.
Excellent emulsifying properties — optimum alkyl chain and EO content depends upon material being emulsified.
Excellent detergency — optimum linear alcohol C12–15 with 6–15 EO. Higher alcohols, e.g. C18, show excellent detergency but need higher EO levels (15–20 EO). The exact EO level required will depend upon the type of soil, the temperature of use and the other components in the formulation.

8. *Disadvantages*. Compared with NPEs, the normal AEs contain free alcohol in appreciable quantities at low levels of EO; but see section 7.1.2.

Applications

1. *Intermediates for sulphation*. The 2 and 3 mole EO products are sulphated for use in detergents and cosmetics (see section 6.5.2). The narrow-range EO products will give different properties to the normal range.
2. *Household products*. Heavy-duty powder detergents; major component of low and medium foam detergents C12–C14 alcohol + 9 EO; heavy-duty liquid detergents, more soluble than LABS for use in high active heavy-duty liquid detergents that are free of or low in phosphates; used with LABS as solubiliser and foam stabiliser — usually C12–15 + 7–9 EO.
3. *Industrial detergents*. Metal cleaning, low-foam products stable to alkali, e.g. benzyl-blocked; machine cleaning of food soils, low foam products, benzyl or propylene oxide tipped.
4. *Textiles*. Wide variety of AEs are used for scouring, depending upon the temperature, e.g. C13 + EO (HLB 10.5) for low-temperature

textile scouring; lubricants and antistatic agents in fibre processing — tallow + 20 EO; also mineral oil and vegetable oils are used as lubricants with AEs as emulsifiers; emulsifiers as dye carriers (better emulsifiers than corresponding NPEs) and for hydrophobic glycerides in blended fibre lubricants; dyeing assistants for wool/synthetic blends — tallow + 20–40 EO; in cotton processing as penetrants, wetting agents and dyeing assistants.

5. *Emulsifiers.* For W/O emulsion in mineral oil, paraffin and chlorinated solvents, C 12–15 + 3 EO; for O/W emulsions of waxes, fats and oils, C 12–15 + 7–8 EO; for solvent emulsifier in dye carriers, C12–15 + 23 EO. The higher alcohols such as oleyl and ceto-stearyl are used as emulsifers in cosmetics to emulsify oils and fats, a combination of a lower (2–3 EO) and a higher (10–20 EO) ethoxylate can give O/W and W/O emulsions; the presence of some free alcohol (i.e. oleyl or stearyl alcohol) is not detrimental to odour as would be the case with the lower alcohols, e.g. lauryl alcohol; castor oil + 5EO emulsifiers and dispersing agents give antifoam properties to chlorinated solvents.

6. *Agriculture.* Excellent emulsifiers for pesticide sprays because they are unaffected by hard water and pH changes; used in emulsifiable concentrates e.g. castor oil + 30 EO; water-soluble emulsifiers used in conjunction with anionics in agricultural herbicides.

7. *Paper industry.* Rewetting agents to improve absorbency, e.g. paper towels; repulping aids for waste paper.

8. *Emulsion polymerisation.* Excellent emulsifier for emulsion polymerisation due to tolerance to inorganic ions and pH changes.

Specification

Active content: usually 100%, but some solid products have water added and are sold as low as 80% active. The reason for adding the water is to supply a liquid product rather than a paste or solid. Small amounts of water are often added to hide the haze caused by polyglycols.

Water content: see % active above

EO content: usually supplied by manufacturer — difficult to check, see hydroxyl number and cloud point.

Cloud point: room temperature to 90°C indicates EO content (see section 7.1.2.). If cloud point is indistinct this may be caused by high polyglycol content.

Polyglycol: very variable, 0.05–5%; low amounts (up to 2%) are found in low EO products and 3–5% found in highly ethoxylated materials.

Sap value: 150 (low EO) to 15 (high EO) for castor oil derivatives

Hydroxyl number: for high amounts of ethylene oxide (< 20 OH number mg KOH/g) the hydroxyl number is difficult to measure accurately.

194 HANDBOOK OF SURFACTANTS

Viscosity: can be used as a rough measure of EO distribution due to the viscosity, preferentially measuring the high molecular part of the distribution.

7.4 Alkanolamides

Nomenclature

Generic: the products descibed in this section are the N-acyl derivatives of monoethanolamine and diethanolamine. The polyethoxylated alkylamides are described in section 7.7 The distinction between these groups of products is shown in Figure 7.9. There are two types of alkanolamides: Type 1:1, the 'superamide', which can be mono- or di-alkanolamide (see Figure 7.9); the word 'superamide' usually refers to the dialkanolamide; and Type 2:1, the Kritchevsky alkanolamide, which consists of the superamide plus free alkanolamine.

$$RCONHCH_2CH_2OH \quad \text{monoalkanolamide}$$

$$RCON{\Large\langle}^{CH_2CH_2OH}_{CH_2CH_2OH} \quad \text{dialkanolamide}$$

$$RCONH(CH_2CH_2O)_nH \quad \text{polyethoxylated monoalkanolamide.}$$

Figure 7.9 Alkanolamides.

Abbreviations
   Amide, alkyl (usually coconut) diethanolamide
   CD — coconut diethanolamide, used in this book
   CE, coconut diethanolamide
   CMEA, coconut monoethanolamide
   LDEA, lauric diethanolamide
   LMEA, lauric monoethanolamide
   LMP, lauric monopropanolamide (made from isopropanolamine)

Note: suppliers often do not differentiate between the superamide and the Kritchevsky alkanolamide, but the difference is shown under Specification, below.

Description

The 'superamide', Type 1:1: for the reaction between one mole of the ester of a fatty acid and one mole of mono- or di-ethanolamine, see Figure 7.10.

NONIONICS   195

R′ is usually Me and MeOH taken off; if R′ = glycerol, i.e. the free glyceride is used, then the glycerol will be left in the alkanolamide. Impurities will include small amounts of free diethanolamine (1–3%), ester amide (see Figure 7.10), soap and possibly glycerol (if free oil is used).

$$RCOOCH_3 + NH_2CH_2CH_2OH \longrightarrow RCONHCH_2CH_2OH + CH_3OH$$

or

$$RCOOCH_3 + NH\big\langle{}^{CH_2CH_2OH}_{CH_2CH_2OH} \longrightarrow RCON\big\langle{}^{CH_2CH_2OH}_{CH_2CH_2OH} + CH_3OH$$

Formation of ester–amide

$$2RCON\big\langle{}^{CH_2CH_2OH}_{CH_2CH_2OH} \longrightarrow RCON\big\langle{}^{CH_2CH_2OCOR}_{CH_2CH_2OH} + NH\big\langle{}^{CH_2CH_2OH}_{CH_2CH_2OH}$$

ester–amide

Figure 7.10 Superamide of Type 1:1.

Kritchevsky, Type 2.1: made by reaction between one mole of the fatty acid or ester and two molecules of diethanolamine. There is no Kritchevsky monoalkanolamide (see Figure 7.11). Impurities include molar amounts of alkanolamine (25–30% is common), ester amide (formed in the same way as for Type 1:1; see Figure 7.10), free fatty acid (at least 5%), dihydroxy ethyl piperazine (formed when two moles of diethanolamine react and split out two moles of water; this occurs when reactions are at high temperature 160–180°C). The reactions indicated in Figures 7.10 and 7.11 are simplifications due to further reactions taking place, involving the esteramide with itself and with fatty acid.

$$RCOOH + 2NH\big\langle{}^{CH_2CH_2OH}_{CH_2CH_2OH} \longrightarrow RCON\big\langle{}^{CH_2CH_2OH}_{CH_2CH_2OH}$$
$$+ NH\big\langle{}^{CH_2CH_2OH}_{CH_2CH_2OH}$$

Figure 7.11 Kritchevsky or Type 2:1 alkanolamides.

General properties

1. *General.* The superamides (Type 1:1) have low free alkanolamine and are therefore more suited for household and personal use. They have poor water solubility at room temperature. The Kritchevsky

196 HANDBOOK OF SURFACTANTS

dialkanolamides have good water solubility at room temperature but high free diethanolamine. Therefore they are more suited to industrial use. The free fatty acid and free diethanolamine are both necessary to obtain the water solubility of the product. The reason for this is that the diethanolamine fatty acid soap forms a co-micelle with the dialkanolamide giving the high solubility in water.

Table 7.18 Solubility of cocomonoethanolamide (I = insoluble, D = dispersible)

1% in water	I	10% in water	I
1% in mineral oil	I	10% in mineral oil	I
1% in white spirit	I	10% in white spirit	I
1% in aromatic solvent	D	10% in aromatic solvent	D
1% in perchlorethylene	I	10% in perchlorethylene	D

2. *Solubility*. Type 1:1 C12 monoethanolamide is solid at room temperature (see Table 7.18). Type 1.1 C18 monoethanolamides is solid (m.p. 80°C), disperses in water, but is not soluble. Type 1:1 C12 diethanolamide is semi-liquid at room temperature (see Table 7.19). Blends with unsaturated acids or lower acids are liquid at room temperature. For Type 2:1 C12–C14 diethanolamide, see Table 7.20.

Table 7.19 Solubility of Type 1:1 cocodiethanolamide (D = dispersible, S = soluble)

1% in water	D	10% in water	D
1% in mineral oil	D	10% in mineral oil	D
1% in white spirit	D	10% in white spirit	D
1% in aromatic solvent	S	10% in aromatic solvent	S
1% in perchlorethylene	S	10% in perchlorethylene	S

Table 7.20 Solubility of Type 2:1 cocodiethanolamide (S = soluble, D = dispersible)

1% in water	S	10% in water	S
1% in mineral oil	D	10% in mineral oil	D
1% in white spirit	D	10% in white spirit	D
1% in aromatic solvent	D	10% in aromatic solvent	S
1% in perchlorethylene	S	10% in perchlorethylene	S

3. *Compatibility with other surfactants*. Type 1:1 dialkanolamides are compatible with anionics and cationics over a wide pH range. Type 2:1 is not compatible with cationics, due to fatty acid.
4. *Chemical stability*. All alkanolamides are unstable to hydrolysis, but Type 1:1 monoethanolamide is more stable to hydrolysis than the corresponding diethanolamide, particularly under alkaline conditions. Ethoxylation of the monoamides gives superior alkaline and

acid stability. The isopropanolamides are more stable in alkaline soution than the corresponding ethanaloamides, but not in acid.

5. *Functional properties.* Foaming properties: good flash foam and stable foam with Type 2:1. Foam stablisation is shown with all the alkanolamides; alkanolamides stabilise foam in anionics when hard water and sebum or soil are present (if alkanolamide is absent, then anionic foam will collapse in hard water/sebum); some alkanolamides boost or increase the volume of foam of an anionic in the presence of soil; alkanolamides do not always increase the amount of foam; in fact, they can decrease it, because some surfactant is often needed to solubilise the alkanolamide. Solubiliser: for anionics, monoethanolamide is a good solubiliser for lauryl sulphate. Emulsification: both Type 1:1 and Type 2:1 are excellent emulsifiers for O/W; type 2:1 C16–C18 are excellent emulsifiers for mineral oil. Excellent lime soap dispersant. Poor wetting with Type 1:1 products, excellent wetting with Type 2:1 products. Poor detergent properties with Type 1:1 products but synergistic with other surfactants which show these properties; excellent detergent properties with Type 2:1

6. *Viscosity effects.* Addition of alkanolamides in themselves to anionic surfactants does not increase the viscosity. It is only when electrolyte (salt), oils or fatty alcohols are present that the viscosity increases. Generally monoethanolamides are superior to diethanolamides in increasing the viscosity of anionics. This is called 'building' the viscosity. The viscosity building effect is inversely proportional to their water solubility, i.e. the lower the water solubility the higher the 'built' viscosity. Type 1:1 dialkanolamides prepared via the methyl ester are better viscosity builders than those prepared from fatty acids (see Milwidsky and Holtzmann, 1972). Monoisopropanolamides are more effective than monoethanolamides or diethanolamides for increasing the viscosity of sodium lauryl sulphate.

7. *Disadvantages of Type 1:1.* Low solubility in water

8. *Disadvantages of Type 2:1 compared with Type 1:1.* High alkanolamine content; fatty acid makes it incompatible with cationics; complex mixture, difficult to analyse and check for quality.

Applications

Type 1:1 applications are:

1. *Household products.* The most common products available are the products based on coconut oil, but some products based on lauric acid are also available. These products were at one time extensively used in detergents and household products for their good detergent properties combined with foam stabilising action on anionics. In

198 HANDBOOK OF SURFACTANTS

Europe they have been largely replaced by nonionics based on ethoxylates of lauryl alcohol and ether sulphates. However mono cocoalkanolamide is still used as a foam stabiliser in powder laundry detergents.

2. *Shampoos and bubble baths.* The most useful property of dialkanolamides is their ability to stabilise the foam and thicken solutions of anionic surfactants, e.g. alcohol sulphates and ether sulphates in shampoos and bath additives. Type 1:1 is more effective than Type 2:1. Recently their use in personal care products has decreased due to problems of nitroso content (see Specification), and replacements are appearing on the market as foam stabilisers and thickeners for ether sulphates.

Type 2:1 applications are:

1. *Detergents.* The C12–C14 alkanolamides can be used as detergents but are more often used for formulating built liquid detergent formulations, including hard surface cleaners, floor cleaners and wax strippers. The addition of dialkanolamides to LABS has a synergistic effect on detergency.

2. *Metal working.* The other major range of products available is the 2:1 based on tall oil fatty acid, soya bean acid or oleins. These products are oil-soluble and are excellent emulsifiers for mineral oil in water combined with corrosion inhibition from the free alkanolamines. These products are mainly used in cutting oils and metal-working fluids. Alkanolamides react readily with boric acid to give 'borate esters'. Such products are used in metal cutting oils as corrosion inhibitors/biocides, but the chemical structure of the 'ester' is now doubtful.

Specification

Typical analysis (for cocodiethanolamide) is shown in Table 7.21.

7.5 Amine oxides, phosphine oxides and sulphoxides

Nomenclature

Generic
    N-Alkyl amidopropyl-dimethyl amine oxides
    N-Alkyl bis (2-hydroxyethyl) amine oxides
    N-Alkyldimethyl amine oxides
    Amine oxides
Examples
    Coco amido propyl dimethyl amine oxide (see Figure 7.12)

NONIONICS 199

Table 7.21 Analysis of alkanolamides[a]

	Type 1:1 (superamide)	Type 2:1 (Kritchevsky)
Appearance	Liquid–paste	Amber or straw liquid
Active (%) (amide content)[b]	80–90	45–66
Free amine (%)[c]	1–6	20–30
Free acid (as soap) (%)	0.1–1	5–10
Amine soap (%)	1–2	8–10
pH (1% soln)	8–10	8–11
Nitroso content	_[d]	_[d]
Ester–amide (%)	1–3	3–5

[a] Alkanolamides are difficult to analyse and discrepancies are often obtained between different laboratories unless the same procedures are followed. The figures for Type 1:1 are via the methyl ester.
[b] The active (amide) content is not usually measured directly but obtained by difference.
[c] Type 1:1 has lower level of free amine (1%) in monoethanolamide than diethanolamides (5%).
[d] Nitroso content: there is concern over the use of alkanolamides in cosmetic products. Some nitrosamines have been shown to be potential carcinogens. N-nitrosodiethanolamine has been detected in very low concentrations in cosmetics in the United States (Fan *et al.*, 1977). One possible source is dialkanolamides. Nitroso compounds have been found at very low levels in alkanolamides. At the time of writing there is considerable work on developing analytical techniques for nitroso compounds. Whether there is real danger to humans is difficult to determine but new additives to shampoos are replacing dialkanolamides, the new additives having claims that there is no possibility of them being precursors to nitrosamines.

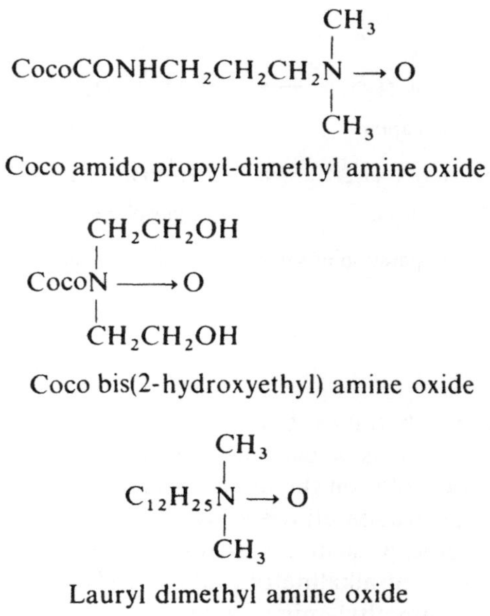

$$
\begin{array}{c}
CH_3 \\
| \\
CocoCONHCH_2CH_2CH_2N \longrightarrow O \\
| \\
CH_3
\end{array}
$$

Coco amido propyl-dimethyl amine oxide

$$
\begin{array}{c}
CH_2CH_2OH \\
| \\
CocoN \longrightarrow O \\
| \\
CH_2CH_2OH
\end{array}
$$

Coco bis(2-hydroxyethyl) amine oxide

$$
\begin{array}{c}
CH_3 \\
| \\
C_{12}H_{25}N \longrightarrow O \\
| \\
CH_3
\end{array}
$$

Lauryl dimethyl amine oxide

Figure 7.12 Amine oxides.

200 HANDBOOK OF SURFACTANTS

Coco bis (2-hydroxyethyl) amine oxide (see Figure 7.12)
Lauryl dimethyl amine oxide (see Figure 7.12)

Description

Amine oxides are prepared by oxidising a tertiary nitrogen group with aqueous hydrogen peroxide at temperatures of about 60–80° C:

$$RR'R''N + H_2O_2 \rightarrow RR'R''N{\rightarrow}O + H_2O$$

R can be: (i) an alkyl chain with R' and R'' methyl groups, to give *N*-alkyl-dimethyl amine oxides; or (ii) an amido propyl alkyl chain $RCONHCH_2$ CH_2CH_2N – with R' and R'' methyl groups to give *N*-alkyl amidopropyl-dimethyl amine oxides. R' and R'' are generally CH_3 but can be any group, e.g. CH_2CH_2OH; thus a primary amine can be reacted with 2 moles of ethylene oxide and then with hydrogen peroxide to give the bis (2-hydroxyethyl) amine oxides. Impurities and by-products include free starting material, e.g. alkyl dimethyl amine

Phosphine oxides and sulphoxides occur frequently in the technical and academic literature, but very few commercial products are available. The sulphoxides can be made by the oxidation of methyl mercaptans (see Webb, 1957), and the phosphine oxides by oxidation of the corresponding methyl phosphines (see Figure 7.13).

$$C_{12}H_{25}SCH_3 + H_2O_2 \longrightarrow C_{12}H_{25}\overset{\overset{\text{O}}{\|}}{S}CH_3 + H_2O$$

Methyl mercaptan · Sulphoxide

$$C_{12}H_{25}P(CH_3)_2 + H_2O_2 \longrightarrow C_{12}H_{25}PO(CH_3) + H_2O$$

Dimethyl phosphine · Phosphine oxide

Figure 7.13 Preparation of sulphoxides and phosphine oxides.

General properties

1. *General.* Similar to betaines, because the substituted amino group is protonated and acts as a cationic surfactant in acid solution. In neutral or alkaline solution the amine oxides are essentially nonionic in character. They are sometimes known as quasi-cationic. Thus, they are weakly cationic at acid pH (cationic below pH 3), and have nonionic properties at alkaline pH (above pH7).
2. *Solubility.* Alkyl dimethylamine oxides are water-soluble up to C16 alkyl chain. Most products are dispersible in mineral oil but insoluble in white spirit, aromatic solvents and perchlorethylene.

NONIONICS 201

3. *Compatibility with other surfactants*. With anionics a 1:1 salt is formed that is more surface-active than either the anionic or amine oxide; thus, in conjunction with anionics amine oxides can replace alkanolamides in shampoos, although they are claimed to be more efficient foam boosters. They can increase viscosity, especially the amido-amine oxides. They also detoxify the effect of anionic surfactants. Above pH 9, amine oxides are compatible with most anionics. At pH 6.5 and below, some anionics tend to interact and form precipitates. A 9/1 ratio of anionic (alkyl and alkyl ether sulphates) to amine oxide gives a clear aqueous solution down to pH 4; at an 8/2 ratio of anionics to amine oxide, amido-amine oxides give a clear solution to pH 4 but alkyl dimethyl only to pH 6. At a 7/3 ratio of anionics to amine oxide, only amido-amines can be used below pH 7 to give clear aqueous solutions. There is an increase in cloud point when a small amount of amine oxide is mixed with a nonionic.
4. *Compatibility with aqueous ions*. Resistant to hard water; good lime soap dispersing properties.
5. *Chemical stability*. Do not show oxidising properties; amine oxides form peroxides with excess hydrogen peroxide.
6. *Surface-active properties*. Surface active properties are lost when the alkyl chain in alkyl dimethyl amine oxides is <10 carbon atoms. The CMC of *n*-dodecyldimethylamine oxide is $1.5–2 \times 10^{-3}$ mol/l (0.03–0.04%) at pH 6.2, but $6–8 \times 10^{-3}$ mol/l (0.12–0.16%) when completely protonated (pH > 7).
7. *Functional properties*. Excellent foaming agents; C12–C14 dimethylamine oxide is an excellent foam booster for LABS and superior on a weight basis to lauric diethanolamide or lauric isopropanolamide; the foam produced has a creamy feel.
8. *Disadvantages*. The main problem in conjunction with anionics is a slight incompatibility at acid pH unless the correct combination of anionic to amine oxide is chosen.

Applications

1. *Shampoos and bath products*. C12 or coco products with anionics give conditioning, reduce eye irritation, impart lubricity, increase viscosity and stabilise foam; these can replace alkanolamides wholly or partly; used in low pH or acid balanced shampoos. C18-dimethyl reduces irritation of zinc pyrithane in shampoos (see Kemp and Gerstein, 1977)
2. *Household products*. Liquid detergents; C12 or coco stabilises foam and increases viscosity. The performance is similar to cocodiethanolamide, but lower concentrations can be used. The foam stabilising properties are most apparent when the anionic in the formulation is

202 HANDBOOK OF SURFACTANTS

only ether sulphate with no LABS. Thickens bleach in conjunction with fatty acid soap (Unilever, (1973) UK Patent 1,329,086) or ether sulphate. The most prominent application of sulphoxides was as a lime soap dispersant; at 5–10% in combination with soap they form scumfree soap bars.

3. *Textiles.* C12 or coco — antistatic softener.
4. *Polymers.* C12 or coco gives foam stabilisation in foam rubber (SBR latex).
5. *Metal treatment.* C12 bis (2-hydroxyethyl) — corrosion inhibitor for non-ferrous metals.
6. *Petroleum production.* Stabilises foam in foam drilling.

Specification

Amine oxide content, 20–40%
Solvent, water, but can sometimes contain isopropanol
Free amine, 1–3% is typical, but shampoos require lower values
Free peroxide, can be up to 0.2%, which can cause problems with colours

7.6 Surfactants derived from mono- and poly-saccharides

Nomenclature

Abbreviations
  APG, alkyl polyglycosides, used in this book.
Generic
  Alkyl glucosides*
  Alkyl *N*-methyl glucosamates
  Alkyl polyglycosides*
  Fatty acid sugar esters
  Glucosamides*
  Glucose esters
  Glucosyl alkyls*
  Methyl glucoside esters*
  Polyhydroxy amides
  Sucrose esters
  Sucroglycerides
  Sugar esters

* Do not confuse these products, which are alkyl derivatives of glucose, with glycerides, which are derivatives of glycerol (see section 7.12)

Description

There is a considerable literature on attempts to make surfactants by using the multifunctional hydroxyl structures in mono- or oligo-saccharides,

NONIONICS

203

sucrose or carbohydrates, which can act as the hydrophilic entity. The interest lies in the ready availability of cheap raw materials which are not derived from petrochemicals; they biodegrade easily and are readily accepted by the consumer. The technical problem is one of joining a hydrophobic group to the multihydroxyl structure. Many useful products have been made on a small scale, e.g. by transesterification between natural glycerides and methyl glucoside. However, in practice there are now at least four different commercial routes each of which gives a range of surfactants. These four routes are:

1. Esterification of sorbitol (or sorbitan) with fatty acids. This method gives a range of products of very limited HLB values (less than 9) and subsequent ethoxylation is needed to make higher HLB products. These products are covered in section 7.11 and will not be discussed in detail in this section.
2. Esterification of sucrose with fatty acids or fatty glycerides. These have been manufactured for many years and have achieved commercial status world wide but there are still very few manufacturers.

$$\text{Sucrose} + n\text{RCOOCH}_3 \rightarrow \text{Sucrose ester with } n \text{ fatty residues} + \text{MeOH}$$

Sucrose esters are prepared by esterification of sucrose with the methyl ester of a fatty acid in a solvent (dimethylformamide) or in aqueous dispersion by emulsion technology. In place of the methyl ester a fatty oil (polyglyceride) can be used and transesterification occurs, leaving glycerol in the product. The mono esters probably have the composition shown in Figure 7.14. Some diesters are formed even when high ratios of sucrose to fatty ester are used. Products made with solvent need the dimethylformamide removed for use in food. Such purification procedures are costly.

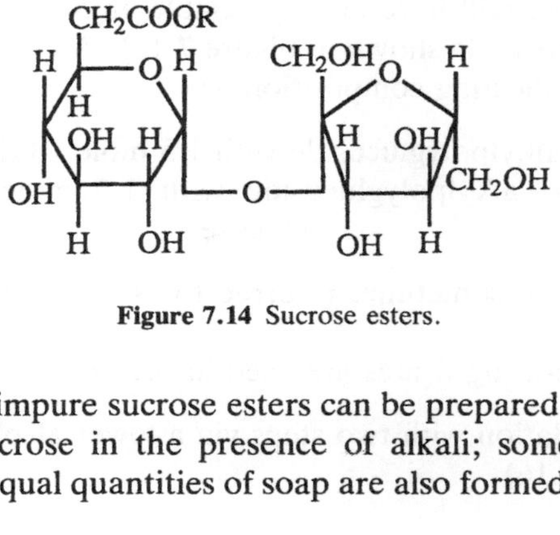

Figure 7.14 Sucrose esters.

Note that impure sucrose esters can be prepared by heating a fatty acid with sucrose in the presence of alkali; some sucrose ester is formed but equal quantities of soap are also formed. The preparation of this type is much easier and cheaper than production of the products mentioned above.

3. Acetalisation by reaction of a fatty alcohol directly with glucose. These are relatively new products but a number of companies are offering the alkyl polyglycosides (APG). The basic raw materials are glucose — this possesses five OH groups and is the hydrophilic part of the molecule — and fatty alcohol R–OH, with R generally being in the region of C12–C16, which can be derived from vegetable oils or petrochemicals. The structure of an alkylpolyglycoside can be described as shown in Figure 7.15. A product with $n=2$ has two glucose residues with four OH groups on **each** molecule and thus has a total of eight OH groups, and thus a product with $n=2$ will be more hydrophilic than that with $n=1$, if R is the same in both cases.

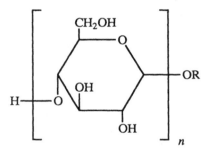

Figure 7.15 An alkyl polyglycoside.

The chemistry is more complex than indicated in this example and commercial products will be mixtures with $n = 1.1$ to 3. In an analogous manner to the addition of ethylene oxide, an alkylpolyglucoside made by the addition of 1.5 mole of alcohol (ROH) to 1 mole of glucose will have an appreciable amount of alkyltetraglucoside ($n=4$ in formula shown in Figure 7.15). A convenient shorthand method for indicating composition is:

APG(1.5) = alkylpolyglucoside with 1.5 mole ROH/mole of glucose
C_{10}PG(1.5) = alkylpolyglucoside with 1.5 mole C_{10}OH/mole of glucose.

The figure 1.5 is sometimes referred to as the degree of polymerisation (d.p.).

Two manufacturing routes are used in practice:

1. Transglucosidation with two steps via a lower alcohol (e.g. butanol) (see Figure 7.16).

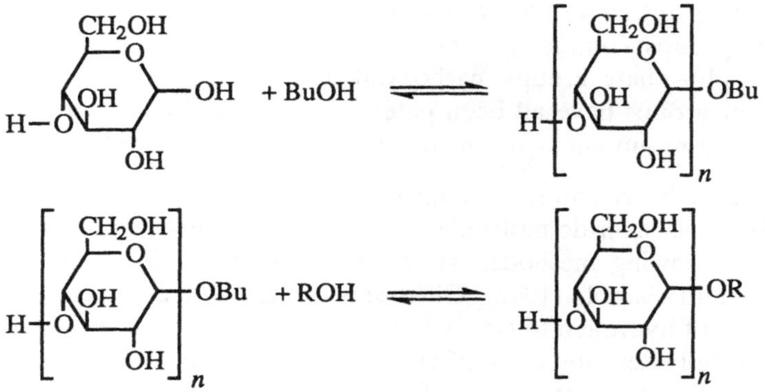

Figure 7.16 Preparation of alkyl polyglycosides.

2. Direct glucosidation where excess fatty alcohol is distilled off. The first stage reaction with butanol is not carried out and the glucose is reacted directly with the fatty alcohol. The reaction does not go as smoothly, higher temperatures are required, and darker coloured material is produced. However, the final products made by both routes are chemically very similar.

In addition to the various isomers with differing values of n, geometric isomers are also formed. These isomers have been shown to differ in surfactant properties (see Bocker and Thiem, 1989, and Matsumura *et al.*, 1990).

The isomers result from stereoisomerism with α- and β-anomers, binding isomerism, where 1,6- and 1,4-interglycosidic linkages are formed, and ring isomerism (furanosides). The details are less important than the facts that an APG(2) can have about 30 possible isomers, an APG(3) several hundred and an APG(4) several thousand isomers. As an APG(1.5) will have APG(2), APG(3), APG(4), etc., present then a commercial product will be an extremely complex mixture. A consistent manufacturing procedure should give consistent products in terms of their function; however, the higher d.p. products, although present in small proportion, may well exert a significant effect in viscosity or other functional effects.

In practice excess alcohol is used to give good yields but this then needs to be removed by distillation or evaporation. This can lead to thermal breakdown and the formation of dark colours. The type of starting carbohydrate can also influence the synthetic route. The less expensive starches will require depolymerisation and then refining to remove by-products before the glucodisation reaction.

Modified alkylpolyglycosides: A considerable number of patents have

206 HANDBOOK OF SURFACTANTS

been taken out covering the modification of the basic APG structure. Amine groups, amide groups, sulphonate groups, sulphosuccinate groups, phosphate groups, carboxyalkylate groups, ether carboxylates and ester groups have all been patented. The APG is thus modified to become more anionic, cationic or amphoteric in character.

4. Amidation by reaction of a fatty acid with an amine group on the glucose or saccharide molecule. The fatty acid amides can be prepared by the following methods, which are a summary of two patents by Procter and Gamble (1992). Glucose is reacted with alkyl amines in the presence of hydrogen to give a N-alkylglucamine. This is an old process and was fully described in 1935 (USP 2,016,962). The glucose amine will then react with a fatty acid to form an amide (USP 1,985,424; 1934). Other patents on a similar reaction scheme have been published over the years; these reactions gave complex mixtures of poor colour but some of the products were quoted as having detergent properties. The reaction scheme is shown in Figure 7.17.

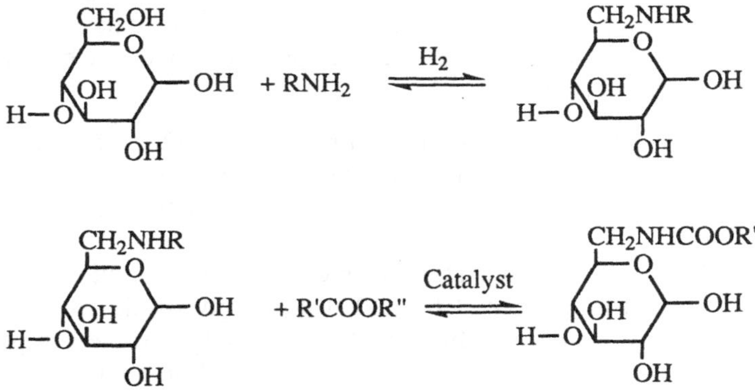

Figure 7.17 Preparation of polyhydroxy amides.

The products contain some impurities such as cyclic glucamide or esteramide (formed from the ester and the – OH groups on the amide) as by-products, but the recent Procter and Gamble (1992) patents claim that the amount of such by-products are kept to below 15 mole % by the use of specific catalysts. The catalysts are inorganic compounds, e.g. sodium carbonate, and the amount recommended is 50–80% of the weight of the reactants. However, if the catalyst is left in the product it can function as a builder in a formulated detergent.

The reaction conditions need careful control to avoid dark colour. When the alkyl group of the ester (R′) is C16 or higher the reactants mix with some difficulty and phase transfer agents need to be used. The latter can be capable of lowering the Krafft temperature of the glucamide.

If R' = C12 and R'' = Me then the glucamide is a solid. However if the reaction is carried out in the presence of an organic hydroxy solvent the products produced are liquid and can be more easily incorporated into detergent formulations.

Thus the polyhydroxy fatty acid amides (PHFA) are different in chemical structure from the alkylpolyglycosides (APG) in at least two characteristics:

1. The hydrophobic group is joined to a reducing sugar ring by an amide group rather than an acetal grouping.
2. The preferred products in the two patents quoted are based on a single reducing sugar ring rather than a mixture of polysaccharides, although there is no reason in principle why the PHFA could not be based on a polysaccharide.

General properties of the four major groups

The basic difference between the four different groups of surfactants based upon saccharides is the solubility in water and the chemical stability of the chemical group that joins the hydrophobic chain to the saccharide. For a comparison, see Table 7.22.

Table 7.22 Comparison of the four groups of surfactants based on saccharides

	Solubility in water	Chemical and heat stability
Sorbitan esters	Limited to short alkyl chains (C12 and below)	Stable over a wide pH range; limited heat stability
Sucrose esters	Excellent water solubility	Stable over reasonable pH range; very limited heat stability
APG (acetals)	Excellent water solubility	Very stable at high pH, unstable at low pH; very limited heat stability
Glucosamides	Excellent water solubility	Stable over a wide pH range; very limited heat stability

General properties of sucrose esters

1. *General.* The monoesters of sucrose with the C12–C18 fatty acids are soluble in warm water due to the large number of free hydroxyl groups remaining. They show general surfactant properties similar to ethoxylated alcohols and alkyl phenols. However, the pure products are solids which are not easy to dissolve in water.
2. *Solubility of sucrose esters.* Sucrose monolaurate, monostearate and

208 HANDBOOK OF SURFACTANTS

mono-oleate are all soluble in warm water, while sucrose distearate and sucrose dioleate are insoluble in water.

3. *Chemical stability*. All sucrose esters show poor heat stability compared with ethoxylated alcohols or ethoxylated nonylphenols. The sucrose esters have the typical stability of esters, i.e. they are more stable in acidic than alkaline solutions.

4. *Surface-active properties*. CMC of sucrose monolaurate is 3.4×10^{-4} mol/l (0.018%) and for sucrose monooleate the CMC is 5.1×10^{-4} mol/l (0.028%). The surface tension of a 1% solution of sucrose monoesters with C12 acid is 33.4 mN/m, and with C18 fatty acid is 33.5 mN/m

5. *Functional properties*. Most products are low HLB and function as emulsifiers.

Applications of sucrose esters

1. *Emulsifiers*. For feeding stuff for calves.
2. *Food additives*. Soluble in water and could replace polysorbates, but they are considerably more expensive than polysorbates.

General properties of alkylpolyglycosides (APG)

1. *General*. In a similar manner to the ethoxylates, the properties of the APG will depend upon the alkyl chain length of the hydrophobic group and the average degree of polymerisation (d.p.). Thus a given HLB value can be obtained by using a short alkyl chain/low d.p. or a long alkyl chain/high d.p. The alkyl glucosides are usually dark brownish liquids with 50–70% solids. They are stable in neutral and alkaline solutions.

2. *Solubility of alkyl polyglycosides*. Alkyl polyglycosides have good solubility in water, a high cloud point (> 100°C), but increased solubility in alkali. They show marked differences in solubility behaviour from ethoxylated alcohols, nonylphenols, fatty acids and fatty amines. The Krafft temperature is about 1°C whereas those of most ethoxylates are well below 0°C; they do not show a cloud point as the temperature rises but rather a distinct separation of separate phases. Rod-like micelles are found at low concentrations but no hexagonal liquid-crystalline phases (see chapter 4) have been identified. At high concentration, lamellar phases appear. The consequence of these differences is that the viscosity of low concentrations of APG will be 1–2 orders of magnitude higher than that of normal ethoxylates, but there is no thick gel phase at intermediate concentrations. Also, as expected, APGs do not show an increase of viscosity by the addition of inorganic salts.

NONIONICS 209

3. *Chemical stability*. The alkyl glucosides are stable in neutral and alkaline solutions but unstable in strong acid solution, due to the acetal linkage
4. *Compatibility of aqueous solutions to ions*. The alkyl glycosides are soluble at all levels of water hardness and extremely tolerant to high concentrations of electrolyte.
5. *Compatibility with other surfactants*. Compatible with all types of surfactants. Addition of small quantities of ether sulphates will improve solubility of the longer chain lengths and avoid coercevation; such solutions can also give high viscosity when electrolyte is added.
6. *Surface-active properties*. Typical CMCs of APGs are shown in Table 7.23.

Table 7.23 CMCs of APGs

Alkyl chain length (R) of RPG (1.5)	CMC* (%)
8–10	0.1
10–12	0.05
12–14	0.009

Data obtained from Balzer (1991)
* There were two CMCs; the ones quoted are the higher CMCs, where the surface tension was constant (30 mN/m); 1% solutions of alkyl glucosides give surface tensions of 27–29 mN/m.

7. *Functional properties*. Alkyl glucosides are available at low and high HLB and show functional properties similar to the alcohol or nonylphenol ethoxylates. Selection of the appropriate product can give good foam and excellent wetting, The foam produced with short alkyl chain APG is comparable to that from LABS and sodium lauryl sulphate. The foaming and wetting ability differs from normal ethoxylates in being reduced by water hardness. Foaming is also reduced by an increase of temperature and/or alkaline solution. Excellent emulsifying properties are found, particularly for polar compounds, e.g. triglycerides. Microemulsions of hydrocarbons can be formed by the addition of sodium chloride but, in contrast to sulphonates, the volume of middle phase emulsion is not dependent upon the sodium chloride concentration; it is also not temperature dependent as for ethoxylated alcohol (see also chapter 4). APG is claimed to give detergency similar to alcohol ethoxylates on organic soil that is easily removed by nonionics (e.g. sebum) but to show superior inorganic soil removal when compared with alcohol ethoxylates. Excellent detergency is found when combined with anionics for

210 HANDBOOK OF SURFACTANTS

hard surface cleaning, and also at low temperature on textiles when combined with ethoxylated nonionics.

Applications of alkylpolyglycosides

1. *Household products*. Laundry detergents, dishwashing detergents, hard surface cleaners.
2. *Cosmetics*. Shampoo base.
3. *Industrial detergents*. Alkali-stable detergents.

Polyhydroxy amides

At the time of writing the available information on properties, applications and specifications is very limited and of doubtful validity, and therefore no entries are made.

Specification (of sucrose esters only)

Solids, 50–70%
Cloud point, >100°C
Free alcohol, 1–4%

7.7 Ethoxylated alkanolamides

Nomenclature

Generic
Alkanolamide ethoxylates
Alkyl monoethanolamide ethoxylates
Amide ethoxylates
Ethoxylated alkanolamides
Ethoxylated monoalkanolamides
Fatty amide polyglycol ethers
Polyoxyethylated alkylamides
Example
Coconut monoethanolamide + 5EO

Description

These products are made by reacting a monoalkanolamide (see section 7.4) with ethylene oxide. In the presence of a basic catalyst the ethylene oxide adds preferentially on to the hydroxyl rather than the hydrogen attached to

NONIONICS 211

the nitrogen (see Figure 7.18). The main commercial products are the coconut and lauryl derivatives.

$$RCONHCH_2CH_2OH + nCH_2{-}CH_2 \longrightarrow RCONH(CH_2CH_2O)_{n+1}H$$

Figure 7.18 Addition of EO to monoethanolamides.

General properties

1. *General.* As the amount of ethylene oxide increases the products change from typical alkanolamides to typical ethoxylates. The addition of the ethylene oxide improves dispersibility or solubility in water.

Table 7.24 Solubility of monoethanolamides (C12) with varying EO contents (S = soluble, D = dispersible, I = insoluble)

	Nil EO	2 EO	5 EO
10% in water	I	D	S
10% in mineral oil	I	I	I
10% in white spirit	I	I	I
10% in aromatic solvent	D	S	S
10% in perchlorethylene	D	S	S

2. *Solubility.* For the solubility of cocomonoethanolamide with varying ethylene oxide contents, see Table 7.24. Water solutions do not form liquid crystalline phases and yield only clear isotropic solutions between the cloud point and solidification temperature (see Brink *et al.*, 1992).
3. *Chemical stability.* Addition of ethylene oxide increases stability to strong alkali. With 6 moles of EO no hydrolysis occurs after refluxing for 2 h with 1 M sodium hydroxide.
4. *Compatibility with other surfactants.* Improves lime soap dispersibility.
5. *Surface-active properties.* Oleyl amide + 5EO, ST of 1% soln = 34 mN/m. Hydrogenated tallow amide + 5EO, ST of 1% soln = 37 mN/m. Hydrogenated tallow amide + 5EO, ST of 1% soln = 46 mN/m.
6. *Functional properties.* Foaming and wetting are at the optimum when 3–4 moles of ethylene oxide are added (see Tagawa *et al.*, 1962).

212   HANDBOOK OF SURFACTANTS

Applications

1. *General*. The properties are similar to either the alkanolamides or the ethoxylated alcohols, but these products are more expensive than either; hence the limited use in practice.
2. *Cosmetics*. Similar to alkanolamides, i.e. thickening and foam stabilising, but with improved dispersibility. Optimum foaming properties at 3–4 moles of EO (see Knaggs, 1965).
3. *Detergents*. Coconut monoethanolamide + 5EO is used in strongly alkaline cleaners where foam stability is required. It is used in car wash and wax–wash formulations where it is claimed to leave a residual wax-like finish on the car.
4. *Emulsion polymerisation*. The polyunsaturated linseed (or soyabean) amide ethoxylates can be used as polymerisable surfactants e.g. in alkyl emulsions, where they react with the alkyd binder in the presence of driers. However, the polyunsaturated amide ethoxylates are sensitive to oxidation of the double bonds which will produce discoloration.

Specification

Appropriate tests are similar to those for ethoxylated alcohols (see section 7.3).

7.8 Ethoxylated long-chain amines

Nomenclature

Generic
    Alkyl polyamine ethoxylates
    Alkyl polyoxyethylene amines
    Ethoxylated amines
    Fatty amine ethoxylates
    Polyoxyethylated fatty amines
    Polyoxyethylene alkylamines
Examples
    Laurylamine ethoxylate, $C_{12}H_{25}N[(CH_2CH_2O)_3H]_2$; usually described as laurylamine + 6EO
    Bis(2-hydroxyethyl)dodecylamine
    $C_{12}H_{25}N[CH_2CH_2OH]_2$

Note: the imidazolines are described in section 8.13.

NONIONICS

213

Description

Whether to call these products nonionics or cationics depends upon the amount of ethylene oxide and the pH at which they are used. At low ethylene oxide levels they are cationic in nature but at high ethylene oxide levels and neutral pH they behave very similarly to nonionics. Prepared by the addition of ethylene oxide to primary or secondary fatty amines (see section 7.1, Ethoxylation). The ethylene oxide preferentially reacts with the amine (like the phenols), so the products have only very small quantities of free amine, unlike the ethoxylated alcohols which have considerable quantities of free alcohol. When ethoxylating a primary amine both hydrogen atoms on the amine react with ethylene oxide before the ethylene oxide adds to the hydroxyl to form the polyethoxylate. No catalyst is needed for the replacement of the hydrogen on the nitrogen, but a basic catalyst is then needed for subsequent reaction of ethylene oxide with the hydroxyl groups (see Figure 7.19).

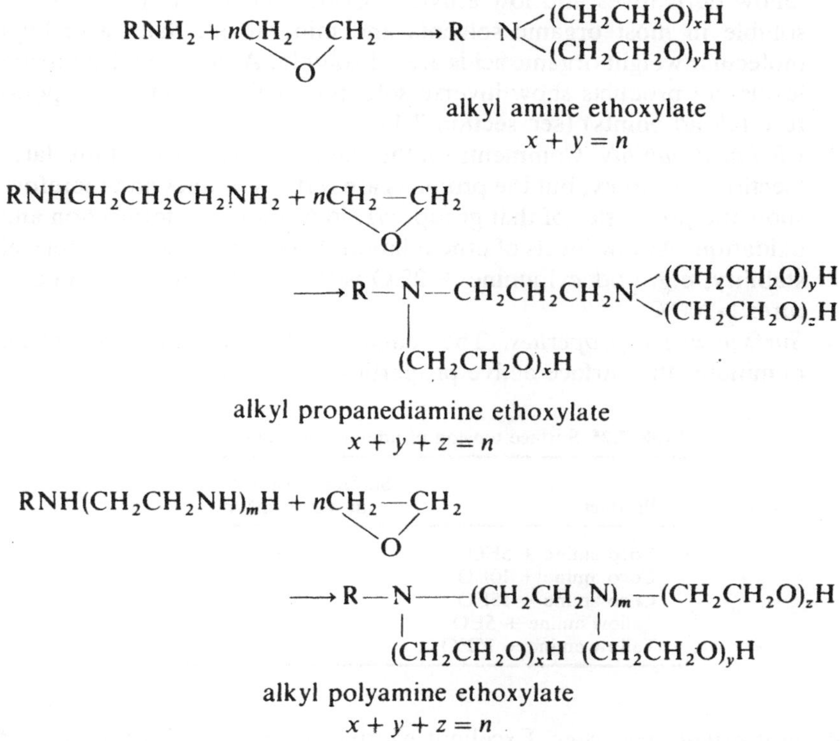

$$RNH_2 + nCH_2-CH_2 \longrightarrow R-N \underset{(CH_2CH_2O)_yH}{\overset{(CH_2CH_2O)_xH}{<}}$$

alkyl amine ethoxylate

$$x + y = n$$

$$RNHCH_2CH_2CH_2NH_2 + nCH_2-CH_2$$

$$\longrightarrow R-N-CH_2CH_2CH_2N \underset{(CH_2CH_2O)_zH}{\overset{(CH_2CH_2O)_yH}{<}}$$
$$|$$
$$(CH_2CH_2O)_xH$$

alkyl propanediamine ethoxylate

$$x + y + z = n$$

$$RNH(CH_2CH_2NH)_mH + nCH_2-CH_2$$

$$\longrightarrow R-N-(CH_2CH_2 N)_m-(CH_2CH_2O)_zH$$
$$|\qquad\qquad\qquad |$$
$$(CH_2CH_2O)_xH\ \ (CH_2CH_2O)_yH$$

alkyl polyamine ethoxylate

$$x + y + z = n$$

Figure 7.19 Common ethoxylated amines.

214   HANDBOOK OF SURFACTANTS

A wide variety of ethoxylated alkyl amines can be prepared depending upon the starting amine. The various amines are described in section 7.13, but the most common ethoxylated amines are ethoxylated primary amines, ethoxylated propanediamines and ethoxylated polyamines (see Figure 7.19). Minor components and by-products include free amine, ethylene oxide and 1,4-dioxane.

General properties

1. *Solubility*. The products become more similar to the corresponding nonionics as the ethylene oxide chain increases and the cationic properties decrease, i.e. the solubility does not change with pH and the incompatibility with anionics diminishes. At low ethylene oxide content, the products are not soluble in water but soluble in acid (mineral or low molecular weight organic) solution. At high ethylene oxide content, the products based on C12 amine are soluble in water but those based on C18 amine are generally insoluble. With the tallow derivatives and low ethylene oxide content the products are soluble in most organic solvents and mineral oil. Salts with high molecular weight organic acids are oil-soluble. At high ethylene oxide levels, all products show inverse solubility with respect to temperature (cloud points) (see section 7.1).
2. *Chemical stability*. Comments on the chemical stability of ethoxylates (section 7.1) apply, but the products are tertiary amines and therefore show the properties of that group, *viz.* ionisation, quaternisation and oxidation. At low levels of ethoxylation they show very good thermal stability, e.g. dodecyl amine + 2EO is thermally stable at temperatures >200°C.
3. *Surface active properties*. The length of the ethylene oxide chain dominates the surface active properties (see Table 7.25).

Table 7.25 Surface tension of amine ethoxylates

Product	Surface tension of 1% in water (mN/m)
Coco amine + 5EO	33
Coco amine + 10EO	38
Coco amine + 15EO	41
Tallow amine + 5EO	33
Tallow amine + 15EO	40

4. *Functional properties*. Excellent wetting properties with products of low EO content. Unsaturated alkyl chains have better wetting properties than saturated alkyl chains. Adsorption on to metal

NONIONICS 215

surfaces occurs when ethylene oxide content is low and/or there are multiplicity of amine groups.

Applications

1. *General.* The ethoxylated alkyl amines have very wide applications. The cationic property has extended the use of these compounds. Most metal, mineral and fibre surfaces are negatively charged in aqueous solution, and the amine group is strongly adsorbed on to such surfaces.
2. *Oilfields and refineries.* Act as corrosion inhibitors. By variation in the alkyl chain of the amine, the ethylene oxide content and the type of organic acid (see Solubility, above) a very wide range of water or oil soluble materials may be made. For cost and logistics reasons the variations on alkyl chain and organic acid are limited, so that the variation in ethylene oxide content becomes the main variable used in practice.
3. *Emulsifying agents (for similar reason to corrosion inhibitors).* Agrochemical emulsions, wax emulsions and two-phase emulsion cleaners.
4. *Textiles.* Processing aid in rayon production in the spinning bath; softener and antistatic agents in processing, because they adsorb on the fibre from aqueous solution; scouring, desizing and dyeing assistants (levelling aid).
5. *Road repairs.* Coating of asphalt aggregates for adhesion to wet surfaces; addition to asphalt or use as emulsifying agent for bitumen emulsions.
6. *Paint.* Dispersing agents for pigments in oil paints.

Specification

Active, usually >99%
Hydroxyl number, can vary over wide range
PEG and non-amine, 2–5%
Free amine, usually very low but can be significant.
Free ethylene oxide, should be very low (<5 ppm)
1,4-Dioxane, should be very low (<25 ppm)

7.9 Ethylene oxide/propylene oxide (EO/PO) copolymers

Nomenclature

Abbreviations
  EO = ethylene oxide

216 HANDBOOK OF SURFACTANTS

PO = propylene oxide
EO/PO or EO_xPO_y =   EO/PO block copolymers
Generic
Ethylene oxide/propylene oxide block copolymers, $(EO)_x(PO)_y$
Meroxapol = reverse Pluronics® (Pluronics®, is a Wyandotte trade name)
Polaxamer = Pluronics®
Polyalkylene oxide block copolymers
Poloxamine
Polyoxyethylated polyoxypropylene glycols $(EO)_x(PO)_y(EO)_x$ = Pluronics®
Polyoxypropylated polyoxyethylene glycols $(PO)_x(EO)_y(PO)_x$ = reverse Pluronics®
Reverse Pluronics®
Examples
All EO/PO copolymers are trade names which do not directly give the composition. Some manufacturers adopt a systematic nomenclature from which it is possible to derive an approximate composition, but they do not give the exact composition.

Description

There are many possible variations in this class of products and more appear very year. However, the major types to achieve significant volume are not so extensive, these are:

Type 1. EO/PO copolymers based on the reaction of a polypropylene glycol (difunctional) with EO (polaxamer range) or mixed EO/PO:

$(EO)_n (PO)_m (EO)_n$, block copolymers, the Pluronic® polyols

$(AO)_n (PO)_m (AO)_n$, where AO = EO or PO, random copolymers

Type 2. EO/PO block copolymers based on the reaction of a polyethylene glycol (difunctional) with PO (Meroxapol range) or mixed EO/PO

$(PO)_n (EO)_m (PO)_n$, block copolymers, the reverse Pluronics®

$(AO)_n (PO)_m (AO)_n$, where AO = EO or PO, random copolymers

A wide range of surfactants can be made using just two basic chemicals, EO and PO, in differing ratios rather than by using different hydrophobes, and therefore logistically this system has great economic attractions for making a wide range of surfactants. However, there are certain disadvantages:

The starting materials are highly toxic and potentially explosive, so that very specialised equipment is needed.

NONIONICS 217

The many variations tend to proliferate product lines.
Slight differences in manufacture can give different properties although the ratio of EO/PO may be correct; this gives problems in quality control.

Type 3. EO/PO copolymers with an alkyl or alkyl/alkyl end group; EO/PO copolymers, both block and random, can be based on any starting material (initiator) with an active hydrogen, but the most common are fatty alcohols ROH. Monofunctional products can be described as follows:

$RO(AO)_n (EO)_m$, where AO = mixed EO and PO to give a random copolymer

$RO(PO)_n (AO)_m$, where AO = EO to give a block copolymer or AO = mixed EO and PO to give a random copolymer

$RO(EO)_n (AO)_n$, where AO = PO to give a block copolymer or AO = mixed EO and PO to give a random copolymer

The products $RO(EO)_n (AO)$, where R = C12–C18, are basically modified alcohol ethoxylates which have reduced foam by addition of the PO.

Functionality. Types 1 and 2 above are difunctional i.e. they have two free hydroxyl groups and two hydrophilic or hydrophobic chains.
Type 3 products are monofunctional, i.e. they have one hydroxyl group, and one hydrophilic chain and one hydrophobic chain.
Trifunctional products are also available, where the starting polypropylene glycol is based on glycerol (see Figure 7.20). These products will have three free hydroxyl groups and three chains of either block or random copolymers.
Tetrafunctional products are available, where the starting material for ethoxylation is ethylene diamine, known as the poloxamine range (see Figure 7.21). These products will have four free hydroxyl groups with four block copolymer chains.

The surfactant properties will be influenced by the hydroxyl group on the end of the chain, the number of chains and the properties of each individual chain. Thus chacterisation by EO content, PO content and EO/PO ratio is generally insufficient in comparing product to product but useful as a quality control on one product.

The great advantage of this product range is that by varying the EO/PO ratio and molecular weight a wide range of products can be made with varying properties. The disadvantage is that the exact composition of a product is very often difficult to determine. The composition will invariably

HANDBOOK OF SURFACTANTS

$$CH_2 - (EO)_x(PO)_yH$$
$$|$$
$$CH - (EO)_x(PO)_yH \qquad \text{block copolymer}$$
$$|$$
$$CH_2 - (EO)_x(PO)_yH$$

$$CH_2 - (PO)_x(EO)_yH$$
$$|$$
$$CH - (PO)_x(EO)_yH \qquad \text{block copolymer}$$
$$|$$
$$CH_2 - (PO)_x(EO)_yH$$

$$CH_2 - (AO)_xH$$
$$|$$
$$CH - (AO)_xH \qquad AO = EO \text{ or } PO$$
$$|$$
$$CH_2 - (AO)_xH \qquad \text{random copolymer}$$

Figure 7.20 Glycerol-based EO/PO copolymers.

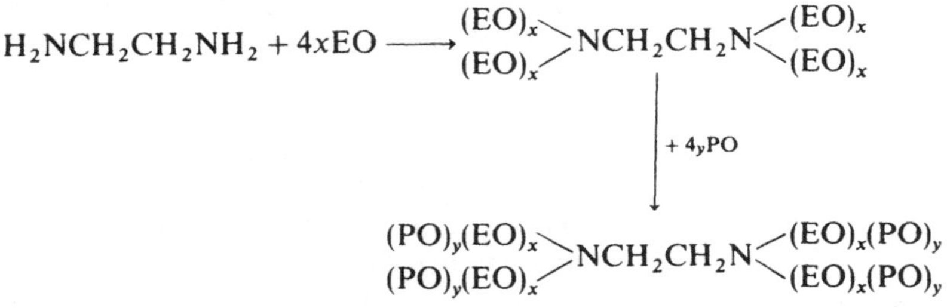

$$H_2NCH_2CH_2NH_2 + 4xEO \longrightarrow \begin{matrix} (EO)_x \\ (EO)_x \end{matrix} NCH_2CH_2N \begin{matrix} (EO)_x \\ (EO)_x \end{matrix}$$

$$\downarrow + 4_yPO$$

$$\begin{matrix} (PO)_y(EO)_x \\ (PO)_y(EO)_x \end{matrix} NCH_2CH_2N \begin{matrix} (EO)_x(PO)_y \\ (EO)_x(PO)_y \end{matrix}$$

Figure 7.21 Ethylene diamine-based products.

be different from that envisaged in the product descriptions in formulae given above. First, there will be the spread of molecular weight normally associated with reactions of ethylene oxide (or propylene oxide). Secondly, traces of water will form polyethylene or polypropylene glycols of different molecular weight from that of the starting initiator, and then these polyglycols will react with the alkylene oxide being added. Thirdly, there may be some random EO/PO chains present where block copolymers were expected. This can occur if the starting initiator (e.g. polypropylene glycol) is made in the same reactor before the addition of the ethylene oxide; when the ethylene oxide is subsequently added there may be some free propylene oxide remaining and so there will some random EO/PO chain in the final product. All these, factors will affect surfactant properties but are difficult to determine analytically.

General properties

A very wide range of properties can be obtained:

- Liquids to solids.
- Water-soluble to water-insoluble.
- Cloud points from room temperature to > 100°C. Some products show two cloud points.
- High foam to no foam.
- Poor wetting to non-wetting.
- Excellent dispersing properties to no dispersing properties.

Some generalisations (for products consisting solely of EO and PO on difunctional initiators, unless otherwise stated) are as follows:

1. *Solubility.* Most products are more soluble in cold water than in hot water, but they often form gels. The reverse products (PO/EO/PO) have a faster rate of solubility in water than the normal products (EO/PO/EO) and rarely form gels. The tetrafunctional products dissolve faster than the difunctional products. Most products are soluble in aromatic solvents or other polar organic solvents. Very short PO chains behave like EO and hydrophobicity only appears when more than five PO units are present.

2. *Cloud points.* EO/PO copolymers show the inverse solubility/temperature relationship normal to ethoxylated nonionics, but the cloud points are often indistinct, and sometimes two cloud points are observed. Increasing EO content increases cloud point, increasing PO content decreases cloud point.

3. *Surfactant properties.* In general the lowest surface tension of aqueous solutions are in the region of 35–45 mN/m at room temperature i.e. significantly higher than ethoxylates where the hydrophobic chain is a paraffinic chain. However surface tensions of 30–35 mN/m are obtained when the temperature is raised. There has been considerable confusion in the literature on whether or not these products form micelles but only the polaxamer range has been thoroughly investigated. Even where a CMC has been identified and measured then varying values of CMC have been given to apparently the same molecule. It is the author's opinion, based on some experience, that the surfactant properties of EO/PO copolymers can differ appreciably for products of apparently similar chemical composition. Thus quoted CMC values are of little use. However the measured CMCs are very sensitive to temperature (see Zhou and Chu 1987, 1988a,b).

4. *Foaming properties.* Maximum foam is at PO/EO ratios of 2/3 for difunctional (EO/PO/EO), 1/1 for difunctional (PO/EO/PO) and 1/2 for tetrafunctional. Low EO content products have low foaming

220 HANDBOOK OF SURFACTANTS

properties, so varying the PO content can control foam. Best antifoam efficiencies are obtained at PO/EO ratios of 4/1 to 9/1. These products are good defoamers for other surfactants if the PO content is high and if the molecular weight is above 2000. Reverse products (PO/EO/PO) give the lowest foam.

5. *Wetting properties (as measured by Draves test — see chapter 4)*. Wetting time decreases as the molecular weight of the hydrophobe increases up to an optimum value; above that the wetting time decreases. Wetting times decrease with a decrease in EO content. High molecular weight products with low EO content have good wetting properties, better than polyethylene glycol fatty esters.

6. *Dispersing properties*. High EO content products have good dispersing properties.

7. *Emulsifying properties*. Good emulsifiers are obtained at molecular weights of the PO portion of 2000–4000.

8. *Disadvantages*. High PO products have poor biodegradability; they are difficult to characterise chemically and there are problems of product replacement; suppliers of many of these products do not give a detailed description of their chemical composition; they may contain traces of free EO

Warning: there are so many products possible that the above generalisations will have many exceptions.

Considerable information is available from the suppliers and also in the technical literature on the properties of these products. For detailed information on the basic properties of the EO/PO block copolymers (Pluronics®, Type 1) see Schick (1967), pp. 309–333. A good summary on the differences between mono-, di-, tri- and tetra-functional products was given by Schmolka (1977).

Some specific properties of interest in formulating mixtures are: some products show an increase in cloud point by addition of anionics, e.g. LABS; the increase can be up to 20°C. Most products form molecular complexes with iodine (the iodophors); most water-soluble products function as lime soap dispersants: products based on tetrafunctional initiators (the best known being ethylene diamine) exhibit excellent cold-water detergency when the EO content is 25–55%; however, they are used more for low foam wetting than for their detergency; they also have the ability to disperse lime soaps in hard water.

Applications

1. *General*. The EO/PO copolymers find many varied applications, generally in small quantities, the main application being foam control. This can be either as a process defoamer or as an additive to

NONIONICS 221

a detergent solution to control the foam. Other major uses are as wetting agents, dispersing agents or emulsifying agents. They are also used as chemical intermediates in the production of polymeric silicone surfactants (see chapter 11).

2. *Defoaming.* Defoaming agents are used in paper manufacture, industrial fermentation, emulsion paints, sugar processsing, textile processing, oilfield chemicals, etc. Most defoaming agents are formulated products with the compositions not disclosed. EO/PO copolymers are widely used in these formulations, but in many cases the products are not standard EO/PO copolymers but tailor-made for the end application. These products are also used for foam control in laundry detergents (low molecular weight with high PO content).

3. *Rinse aids in machine dishwashing.* Low molecular weight with low EO (2000) (10%) content as defoamer constituent; blended with molecular weight 2500 diol with an EO content of 30% as low foam wetter.

4. *Disinfectants.* Used in the manufacture of iodophors.

5. *Textile industry.* Defoamer components in defoaming, dyeing and finishing processes. Wetting and emulsifying agents in lubricants and spin finish formulations.

6. *Emulsion polymers.* Low molecular weight (<2000) products are used as primary emulsifiers. High molecular weight products with high (>70%) EO content are used as post stabilisers.

7. *Emulsion paints.* Dispersants for pigments in emulsion paints (high molecular weight with high PO content); defoamer component.

8. *Agriculture.* Emulsifier for herbicides and insecticides — usually used in conjunction with other surfactants (mixtures of anionic and nonionic); spray additives for herbicides to give good wetting and improve penetration but without promoting foam.

9. *Water treatment.* Scale removal in boilers — high molecular weight with high EO content.

10. *Dispersing agents.* Used as pigment dispersants in aqueous and nonaqueous solvents

Specification

Water content: usually low (<0.5%).
Cloud point of 1% aq solution: Can obtain double cloud points or indeterminate cloud points as they are susceptible to small amounts of polyglycol. But cloud point is important, as a small difference in EO level can affect cloud point appreciably. Thus it can be a simple sensitive test for EO content.

222   HANDBOOK OF SURFACTANTS

Viscosity: for high molecular weight products viscosity can be a useful quality control.

EO and PO content: nice to know but difficult to measure, and no guarantee of consistency

Hydroxyl number: useful for quality control but not a guarantee of consistency.

Functional test: wetting, foaming, defoaming, solubility, etc., can be often more discerning in checking batch-to-batch variation or product replacement.

7.10 Fatty acid ethoxylates

Nomenclature

Abbreviation
  PEG esters, polyethylene glycol esters, used in this book
Generic
  Fatty acid ethoxylates
  Long chain carboxylic acid esters
  Polyethylene glycol esters
  Polyoxyethylene fatty acid esters
  Polyoxyethylene esters
Example
  PEG(400) monolaurate — this indicates that a polyethylene glycol of molecular weight 400 has one hydroxyl group replaced by an ester grouping formed with lauric acid.

Description

Monoesters can be made in two ways:

  1. Reaction of ethylene oxide with a fatty acid

$$RCOOH + nEO \rightarrow RCOO-(CH_2CH_2O)_nH$$

  2. Reaction of the acid with a polyglycol

$$RCOOH + HO(CH_2CH_2O)_nH \rightarrow RCOO-(CH_2CH_2O)_nH + H_2O$$

The most common fatty acids used are coconut (C12), tallow (C18) or oleic (C18 unsaturated). The usual practice in naming is to give the molecular weight of the ethylene glycol; 200, 400 and 600 are common.

Reaction 2 is basically an esterification and so can be performed on simple equipment, but reaction 1 demands specialised equipment to handle ethylene oxide. The products look identical, but reaction 1 will give a spread of n (see section 8.2) whilst reaction 2 will give a mixture of the

NONIONICS 223

monoester (as shown above) plus some diester (shown below); also it is an equilibrium reaction, and therefore some free fatty acid and polyglycol are present.

$$RCOO-(CH_2CH_2O)_n-OCOR \text{ is the diester}$$

Up to 30% of the diester is very common.

In reaction 1 the carboxyl group is highly ionised (see section 7.2) and therefore the first step in the ethoxylation is the reaction of the fatty acid with one molecule of ethylene oxide. Only when all the carboxyl groups have disappeared will the ethylene oxide react with the hydroxyl group which is formed. The result is that ethoxylated fatty acids made by reaction 1 are quite free of fatty acid, unlike those from reaction 2. However, during the ethoxylation reaction in the presence of an alkaline catalyst transesterification occurs:

$$2RCOO-(CH_2CH_2O)_nH \leftrightarrow RCOO-(CH_2CH_2O)_n-OCOR$$
$$+ HO-(CH_2CH_2O)_n-H$$

This gives a mixture of polyglycol, monoester and diester, generally in the proportions of 1:2:1. The relative proportions of polyglycol, monoester and diester in reaction 2 will depend upon the ratio of the reactants. An equimolar ratio of fatty acid and polyglycol results in a mixture which is predominant in monoester and similar to that produced in reaction 1. In the preparation of monoesters by reaction 2, an excess of the polyglycol is usually used to react with the fatty acid to ensure a high conversion to the monoester.

Minor components and impurities depend upon route. For reaction 1 (EO route) these include: traces of EO, polyglycol (from water), diester, sodium methoxide (catalyst for ethoxylation), acetic acid (used in the neutralisation of the sodium methoxide), and sodium acetate. For reaction 2 (polyglycol route) these include: diester, fatty acid and polyglycol. The diesters are made by the acid/polyglycol route with two moles of acid to one mole of polyglycol. This reaction is again an equilibrium, with fatty acid and polyglycol present plus some monoester.

General properties

1. *Solubility.* The longer the chain length of the fatty acid, the less soluble the product in water. The larger the amount of ethylene oxide, the better the solubility in water. Addition of 1–8 moles of ethylene oxide gives oil-soluble products; at 12–15 moles of EO, water dispersibility or solubility occurs; 8 moles of EO = 372 molecular weight; 12 moles of EO= 528 molecular weight. Solubility in water depends upon whether the product is a monoester or a diester. The monoesters are much more soluble in water than the

224 HANDBOOK OF SURFACTANTS

diesters, which are only dispersible in water (at PEG molecular weight <1000).

2. *Compatibility with aqueous ions*. Excellent unless fatty acid is present.
3. *Chemical stability*. Readily hydrolysed by hot alkali.
4. *Surface-active properties*. Lowest surface tension at 10 moles of EO with most acids at 30–36 mN/m. Lowest surface tension (30 mN/m) with C10–C12 acids and 8–10 moles of EO. HLB values (see chapter 4) can be calculated from the relation HLB = (ethylene oxide content in % + propylene oxide content in %)/5.
5. *Functional properties*. Outstanding emulsifying properties, better than AE or NPE. By choice of the fatty acid, the ethylene oxide content and mono/diester content a very wide range of HLB values can be obtained (3 to 20). Lower flash foam and poorer foam stability than other nonionic types, and the PEG diesters are often used as defoamer components. Increase in temperature has no marked effect on the foaming properties of the majority of the products. Tall oil derivatives (with rosin) have lower foaming properties than corresponding fatty acid derivatives. The PEG esters are poor wetting agents compared with alcohol ethoxylates or alkyl phenol ethoxylates. The best are the lower chain length acids (C 12) with low amounts of EO (10 moles = PEG (400)) monolaurate. Detergency is not outstanding.
6. *Disadvantages*. Poor chemical stability to alkali; generally poor wetting properties.

Applications

1. *Emulsifying agents for O/W and W/O emulsions*. Detergents, cosmetics, solvent cleaners, degreasers, leather industry, oil industry, textile industry, agriculture. Monoesters are usually used as blends of different HLBs. For some typical HLB values, see Table 7.26.

Table 7.26 HLB values of fatty acid PEG esters

Monoesters	HLB
Oleic monoester with EO $n = 1$	
(ethylene glycol)	3.5
PEG 200 monolaurate (C12)	9.8
PEG 400 monolaurate (C12)	12.8
PEG 600 monolaurate (C12)	14.6
PEG 200 monostearate (C18)	8.0
PEG 400 monostearate (C18)	11.2
PEG 600 monostearate (C18)	13.2

NONIONICS 225

Diesters are generally more oil-soluble with much lower HLBs so used as emulsifiers for oil particularly where foam has to be low. For typical HLBs, see Table 7.27.

Table 7.27 HLB values for fatty acid PEG diesters

Product	HLB
PEG 200 distearate	5.0
PEG 400 distearate	7.8
PEG 600 distearate	10.6
PEG 200 dilaurate	7.4
PEG 400 dilaurate	10.0
PEG 600 dilaurate	11.5

2. *Shampoos*. Ethylene glycol monostearate (EGS) (can be up to 30% diester) is used as a pearling agent in other surfactants (e.g. shampoos); one needs to add the solid EGS to the heated shampoo, then stir and cool. It is, difficult to get a pearling effect with sulphosuccinates but easy with AOS, LABS, AS and AES.
3. *Textiles*. Lubricant components and emulsifiers (particularly for fatty acid esters); also give antistatic properties with high EO contents; dyeing assistants.
4. *Cutting oils*. Emulsifier for both O/W and W/O emulsions.
5. *Paints*. Defoamers and levelling agents in emulsion paints.

Specification

See Table 7.28.

Table 7.28 Properties of PEG esters (S = soluble; D = dispersible)

Typical properties (PEG 400 laurate)	Monoester	Diester
Appearance	Liquid-paste	Liquid-paste
Active content	100	100
Acid value (mg KOH/g)	0.05–0.5%	—
Water content (%)	0.5–5%[a]	0.1–1.0%
Sap. value (mg KOH/g)[b]	80–90	130–140
Hydroxyl value (mg KOH/g)[b]	80–90	10–20
Solubility in water (10%)	D	D
Solubility in white spirit	D	S

[a] Some manufacturers add up to 5% water to keep products liquid. If water is >5% then there is danger of corrosion of mild steel drums.
[b] Sap. value will depend upon whether monoester or diester is used. The diester will always have higher sap. value than the monoester at the same molecular weight of the PEG portion. The higher the molecular weight of the PEG part of the molecule, the lower the sap. number. The higher the molecular weight of the fatty acid the lower the sap. value. Sap. value should equal the hydroxyl value for monoesters but be higher (approximately twice) for diesters.

226   HANDBOOK OF SURFACTANTS

7.11   Sorbitan derivatives

Nomenclature

Two groups of products are covered in this section, the fatty acid esters of
sorbitan and their ethoxylated derivatives.
Generic — esters
  Anhydrohexitol esters
  Sorbitan esters
  Sorbitan fatty acid esters
  Spans (Atlas trade name)
Examples
  Sorbitan monolaurate
  Sorbitan trioleate

Generic — ethoxylated esters
  Polyoxyethylene sorbitan esters.
  Sorbitan ester ethoxylates
  Tweens (Atlas trade name)
Example
  Polyoxyethylene (20) sorbitan monolaurate — this indicates that 20
  moles of EO have been reacted with sorbitan monolaurate

Atlas were the first company to commercialise these products and as such
they are often called by the Atlas trade names, Spans (the sorbitan esters)
and Tweens (the ethoxylated sorbitan esters). However, care should be
taken not to infringe trade marks.

Description

The two groups of products, the sorbitan esters and the ethoxylated
sorbitan esters, have been grouped together because they are so often used
together.
  The esters. Sorbitol is reacted with a fatty acid at $> 200°C$. The sorbitol
  dehydrates to 1,4-sorbitan and then esterification takes place. The
  resulting product is a mixture, with esterification taking place primarily
  on the primary OH (see Figure 7.22). If one mole of fatty acid is reacted
  with one mole of sorbitol then the product is principally the monoester,
  but some diester is formed.
  The ethoxylates. Ethylene oxide can react on any hydroxl group
  remaining on the sorbitan ester group; in addition there can be ester
  interchange and some polyoxyethylene groups can get between the
  sorbitan and the acid radical. If sorbitol is first reacted with ethylene
  oxide and then esterified, an ethoxylated sorbitan ester is obtained,
  which will have different surfactant properties to the Tweens.

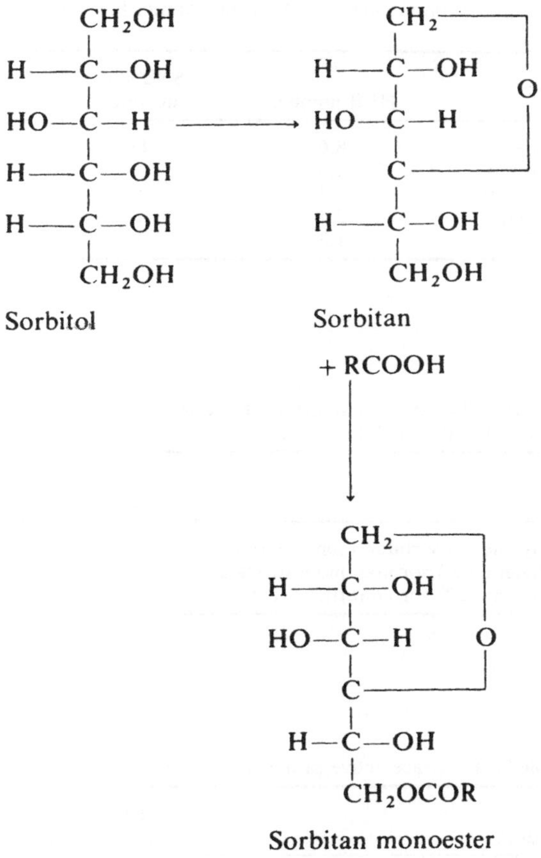

$$\text{NONIONICS} \qquad \qquad 227$$

Figure 7.22 Sorbitan esters.

General properties

1. *Solubility*. The esters are insoluble in water but soluble in most organic oils (see Table 7.29). Depending on how much ethylene oxide is added ethoxylated derivatives can be made soluble or dispersible in water to give a high HLB value (see Table 7.30).
2. *Surface active properties*. See Tables 7.31 and 7.32.
3. *Chemical stability*. Unstable at higher temperatures.
4. *Functional properties*. Esters give poor wetting, foaming, dispersing and detergency, but they are excellent emulsifiers, particularly when used in conjunction with the ethoxylates. They are also excellent lubricants for fibres. Ethoxylates, depending upon EO content, can be excellent foamers, dispersing agents, wetting agents and deter-

228 HANDBOOK OF SURFACTANTS

Table 7.29 Solubility of sorbitan esters (1–10% concentration) (S = soluble, D = dispersible, I = insoluble)

Product	HLB number	Solubility in water	Solubility in mineral oil
Monolaurate (Span 20)	8.6	D	S
Monostearate (Span 60)	4.6	I	D (at 50°C)
Mono-oleate (Span 80)	4.2	I	S
Tristearate (Span 65)	2.1	I	D (at 50°C)
Trioleate (Span 85)	1.8	I	S

Table 7.30 Solubility of sorbitan ester ethoxylates (1–10% concentration) (S = soluble, D = dispersible, I = insoluble)

Product	Solubility in water	Solubility in mineral oil
Polyoxyethylene (4) sorbitan monostearate	D	I
Polyoxyethylene (20) sorbitan monostearate	S	I
Polyoxyethylene (20) sorbitan tristearate	D	I

Table 7.31 Surface-active properties of the esters

Product	Surface tension of 1% in water (mN/m)
Monolaurate (Span 20)	28
Monostearate (Span 60)	46
Mono-oleate (Span 80)	30
Tristearate (Span 65)	48
Trioleate (Span 85)	32

Table 7.32 Surface-active properties of the ethoxylates

Product	Surface tension of 1% solution in water (mN/m)	CMC (%)
Polyoxyethylene (4) sorbitan monolaurate	—	0.0013
Polyoxyethylene (4) sorbitan monostearate	38	—
Polyoxyethylene (20) sorbitan monostearate	43	0.0027
Polyoxyethylene (20) sorbitan tristearate	31	—

gents. However, in the majority of cases other nonionics (ethoxylated alcohols or ethoxylated alkyl phenols) are more cost-effective.
5. *Disadvantages*. High cost.

Applications

1. *General*. The sorbitan esters and ethoxylated derivatives have been used and approved as food additives in most countries throughout the world. Therefore they are used in many applications where they may be less cost-effective than other surfactants but governmental regulations inhibit the use of the competitive materials. They are also a family of surfactants where there is long experience of their use as emulsifying agents.
2. *Food*. There are many applications where esters are used as oil-soluble emulsifiers which can change the wetting, dispersing and the physical properties of oil or fat mixtures. This can give improved palatability or shelf storage properties. They can also function as water-in-oil emulsifiers to promote air retention and other physical properties of food mixtures where the water content is low, e.g. cakes.

   Ethoxylates have many applications where they can function in a similar manner to the esters, but acting more as a oil-in-water emulsifier for products with a high water content, e.g. ice creams. Mixtures of esters and ethoxylates are commonly used in many foods in order to obtain the benefits of both sets of products.
3. *Shampoos*. Ethoxylated products are used with anionics (ether sulphates) for the production of baby shampoos (low eye irritation).
4. *Cosmetics*. Ethoxylates are used as solubilisers for oils and fragrances. Perfume oils are solubilised for use in products such as colognes, bath oils and after-shave lotions. Lotions and creams of the oil-in-water type are prepared from mixtures of esters and ethoxylates. Creams of the water-in-oil type, e.g. cold creams, are generally prepared with the esters, but small quantities of the ethoxylates make a more easily prepared cream.
5. *Pharmaceuticals*. Both esters and ethoxylates are used in pharmaceutical preparations as solubilisers, emulsifiers and dispersants. Mutually incompatible ingredients can be used as ointments or lotions by emulsification or dispersion, particularly in aqueous solution. Increased biological activity is sometimes obtained by administration in solubilised form.
6. *Textiles*. Low HLB products are used as lubricants for fibres. The ethoxylates were commonly used as emulsifying agents for the esters and mineral oils, but are now being replaced by more cost-effective nonionics.

230 HANDBOOK OF SURFACTANTS

7. *Metal protectants and metal working*. Sorbitan oleates were at one time extensively used in anticorrosion compositions in lubricating oils. Removable protective coatings or temporary protective coatings often incorporate the esters, usually the oleates.

The ethoxylates are used as emulsifying agents in cutting oils and give some degree of corrosion inhibition as well. However, more cost-effective formulations based on oil-soluble sulphonates and fatty acids are replacing the sorbitan based products.

8. *Oil-slick dispersants*. The excellent emulsifying properties of the ethoxylates plus extensive knowledge of their toxicity has lead to their use in oil-slick dispersing agents.

9. *Explosives*. The sorbitan esters are used as low HLB emulsifiers for high internal phase ratio W/O emulsions (see chapter 4) in the form of concentrated solutions of nitrates in oil. These emulsions are used as liquid explosives in mining operations.

Specification

See Tables 7.33 and 7.34.

Table 7.33 Analysis of esters

	Monoester	Triester
Acid value (mg KOH/g)	2–10	2–15
Sap. value [a]	145–170	170–190
Hydroxyl value[b]	200–350	50–90

[a] Sap. values will depend upon the hydrophobe (the lower the chain length of the hydrophobe the higher the sap. value) but also upon the amount of diester formation. The difference between mono- and tri-esters may not be large.
[b] The hydroxyl value indicates more readily the difference between mono- and tri-esters.

Table 7.34 Analysis of ethoxylates

Acid value (mg KOH/g)	1–5
Sap. value (mg KOH/g)[a]	20–120
Hydroxyl value (mg KOH/g)[a]	40–250

[a] Depends upon degree of ethoxylation — low values for high degress of ethoxylation.

The following combinations of sap. value and hydroxyl value are useful indicators to composition:

High (150–170) sap. and high (200+) OH value = monoester

NONIONICS 231

High (150+) sap. and low (<80) OH value = triester
Low (<100) sap. and low (<60) OH value = ethoxylated triester

7.12 Ethylene glycol, propylene glycol, glycerol and polyglyceryl esters plus their ethoxylated derivatives

The esters are grouped together because they are all formed by a hydrophobic group or groups attached by an ester group to a multihydroxyl (two or more hydroxyl groups) compound. Most products are synthesised, but some occur naturally. There are very similar products which are derived from carbohydrates (section 7.6). The fatty acid ethoxylates are the esters of polyethylene glycol but these are described in section 7.10.

The ester ethoxylates are those esters with free hydroxyl groups remaining after esterification which have been reacted with ethylene oxide to increase water solubility. Again, the sorbitan ester ethoxylates strictly belong to this group but they are described in section 7.11.

Nomenclature

Nomenclature can be confusing in these categories.
Generic: esters (see Figure 7.23 for structural formula)
    Ethylene glycol esters or glycol esters.
    Glycerol esters–mono- or di-glycerides
    Glyceryl esters — same as glycerol esters.
    Polyethylene glycol esters — not described here, see section 7.10
    Polyglyceryl esters
    Polyol monoester — careful, this can mean polyglyceryl or polyoxyethylene or even a polyoxyethylene polyglyceryl ester. Ask the supplier what it means
    Propylene glycol esters
Examples: esters
    Glycerol (or glyceryl) monostearate
    1-Monolaurin = glycerol monolaurate
    Triglycerol monostearate = a polyglyceride (3 moles of glycerol polymerised then esterified with 1 mole of stearic acid)
    Decagylcerol tristearate = a polyglyceride (10 moles of glycerol polymerised then esterified with 3 moles of stearic acid.)
Generic: ethoxylated esters (see Figure 7.23 for structural formula)
    Polyoxyalkylene glycol esters
    Polyoxyalkylene propylene glycol esters
    Polyoxyalkylene polyol esters = probably ethoxylated polyglyceryl esters.
    Polyoxyalkylene glyceride esters
Example: ethoxylated esters

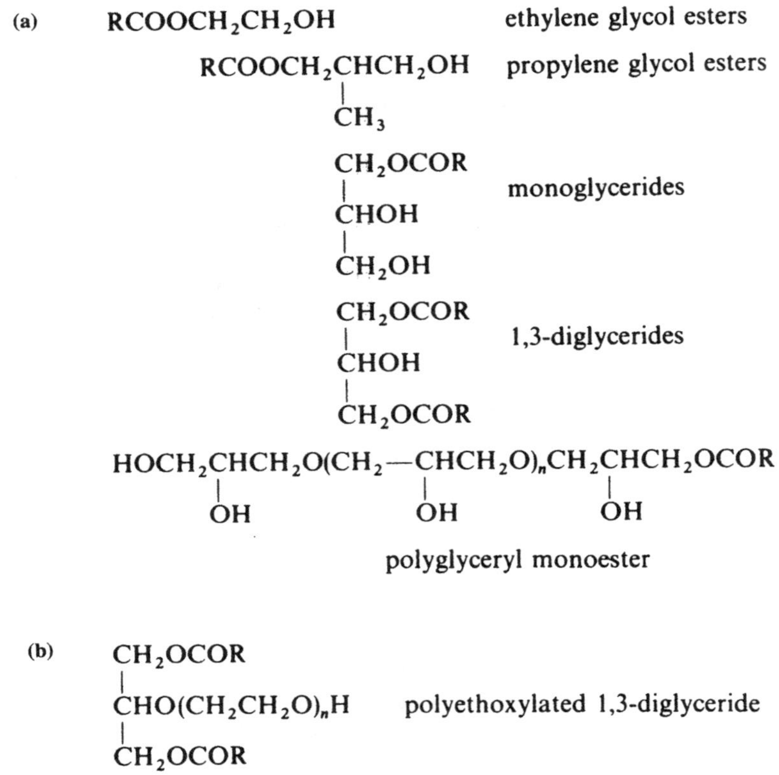

232   HANDBOOK OF SURFACTANTS

(a) $RCOOCH_2CH_2OH$ ethylene glycol esters

$RCOOCH_2CHCH_2OH$ propylene glycol esters
$|$
CH_3

CH_2OCOR
$|$ monoglycerides
$CHOH$
$|$
CH_2OH

CH_2OCOR
$|$ 1,3-diglycerides
$CHOH$
$|$
CH_2OCOR

$$HOCH_2CHCH_2O(CH_2-CHCH_2O)_nCH_2CHCH_2OCOR$$
$$\underset{OH}{|}\qquad\qquad\underset{OH}{|}\qquad\quad\underset{OH}{|}$$

polyglyceryl monoester

(b) CH_2OCOR
$|$
$CHO(CH_2CH_2O)_nH$ polyethoxylated 1,3-diglyceride
$|$
CH_2OCOR

Figure 7.23 Structural formulae of (a) esters and (b) ethoxylates.

Polyoxyethleneglycol(400)triglycerol monostearate. In this case the number 400 indicates molecular weight of the polyethyleneglycol

Description

1. *Glycol esters* (ethylene or propylene glycol monoesters). Preparation by reaction of one mole of glycol with one mole of fatty acid gives a mixture of mono- and di-esters. Commercial glycol ester surfactants usually consist of such mixtures. An excess of glycol will give an ester with a high monoester content. Both hydroxyl groups of ethylene glycol are of equal reactivity, being primary, but the two hydroxyl groups of 1,2-propylene glycol have not the same reactivity, with the 1-hydroxyl being a primary hydroxyl group and the 2-hydroxyl group being a secondary hydroxyl group. The primary group esterifies at a faster (about three times) rate than the secondary hydroxyl group.

NONIONICS 233

2. *Glyceryl esters* (mono- and di-esters of glycerol). Method 1: directly from animal fats or vegetable oils. When a triglyceride (an animal fat or vegetable oil) is reacted with glycerol in the presence of an alkaline catalyst transalcoholysis takes place, and mono-, di- and tri-esters of glycerol are formed, the proportions depending upon the ratio of starting materials. Method 2: partial hydrolysis with alkali gives a mixture of soap, mono-, di- and tri-glycerides of varying composition depending upon the extent of saponification. The products are easy to make in simple equipment but are ill-defined and not easy to reproduce. Method 3: from purified fatty acids and glycerol. An alternative method of manufacture of glyceryl esters is to esterify the fatty acid (which can be a fractionated one) with glycerol; much better defined products are obtained but they are more expensive. The usual products available are described as monoesters, e.g. glyceryl monolaurate, indicating two free hydroxyl groups, or tri-esters indicating no free hydroxyl groups. However, most commercial monoesters will have some diester present. The number of free hydroxyl groups is important as they can influence solubility, i.e. the greater the hydroxyl value the more water-soluble. There is considerable complexity in these reactions and most of the reactions are reversible. The monoesters can be separated from the mixtures by molecular distillation and can be obtained in a pure stable state.
3. *Polyglycerides*. When glycerol is heated in the presence of an alkaline catalyst dehydration occurs with the formation of polyglycerols. The polyglycerols can be esterified with fatty acids. The degree of polymerisation of the glycerol and the degree of esterification can give a large number of products with widely varying surfactant properties.
4. *Polyoxyalkylene glyceride esters*. These are formed by reaction of ethylene oxide with free hydroxyl groups in monoglycerides or diglycerides or polyglycerol esters.

General properties

1. *Solubility*. See Tables 7.35-7.37.
2. *Solubility* (general). Practically all the esters are insoluble in water unless there is a high proportion of hydroxyl groups, e.g. polyglyceryl esters, when the products can disperse to form hazy solutions. Monoglycerides can absorb up to 50% water and form liquid-crystalline phases; those with C20 or C22 alkyl chains form highly viscous hexagonal or cubic phases, (see Chapter 4). Di- and tri-glycerides alone do not form liquid crystals in water although they show some solubility in aqueous solutions of monoglycerides. The phase diagrams of unsaturated monoglycerides resemble those of the

234 HANDBOOK OF SURFACTANTS

Table 7.35 Solubility of monoesters[a] (S = soluble, D = dispersible, I = insoluble)

	Monolaurate			Mono-oleate		Monostearate	
	EG	PG	G	EG	G	EG	G
HLB	–	–	3	–	3	–	3
1% in water	I	I	I	I	I	I	I
1% in mineral oil	S	S	S	S	D	S	I
1% in white spirit	—	—	S	—	D	—	I
1% in aromatic solvent	D	D	D	D	S	D	I
1% in perchlorethylene	—	—	D	—	S	—	—

[a] EG = ethylene glycol monoester, PG = propylene glycol monoester, G = glycerol monoester.

Table 7.36 Solubility of glycerol trioleate (HLB 0.8) (S = soluble, D = dispersible, I = insoluble)

1% in water	I	10% in water	I
1% in mineral oil	S	10% in mineral oil	S
1% in white spirit	S	10% in white spirit	S
1% in aromatic solvent	S	10% in aromatic solvent	S
1% in perchlorethylene	S	10% in perchlorethylene	S

Table 7.37 Solubility of polyglyceryl esters[a] (S = soluble, D = dispersible, I = insoluble)

	TG/MS	DG/MS	DG/TS
1% in water	D	D	D
1% in mineral oil	D	D	S
1% in aromatic solvent	S	S	S

[a] TG/MS = triglycerol monostearate, DG/MS = decaglycerol monostearate, DG/TS = decaglycerol tristearate.

saturated, but with the phase boundaries moved to lower temperatures. Conversely most esters are soluble in mineral oil unless there is a high proportion of hydroxyl or ether groups. Ethoxylation improves water solubility and decreases mineral oil solubility.

3. *Chemical stability.* All these products contain an ester group and hence will be hydrolytically unstable at high temperatures and in the presence of strong acid or even more in the presence of alkali. At room temperature and neutral pH they can be considered stable.

4. *Functional properties.* Wetting, foaming, dispersing and detergent properties are poor compared with the alcohol or alkyl phenol

NONIONICS 235

ethoxylates. This is because the molecular weight has to be high before water solubility is achieved. Emulsification properties, particularly water in oil, are excellent and the relatively high molecular weight gives very stable emulsions. This group of products is one example of 'polymeric' surfactants (see chapter 11).

Applications

1. *General.* Confined to emulsification and/or applications where the low toxicity properties are utilised, e.g. food additives. The ethoxylated sorbitan derivatives are used in very similar applications, but these are discussed in section 7.11
2. *Food.* Many foods contain a water-in-oil emulsion and/or oil-in-water emulsion inside a fat or oil. The oil-soluble surfactants (often used with water-soluble surfactants, e.g. ethoxylated sorbitan esters; see section 7.11) play an important role in entrapping air and holding small bubbles of air. They can also affect the physical properties of the mixtures and also the physical stability over a period of time of such mixtures. For many years glyceryl monostearate (with some distearate), was the standard anti-staling agents in bread products. Addition of about 0.35% surfactant (based on the weight of flour) retards the crystallisation of amylopectin and the resultant release of moisture thus delaying the firming of the bread. Propyleneglycol monoesters are used in conjunction with glyceryl monoesters in high volume cakes and whipped toppings. Other applications of emulsifiers are in ice cream, margarine and synthetic cream.
3. *Pharmaceuticals.* Emulsifying agents in ointments and lotions. Again, often used in combination with the ethoxylated sorbitan derivatives, (see section 7.11)
4. *Defoamer components.* The type of esters described in this section are excellent defoamers for a variety of aqueous systems, particularly where toxicity is an important consideration. They are usually used in conjunction with other surfactants of a higher HLB value.
5. *Cosmetic emulsifiers.* Lotions, creams, gels and other vehicles are used in cosmetics, many of them being water-in-oil emulsions. The esters can either be the oil base for the emulsion or be used as one of the emulsifiers for the system. Ethylene glycol, propylene glycol and glycerol monoesters (particularly stearates) are used to thicken creams and lotions and opacify solutions of soap and more hydrophilic materials.
6. *Textiles.* Used as emulsifiers and/or lubricants in spinning oils.
7. *Polymers.* Mono- and di-glycerides are applied to the surface of food packaging to give anti-fogging properties.

236 HANDBOOK OF SURFACTANTS

Specification

Appearance, usually viscous liquid or pasty solid
Sap. value (mg KOH/g), C16–18 monoglycerides 150–170.
Free glycol, Can be 5–10% for mono esters but purified products can be as
low as 0.5%
Acid value (mg KOH/g) 0.2–4
Soap, 1–6%
Iodine value, check on unsaturation but not applicable to ethoxylates.
Hydroxyl value, useful for ethoxylates (see section 7.10) or polyglyceryl-
esters (100–350 mg KOH/g)

Safety and toxicity

This group of products is more widely accepted as food additives than any
other group of surfactants (see also section 7.11). One reason for this is
that the chemical structures are very similar to those of naturally occurring
oils and fats. Also, glycerides are prepared from naturally occurring glycerol and fatty acids or even direct from the naturally occurring glycerides.
Members of this group appear in lists of approved additives in the USA
and the EC. Suppliers will give full details of whether or not their products
comply with Food Regulations.

7.13 Alkyl amines and alkyl imidazolines

Nomenclature

Generic (see Figure 7.24 for formulae)
  Alkyl amidopropyl dimethylamine
  Alkyl primary amines
  Alkyl secondary amines
  Alkyl tertiary amines
  Alkyl propanediamines
  Alkyl ethylene diamines
  Alkyl diethylene triamines
  Alkyl triethylene tetramines
  Alkyl tetraethylene pentamines
  Alkyl polyamines
  3-alkyloxypropylamines
  Diamines
  Fatty primary amines
  Imidazolines
  Alkylimidazoline hydroxyethylamines
  Alkylimidazoline ethylenediamine

NONIONICS　　　　　　　　　　　　　　237

RNH$_2$　　　　　　　primary amine

$\underset{R'}{\overset{R}{>}}$NH　　　　　　secondary amine

$\underset{R''}{\overset{R}{\underset{R'}{>}}}$N　　　　tertiary amine

RNHCH$_2$CH$_2$CH$_2$NH$_2$　　　　　　propane diamine

RNHCH$_2$CH$_2$NH$_2$　　　　　　ethylene diamine

RNHCH$_2$CH$_2$NHCH$_2$CH$_2$NH$_2$　　diethylene triamine

RNH(CH$_2$CH$_2$NH)$_x$H　　　　　alkyl polyamine

RCONHCH$_2$CH$_2$CH$_3$N$\underset{CH_3}{\overset{CH_3}{<}}$　　N,N-dimethyl-n(3-alkylamido propyl)amine

Figure 7.24 Alkyl amines.

N,N'-dimethyl-N-(3-alkyl amidopropyl)amines
Polyamines
Examples
Laurylamine, C$_{12}$H$_{25}$NH$_2$
N,N-dimethyl-N-(3-lauryl amidopropyl)amine = laurylamidopropyl dimethylamine, C$_{12}$H$_{25}$CONHCH$_2$CH$_2$CH$_2$N(CH$_3$)$_2$

Description

The major commercial products first available were based on natural fatty acids which have been converted to primary amines via the amide and hydrogenation see Figure 7.25. From the primary amine formed, secondary, tertiary and alkyl propanediamines can be made see Figure 7.25. The most common alkyl groups are derived from coconut oil, tallow or soyabean oil and may be fractionated or not. In more recent years alternative routes to the primary amines and tertiary amines have been developed to avoid the costly route shown in Figure 7.25. The three principal routes are:

1. From the alkyl halide by reaction with ammonia
2. From the olefin by reaction with ammonia
3. From aliphatic alcohols by reaction with acrylonitrile and hydrogenation (this route gives the 3-alkyloxypropylamines)

The importance to the formulator of the different routes via alkyl

238 HANDBOOK OF SURFACTANTS

Primary amines

$$RCOOH + NH_3 \longrightarrow RCONH_2 \longrightarrow RCH_2NH_2$$

Propane diamines

$$RNH_2 + CH_2{=}CHCN \longrightarrow RNHCH_2CH_2CN \xrightarrow{+H_2} RNHCH_2CH_2CH_2NH_2$$

Secondary amines

$$2RNH_2 \xrightarrow[\text{nickel}]{\text{Raney}} HN{<}^{R}_{R} + NH_3$$

Tertiary amines (alkyl dimethyl)

$$RNH_2 + 2HCHO + 2H_2 \longrightarrow RN{<}^{CH_3}_{CH_3} + 2H_2O$$

Figure 7.25 Manufacture of amines.

alcohol, via alkyl halide and via the olefin compared to the fatty acid route is as follows. (i) The distribution of the alkyl chain will be different because the fatty acid is from natural sources and the materials of the other route are from petrochemicals. Even if the petrochemical route is via ethylene polymerisation the distribution of chain lengths may be different. (ii) The minor components will be different; this may be very important or not at all. Thus, care should be taken in changing raw materials. Very few amines are used as such, but most are changed into ethoxylated products, betaines, amine oxides and quaternaries. The by-products can sometimes interfere with the conversion process or even pass through to the finished derivative.

There is much interest in developing new routes to tertiary amines because the alkyl dimethyl tertiary amines are the raw materials for amine oxides (section 7.5), quaternaries (section 8.2) or betaines (section 9.2). The principal tertiary amines available are imidazolines, alkylamidodimethyl propylamines and alkyl dimethylamines.

Imidazolines are made by reacting a fatty acid with a substituted ethylene diamine and then cyclising by heating at 220–240°C) (see Figure 7.26). R is the hydrophobe, usually in the range C12–C18, derived from a fatty acid. The group R' can be practically any group although if it contains a reactive group capable of reacting with a COOH group, e.g. a –OH group, then the resulting products can be very complex. Typical commercial products are: R' = CH_2 CH_2 NH_2 aminoethyl imidazoline, R' = CH_2 CH_2OH, hydroxyethyl imidazoline. The imidazoline ring has a tertiary

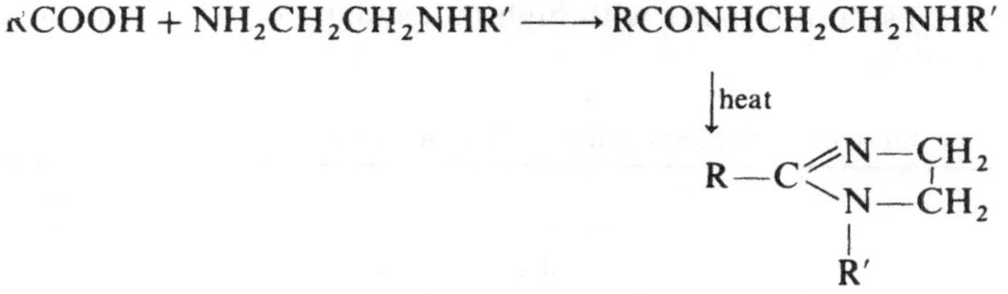

NONIONICS 239

$$\text{RCOOH} + \text{NH}_2\text{CH}_2\text{CH}_2\text{NHR}' \longrightarrow \text{RCONHCH}_2\text{CH}_2\text{NHR}'$$

$$\downarrow \text{heat}$$

$$R-C \underset{N-CH_2}{\overset{N-CH_2}{\lessgtr}} \quad \begin{array}{c} | \\ R' \end{array}$$

R' = CH$_2$CH$_2$NH$_2$ = alkyl aminoethyl imidazoline
R' = CH$_2$CH$_2$OH = alkyl hydroxyethyl imidazoline

Figure 7.26 Manufacture of imidazolines.

nitrogen group which can be quaternised. However, if the group R' contains amino groups then these can also act as cationic groups when acidified. The imidazoline ring readily splits open on hydrolysis (see section 9.3).

Alkylamido dimethyl propylamines (see Figure 7.27) are now available principally to make amine oxides or amidopropyl betaines.

$$\text{RCOOH} + \text{NH}_2\text{CH}_2\text{CH}_2\text{CH}_2\text{N} \underset{CH_3}{\overset{CH_3}{\big\langle}} \longrightarrow \text{RCONHCH}_2\text{CH}_2\text{CH}_2\text{N} \underset{CH_3}{\overset{CH_3}{\big\langle}}$$

Figure 7.27 Alkylamido dimethyl propylamines.

In alkyl dimethylamines, the alkyl group most often found is C12–C14. The products are used to make amine oxides (section 7.5), quaternaries (section 8.2) and betaines (section 9.4). There are now several ways of preparing these products via the primary amine, via olefins or via alkyl alcohols. There can be small differences in the by-products depending upon the manufacturing route.

General properties

1. *Solubility*. Generally insoluble in water but soluble in strong acid solutions (see Table 7.38). Solubility of alkyl amino ethyl imidazolines is shown in Table 7.39. Some salts are compatible with anionics e.g. the propionic acid salt of cocoamidopropyldimethylamine is compatible with lauryl sulphate and lauryl ether sulphate and small amounts (1–5%) will actually give an increase in flash foam. Most alkyl polyamines are soluble in acid (mineral or low molecular weight

240 HANDBOOK OF SURFACTANTS

organic) solution. Salts with high molecular weight organic acids are oil-soluble.

Table 7.38 Solubility (%) of amines and salts (I = insoluble)

	In water		In ethanol	
	20°C	40°C	20°C	40°C
$C_{12}H_{25}NH_2$	I	I	53%	87%
$C_{12}H_{25}NH_2$ + HCl	0.3%	36%	13%	34
$C_{12}H_{25}NH_2$ + HOAc	25%	28%	45%	72%

Table 7.39 Solubility of amino ethyl imidazolines (S = soluble, D = dispersible, I = insoluble)

	Coco	Oleic	Tall oil	Stearic
10% in water	D	D	I	I
10% in mineral oil	I	S	D	D
10% in white spirit	I	S	S	S
10% in aromatic solvent	I	S	S	S
10% in perchlorethylene	I	S	S	S

2. *Chemical stability.* The amines are all reactive, the principal reactions being: salt formation with mineral or low molecular weight organic acids for *p*-, *sec*- and *t*-, form carbamate salts with carbon dioxide, which reaction is reversible by heating for the *p*- and *sec*-; reaction with aldehydes and ketones, *p*-; reaction with alkyl halides, *p*-, *sec*-; reaction with ethylene oxide, *p*-, *sec*-or *t*-; *p*- and *sec*- decompose on heating at above 90°C; *t*-amines are more stable. *N,N*-Dimethyl-*N*-(3-laurylamidopropyl) amine hydrolyses at temperatures of 90°C but acid hydrolysis will be faster.

3. *Surface active properties* The fatty amines are insoluble in water and therefore show little or no surface-active properties. When neutralised by acid (e.g. hydrochloric or acetic acid) they give high surface activity (see section 8.3).

4. *Functional properties.* Adsorption on metallic surfaces to give water repellency; emulsifying properties for O/W emulsions at low pH.

Applications

The major applications are as chemical intermediates for amine salts, quaternary ammonium salts, ethoxylated derivatives or betaines. However, there are a large number of applications of the amines or imidazolines themselves.

NONIONICS 241

1. *Primary amines.* Cationic emulsifying agents below pH 7. Corrosion inhibitor for fuels, lubricating oils and for metal surfaces. Anticaking agents for fertilisers, normally $C_{18}NH_2$, but liquid formulations are used composed of amine in mineral oil and sometimes in conjunction with a fatty acid; solid formulations are made by coating kaolin or talc with the amine. Adhesion promoter for painting damp surfaces. Ore flotation collector (the largest application for the 3-alkyloxypropylamines is as the collector for silicate gangue from iron oxide).
2. *Diamines, polyamines and imidazolines.* Uses as above, but the main uses are as adhesion promoters for bitumen coating of damp road surfaces. Textile softener. Pigment coatings. Oilfield chemicals as corrosion inhibitors; by variation in the alkyl chain of the amine and the type of organic acid used to neutralise the amine a very wide range of water-soluble and oil-soluble materials can be prepared.
3. *Tertiary amines.* For manufacture of quaternaries, amine oxides and alkyl betaines.
4. *Imidazolines.* Can function as oil-soluble emulsifiers producing cationic O/W emulsions; if they are neutralised below pH 8 they become hydrophilic and can act as emulsifiers for polar organic solvents, e.g. toluene, pine oil or triglyceride. Used in water-displacing solvents as they adsorb at metal surfaces in place of water. Addition to oils, waxes or bitumen to improve adhesion to substrates. Addition to paints to improve adhesion.

Specification

Amine number or neutralisation equivalent[a,d]
Amine value[b,d]
Acid value (mg KOH/g)[d]
Iodine value[d]
Impurities[c,d]:

p-amines	up to 4% sec	up to 1% *t*-
sec-amines	up to 8% *p*-	up to 6% *t*-
t-amines	up to 1% *p*-	up to 4% *sec*-

[a] The amount of hydrochloric acid (in mg) needed for the neutralisation of 1 g of sample = amine value × 0.6498. The fatty amines are insoluble in water, so isopropanol is used as solvent and titration is carried out with isopropanol/hydrochloric acid.
[b] The amount of KOH (in mg) that would be equal to the number of milliequivalents of amine present in 1 g of sample = amine number × 1.539.
[c] Most primary amines contain small amounts of *sec*- and *t*-amines. Likewise secondary amines contain small amounts of *p*- and *t*-amines. The

242   HANDBOOK OF SURFACTANTS

impurities can often be important. The diamines and polyamines, however, can have equal quantities of *p-* and *sec-*. Thus the relative amount of *p-*, *sec-* and *t-*amine is a useful identifier for unknown products.

d The values vary so widely in this product range that there is no point in giving ranges. However, these values are useful in characterising product quality.

7.14   Ethoxylated oils and fats

Nomenclature

Generic
  Ethoxylated lanolin
  Ethoxylated castor oil
Chemically these products could be classified as ethoxylated esters and this would be a subgroup of section 7.12. However, in the case of section 7.12 the hydrophilic properties are derived from multiple–OH groups, whilst in this section the hydrophilic properties are due to the polyethylene oxide chain.

Description

A number of natural oils or fats have been ethoxylated to give surfactant properties. The two principal raw materials are wool fat or lanolin and castor oil.

Raw wool consists of a mixture of fatty acid esters of cholesterol, isocholesterol, other higher fatty alcohols and terpene derivatives. Normally the wool fat undergoes chemical fractionation and the ethoxylation is carried out on the lanolin alcohol, which is still a mixture of aliphatic alcohols and sterols. The initial site for ethoxylation is the hydroxyl group on the ricinoleic acid chain. However, the ester groups may provide additional reaction sites under the effect of alkaline catalyst and heat. Thus the composition of the products will depend upon the conditions of ethoxylation.

General properties

The properties follow all the normal ethoxylate properties, as the main variable is the polyoxyethylene content. Specific properties are summarised below:

1. *Lanolin ethoxylate.* With >55% ethylene oxide, the products are water-soluble and give solubilisation and emulsification properties.

NONIONICS 243

2. *Castor oil ethoxylate.* With >60% ethylene oxide the products make excellent emulsifiers.

Applications

1. *General.* The high molecular weight of most products (e.g. castor oil + 40 moles of ethylene oxide, molecular weight = 2800) restricts their use to emulsifiers, demulsifiers, defoamers and dispersing agents.
2. *Cosmetics* Numerous ethoxylated natural products have been made for the cosmetic industry, mainly to be able to claim that a chemical had a 'natural' base.

Specification

Similar to most ethoxylates in cloud point, hydroxyl number and acid number.

7.15 Alkyl phenol ethoxylates

Nomenclature

Abbreviations
  APE, alkyl phenol ethoxylates, used in this book
  NPE, nonyl phenol ethoxylate, used in this book
  APEO, APO, PEO, NPO
Generic
  Alkylphenol ethoxylates
  Alkylphenol polyglycol ethers
  Polyoxyethylates alkylphenols
  Polyoxyethlylene alkylphenols
Examples of names for the product $C_9H_{19}C_6H_4O(CH_2CH_2O)_9H$
  Nonyl phenol + 9EO
  Nonyl phenol ethoxylate (9EO)
  Ethoxylated (9EO) nonyl phenol
  Nonoxynol 9

Description

These surfactants are prepared by reaction of ethylene oxide with the appropriate alkyl phenol. The principal products commercially available are ethoxylates of *para*-nonylphenol, *para*-dodecylphenol, dinonylphenol (mixture of isomers) and *para*-octylphenol (once the major product).

244 HANDBOOK OF SURFACTANTS

The most common APEs are the nonyl phenol ethoxylates which are made using distilled nonyl phenol which is predominantly (usually > 90%) *para*. The nonyl group is derived from propylene trimer and therefore the alkyl group is a branched chain. Nonyl phenol ethoxylates (at the time of writing) are the only large volume surfactant on the international market with a branched hydrophobic group. The reason is their low price and excellent properties, but the biodegradability is suspect and therefore their long-term future is in doubt. However, exactly the same statement has been made for twenty years, and they are still bought and sold, but mainly for use in industrial products rather than domestic detergents. They are likely to be used whilst they are cheaper than the similar alcohol ethoxylates.

The ethylene oxide distribution is described in section 7.1.1, showing that there is no free alkyl phenol remaining in ethoxylated products, even those with low amounts of ethylene oxide, unlike alcohol ethoxylates. However, newer procedures for ethoxylating alcohols can now give low free alcohol and an ethoxylate distribution similar to the APE (see section 7.1.1). Impurities include polyethylene glycols (which often show as a haze), but a small proportion of water gives clear solutions. The presence of polyglycols can often increase products' tendency to gel. Catalyst residues, sodium hydroxide or sodium methoxide are usually neutralised with acetic acid. 1,4-Dioxane can be easily removed from APEs. Free ethylene oxide may also be present.

General properties

1. *Solubility*. The solubility of NPEs is shown in Table 7.40.

Table 7.40 Solubility of NPEs[a] (S = soluble, D = dispersible, I = insoluble)

	NP+1EO	NP+4EO	NP+6EO	NP+8EO	NP+11EO	NP+20EO	DNP+10EO
HLB	4.5	9	11	12	14	16	11
Cloud point[b]	<RT	<RT	<RT	20–30	60–70	>100	—
10% in water	I	I	I	D	S	S	I
10% in mineral oil	S	S	I	I	I	I	S
10% in white spirit	S	S	S	S	I	I	S
10% in aromatics	S	S	S	S	S	I	S
10% in perchlor-ethylene	S	S	S	S	S	I	S

[a] NP = nonyl phenol, DNP = dinonyl phenol; RT = room temperature.
[b] in °C of a 1% soln. in water (DIN 53917). For low EO contents a 2% solution in 25% diethylene glycol monobutyl ether can be used. For high EO contents a 1% solution in 10% sodium chloride can be used.

In comparison with the alcohol ethoxylates, NPEs show better solubility in organic solvents at similar HLB values. Note the

NONIONICS 245

improved solubility of dinonyl phenol ethoxylate + 10 EO in white spirit over nonyl phenol + 6EO, although they have the same HLB value of 11. NPEs are more liquid and have lower viscosities than equivalent AEs. Aqueous solutions of many NPEs show a very large increase in viscosity, or even gels, at about 50–60% concentration with lower viscosities below or above this concentration. The solubility decreases with increasing temperature (see Cloud point in section 7.1.2).

2. *Compatibility with aqueous ions.* Excellent compatibility with all aqueous ions. Lime soap dispersion index for most products is in the range of 4–8% of NPE relative to sodium oleate in 333 ppm hard water. However, at 6EO (optimum surface active activity at room temperature), the index is 10–15 (GAF Corp., 1965).

3. *Chemical stability.* Stable to heat for short periods but show a slow increase in colour when held for extended periods above 80°C in air. Addition of antioxidants (e.g. hindered phenols) can improve stability to well above 100°C. NPEs are stable to hot dilute acid, and to alkali (except yellowing in solid powders), but end-blocking improves the stability to alkali. Not entirely stable to oxidising agents, e.g. hypochlorite, peroxide and perborate (see section 7.1.1).

Table 7.41 CMC values for ethoxylated alkylphenols

Product `	CMC (mmoles/1)	CMC (%)
Octyl phenol + 8.5EO	180–230	0.01 –0.013
Nonyl phenol + 9.5 EO	78–92	0.0049–0.0057
Nonyl phenol + 10.5 EO	75.90	0.0050–0.0060
Nonyl phenol + 15 EO	110–130	0.0095–0.011
Nonyl phenol + 20 EO	135–175	0.015 –0.019
Nonyl phenol + 30 EO	250–300	0.038 –0.045
Nonyl phenol + 100 EO	1000	0.46

4. *Surface-active properties.* CMC values are shown in Table 7.41. Minimum surface tension in water of 28 mN/m occurs with nonyl phenol at the 6EO level, where it only disperses rather than dissolves. Maximum surface activity is near the cloud point.

5. *Functional properties (for NPEs).* Excellent wetting agents with lowest Draves wetting times near cloud point; thus the optimum EO content will depend upon the temperature, but higher temperatures and hence higher cloud points give better wetting. Excellent detergent for oils and greases; EO = 9 for optimum, but this depends upon formulation. Solubiliser for oils and perfumes, EO = 3–5 for optimum. Foaming agent, EO = 15 for optimum, but low foam

compared to anionics; depends upon temperature in a similar way to wetting. Emulsifier, W/O, EO = 1–5 for optimum, O/W, EO = 8–40 for optimum; for waxes, oils and fats, EO = 4–6 for optimum. Defoaming, EO = 1–3 for optimum.

6. *Disadvantages*. Doubts on large scale availability in the future due to the biodegradation properties.

Applications

1. *Household products and industrial cleaning*. Not used in the Western world because of biodegradation properties, but these products give excellent cleaning compounds especially in detergent sanitisers. Most usual products NPE with 10–12 EO (compatible with iodophors, quaternaries and phenolics), but in the presence of strong electrolyte and high temperatures EO = 15 is optimum (particularly for bottle cleaners, metal cleaners and heavy duty alkaline cleaners). In formulated conveyor lubricants, gives both good detergency and lubrication (not found with AE). Make excellent heavy-duty solvent-type cleaners for floors and general cleaning, containing >25% kerosene with 10% of NPE with 9 EO. Now superseded by aqueous dispersions wherever possible.
2. *Textiles*. Detergent, wetting agents and emulsifiers for processing wool, cotton and synthetics (scouring, bleaching, kier boiling, warp desizing), NPE with 8–9 EO at low temperatures and 20 EO at high temperatures; antistatic agent; dye retarders.
3. *Agriculture*. Used as emulsifiers in self-emulsifying herbicides and insecticides.
4. *Concrete*. Foam entrainment for frost protection, NPE with EO >15.
5. *Emulsion polymers*. Emulsifiers and stabilisers, NPE with EO >20.

Specification

Water content, 0.1–20%
Polyglycol content, 0.1–5%
Melting point, a more realistic measure of EO content than cloud point for high EO content materials
Refractive index, a quick method for checking EO content.
Cloud point, a quick method for checking EO content but can be misleading if polyglycols present.

References

Balzer, D. (1991) *Tenside Surfactants Detergents*, **28**(6), 419–27.
Blease, T.G. *et al.* (1992) *Proc. 3rd International Surfactants Congress*, London, Vol. C, pp. 275–280.

NONIONICS 247

Bocker, Th. and Thiem, J. (1989) *Tenside Surfactants Detergents*, **26**, 318.
Brink, C. *et al.* (1992) *Proc. 3rd CESIO Congress*, London, vol. D, p. 211.
Crass, G. (1992). *Seifen, Oele, Fette, Wachse*, **118**(15), 921, 923–4.
Dillan, K.W. *et al.* (1985) *J. Am. Oil Chem. Soc.* **62**(7), 1144.
Fan, T.Y., Goff, U., Song, L., Fine, D.H., Arsenault, G.P. and Biemann, A. (1977) Nitrosomamines in cosmetics, lotions and shampoos, presented at the American Chemical Society Meeting, New Orleans, Los Angeles.
GAF Corp. (1965), *Technical Bulletin* 7543–002.
Gao, Z. *et al.* (1992). *Shiyau Huagong*, **21**(4) 242–8.
Henderson, G. and Newton, J.M. (1966) *Pharm. Acta Helv.* **41**, 228.
Henderson, G. and Newton, J.M. (1969) *Pharm. Acta Helv.* **44**, 129.
Hugo, W.B. and Newton, J.M. (1963) *J. Pharmacol.* **15**, 731.
Kassem, T.M. (1984). *Tenside, Surfactants, Detergents*, **21**(3), 144.
Kemp, R.A. and Gerstein, T. (1977) USP 4,033,895.
Kemp, R.A. (1989), Eur. Pat. Appl. EP 398,450.
Kemp, R.A. (1992) USP 5,102,849.
Knaggs, E.A. (1965) *Soap Chem. Specialities*, **41**(1), 64.
Matsumura, S. *et al.* (1990) *J. Am. Oil Chem. Soc.* **67**, 996.
Meguro, K., Ueno, M. and Esumi, K. (1987) *Nonionic Surfactants — Physical Chemistry*, Marcel Dekker, New York, p. 150.
Milwidsky, B. and Holtzmann, S. (1972). Effects of regular amides and superamides on the foaming and viscosity of detergents, paper presented at the VIth International Congress of Surface Active Substances, Zurich, Switzerland.
Nakagawa, T. (1967) In *Nonionic Surfactants*, ed. M.J. Schick, Marcel Dekker, New York, p. 599.
Procter and Gamble (1992) International Patent Applications WO 92/06070 and WO 92/08687.
Raths, H.-C. (1992) *Proc. 3rd International Surfactants Congress*, London, Vol. C, pp. 291–300.
Sanracesaria, E. *et al.* (1992) *Proc. 3rd. International Surfactants Congress*, London, Vol. C, pp. 281–290.
Schick, M.J. (1967) In *Nonionic Surfactants*, ed. M.J. Schick Marcel Dekker, New York, p. 612.
Schmolka, I.R. (1977) *J. Am. Oil Chem. Soc.* **54**, 110–116.
Tabata, Y., Ueno, M., and Meguro, K. (1984). *J. Am. Oil Chem. Soc.* **61**, 123.
Tagawa, T., Iino, S., Sonoda, T. and Oba, N. (1962) *Kogyo Kagaku Zasshi*, **65**, 953.
Unilever (1973) UK Patent 1,329,086.
Webb, I. (1957) USP 2,787,595.
Wieder, P.R. (1992) USP 5,102,849.
Wimmer, I. (1991) German Patent DE 4,012,725.
Wu, Z. and Cheng, L. (1992) *Riyong Huaxue Gongye* **1**, 1–9.
Zhou, Z. and Chu, B. (1987) *Macromolecules*, **20**, 3089.
Zhou, Z. and Chu, B. (1988a) *Macromolecules*, **21**, 2548.
Zhou, Z. and Chu, B. (1988b) *J. Colloid Interface Sci.* **126**, 171.

8 Cationics

8.1 Cationics (general)

Nomenclature

The cationics are named after the parent nitrogenous phosphorus or sulphur starting material. Thus, the quaternary ammonium compound dodecyltrimethyl ammonium chloride is formed from the starting material dodecyldimethylamine reacted with methyl chloride.

Description

Cationic surfactants are those surfactants where the ionic group on the hydrophobic group would go to the cathode (negatively charged) and hence have a positive charge. With very few exceptions, commercially available cationics are based on the nitrogen atom carrying the positive charge. The only other products are based on phosphorus and sulphur; these, however, are not commercially available except for sulphobetaines. $C_{12}H_{25}NH_2$ is a primary amine and in neutral solution is uncharged, and hence is strictly **not** cationic. However, as a salt of, for example, acetic acid, $C_{12}H_{25}NH_3^+ CH_3COO^-$, the amine is now a cationic surfactant. The largest volume products are the products used for fabric softening in domestic use. These are quaternary ammonium products with two long alkyl chains.

General properties

Non-quaternary cations are sensitive to high pH, polyvalent ions and high concentrations of electrolyte, whereas quaternary cations are not sensitive to high pH, polyvalent ions or high concentrations of electrolyte. Cationics differ from anionic and nonionic surfactants in their high degree of substantivity. This term substantivity encompasses the uptake of surfactant from solution on to the surface of a wide variety of negatively charged surfaces: fibres, cellulosics such as paper and cotton, protein such as wool, and synthetic fibres such as polyamide and acrylic; plastics, polyvinyl chloride and polyvinyl acetate; silicates; metals; pigments. Depending upon the chemical structure of the cationic surfactant, it is possible to make a hydrophilic solid behave as if it was hydrophobic or (less usual) make a

CATIONICS 249

hydrophobic solid behave as if it were hydrophilic. Thus the surface properties of solids can be modified by using cationic surfactants.

1. *Bactericidal action*. The long chain fatty amines and their salts, the quaternaries and the imidazolines have the ability to kill microorganisms or restrict their growth. The chemical structure for optimum effectiveness varies depending upon the type of cationic. However, the imidazolines do not appear to be used very often as biocides.
2. *Complex formation with anionic surfactants*. Cationics will generally, but not always, form insoluble complexes with anionic surfactants which are insoluble in water and lose their surfactant properties, e.g. the ability to foam or wet. However, in organic solvents and mineral oil, anionic/cationic surfactant complexes can show substantivity, wetting and corrosion resistance, suggesting that they are surface-active in such an environment. They can also give very low surface tensions (below 25 mN/m).
3. *Disadvantages*. More expensive than anionics; poor detergency; poor suspending power for solids (e.g. carbon); readily adsorbed on surfaces from aqueous solution and therefore the solution is depleted of surfactant.

Applications

Most solid surfaces are negatively charged, and thus cations will adsorb on to solid surfaces. Examples of this application are: softeners and antistats for textiles, anticaking agents in fertilisers, emulsifiers for bitumen, corrosion inhibitors in oilfields, dispersing agents for pigments, and flotation agents in mineral processing.

8.2 Quaternary ammonium

Nomenclature

These compounds are substituted ammonium salt where none of the four substituent groups are hydrogen

$$R'R''R'''R''''N^+X^-$$

X^- is usually the chloride ion Cl^- or the ethyl sulphate ion $C_2H_5SO_4^-$, e.g. N-alkyltrimethylammonium chloride, R'=alkyl, R''=R'''=R''''=methyl, X=chloride; N-ditallowdimethylammonium chloride; R'=R''=alkyl chain with same length as tallow fatty acid, R'''=R''''=methyl, X=chloride; bis (hydrogenated tallow alkyl) dimethylammonium chloride (see Figure 8.1);

250   HANDBOOK OF SURFACTANTS

benzalkonium chloride, (see Figure 8.1); Quaternised polyoxyethylene fatty amines (see Figure 8.1).

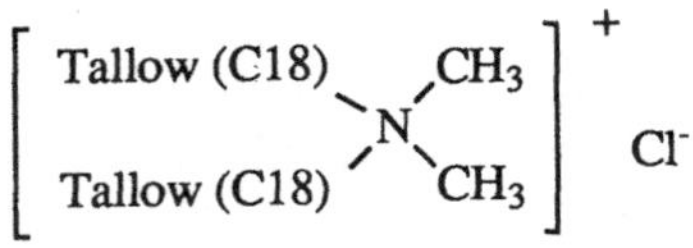

$$\left[\begin{array}{c} \text{Tallow (C18)} \\ \text{Tallow (C18)} \end{array} \hspace{-0.3em} N \hspace{-0.3em} \begin{array}{c} CH_3 \\ CH_3 \end{array} \right]^{+} Cl^{-}$$

bis (hydrogenated tallow) dimethylammonium chloride

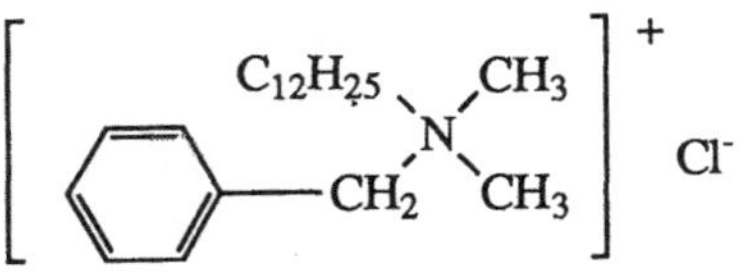

benzalkonium chloride

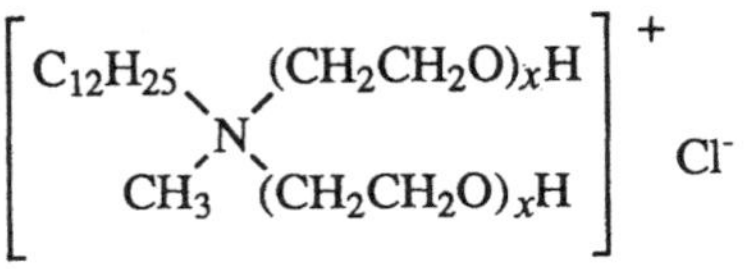

dodecyl methylpolyoxyethylene ammonium chloride

Figure 8.1 Quaternary ammonium compounds.

Abbreviation: QAC, quaternary ammonium compound, used in this book

Description

Quatenaries can be made in simple equipment under mild conditions by reacting an appropriate tertiary amine with an organic halide or organic sulphate (Figure 8.2). The products made in largest volume are the textile

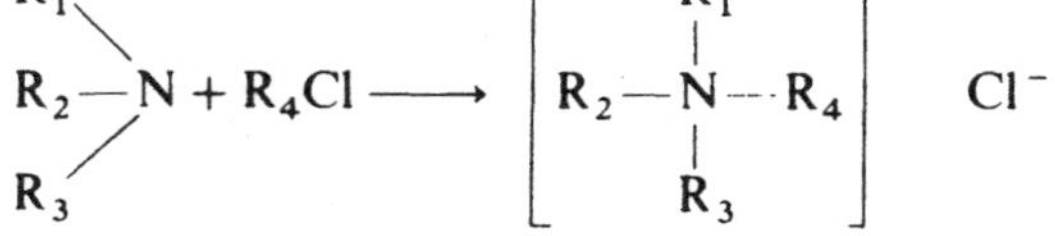

$$\begin{array}{c} R_1 \\ R_2 \!-\! N + R_4Cl \longrightarrow \\ R_3 \end{array} \left[\begin{array}{c} R_1 \\ | \\ R_2 \!-\! N \!-\! R_4 \\ | \\ R_3 \end{array} \right]^{+} Cl^{-}$$

Figure 8.2 Preparation of quaternary ammonium compounds.

softeners for household use, where there are usually two long alkyl groups and two short alkyl groups; the anion can be chloride or sulphate. An example would be bis (hydrogenated tallow alkyl) dimethylammonium chloride (see Figure 8.1).

The best known quaternary is benzalkonium chloride. It is alkyl dimethyl benzyl ammonium chloride where the alkyl group is usually derived from coconut fatty acid or stripped coconut fatty acid and contains C12, C14 and C16. It is a white powder soluble in water and alcohol. Aqueous solutions are also on the market, with products of more than 40% containing cosolvent. Imidazolines can form quaternaries; the most common product is the ditallow derivative quaternised with dimethyl sulphate (see Figure 8.3).

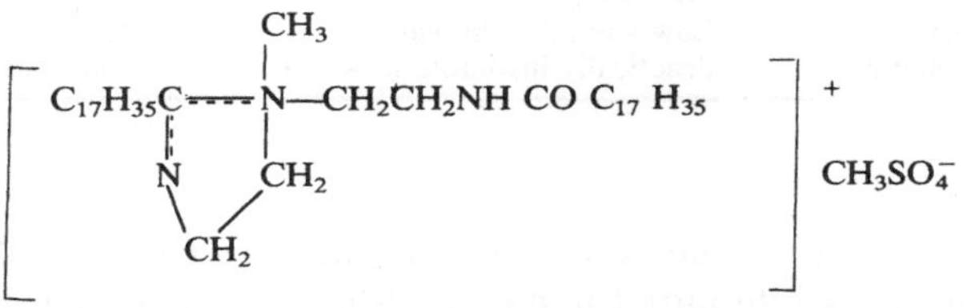

$$\left[\begin{array}{c} CH_3 \\ | \\ C_{17}H_{35}C\text{----}N\text{---}CH_2CH_2NH\ CO\ C_{17}\ H_{35} \\ | \quad\quad | \\ N \quad\ CH_2 \\ \backslash \ \diagup \\ CH_2 \end{array} \right]^{+} \quad CH_3SO_4^-$$

Figure 8.3 Imidazoline-based quaternary compounds.

General properties

1. *Solubility*. Generally water-soluble when there is only one long hydrophobe, with corresponding insolubility in mineral oil, white spirit and perchlorethylene. When there are two or more long-chain hydrophobes the products become dispersible in water and soluble in organic solvents such as white spirit and perchlorethylene. The benzyldimethylalkyl ammonium chlorides show excellent solubility in water and isopropanol.
2. *Compatibility with aqueous ions*. Compatible with most inorganics and hard water, but incompatible with metasilicates and highly condensed phosphates. Incompatible with protein-like organic matter, which is precipitated. Incompatible with substituted phenolic ions when the quaternary molecule is large.
3. *Chemical stability*. Stable to pH changes, both acid and alkaline, especially acid (even HF), but hot alkali usually causes separation. Decomposes when heated to over 100°C.
4. *Compatibility with other surfactants*. Incompatible with anionics. Compatible with nonionics except alkanolamides with high soap content, polyols with high propylene oxide content, and hydrophobic ethoxylates (e.g. nonyl phenol + 4 EO).

252 HANDBOOK OF SURFACTANTS

5. *Surface-active properties*. CMC of benzalkonium chloride (mol. wt 340) = 5×10^{-3} mol/l (0.17%); 0.1% solution of benzalkonium chloride has a surface tension of 34 mN/m.
6. *Functional properties*. The functional properties depend upon the water solubility, with poor water solubility giving high adsorption and maximum surface activity. This is illustrated by the difunctional products, which show a change in properties as the chain length increases (see Table 8.1).

Table 8.1 Properties of quaternaries and chain length

2 C8 chains	Very soluble in water	Mild germicide
2 C10 chains	Soluble in water	Strong germicide
2 C12 chains	Poor solubility in water	Weak germicide
2 C14 chains	Low solubility in water	Antistatic
2 C16–18 chains	Practically insoluble in water	Softener and antistatic

Substantive to negatively charged surfaces, e.g. conditioner for hair, corrosion inhibitor for metals. Moderate foaming properties. Poor wetting properties. Good emulsification properties. Detergency poor to moderate and at best comparable with nonionics, but more expensive. Germicidal properties present at low concentrations and therefore cost-effective germicidal detergents are obtained with the maximum amount of nonionic. However, small amounts of nonionics will not adversely affect germicide properties but a high nonionic/cationic ratio can have reduced germicidal activity. Monoalkyl has more effective germicidal properties than bis(long-chain alkyl); reaction with ethylene oxide reduces germicide effect; chlorination of aromatic ring increases germicide effect. The most common germicides with optimum chain lengths are shown in Figure 8.4.

The general word 'germicide' has been used to describe quaternaries because of the different properties of the many quaternaries. A quaternary can be a bactericide (an agent which kills bacteria) or a bacteriostat (an agent that prevents growth of bacteria).

Quaternaries are adsorbed on to soil and protein, so the performance of quaternaries can be adversely affected by dirt or blood. Bactericidal properties appear with the C12 isomer and continue up to C16–18. As the cations cannot differentiate between bacteria and associated protein the strongly adsorbing higher isomers, C16-C18, are not available to the bacteria. The performance of the C12 isomer is inferior to the C16 isomer in distilled water but superior in the presence of significant contamination.

Quaternaries are moderate foaming agents

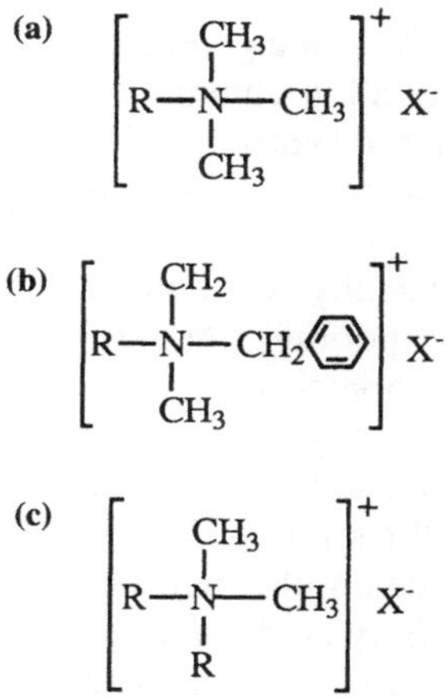

CATIONICS 253

(a)

$$\left[\begin{array}{c} CH_3 \\ | \\ R-N-CH_3 \\ | \\ CH_3 \end{array} \right]^+ X^-$$

(b)

$$\left[\begin{array}{c} CH_2 \\ | \\ R-N-CH_2\!\!\bigcirc \\ | \\ CH_3 \end{array} \right]^+ X^-$$

(c)

$$\left[\begin{array}{c} CH_3 \\ | \\ R-N-CH_3 \\ | \\ R \end{array} \right]^+ X^-$$

Figure 8.4 Quaternaries as bactericides.(a) and (b) R = C12–C16; (c) R = C10.

7. *Disadvantages*. As a germicide, readily deactivated and cannot kill spores (certain bacteria and fungi); thus, cannot be used for sterilisation.

Applications

1. *Biocides*. Benzalkonium chloride BP is C_6H_5 CH_2N $(CH_3)_2$ R^+ Cl^-, where 'R = mixture of alkyls from C_8H_{17} to $C_{18}H_{37}$. There is a difference for the alkyl chain distribution for the British Pharmacoepia and US Pharmacoepia specifications. The USP contains no C8, C10 or C18 alkyl chains. Cetrimide BP is $R–N$ $(CH_3)_3^+$ Br^-, where R = mixture of C12–C16, mainly C14, and is a germicidal quaternary that has some detergent properties, and is usually used in conjunction with chlorhexidine digluconate. Benzalkonium chloride and *N*-benzyl-*N*-alkyldimethylammonium halides are used as bactericides against Gram-positive bacteria, but are less effective against Gram-negative bacteria in hard water. They are used as germicides, disinfectants and sanitisers, are compatible with alkaline inorganic salts and nonionics, and are used with them in alkaline detergent–sanitisers for dishwashing in pubs, restaurants, etc.; they are deactivated by anionics. Hard water tolerance can be improved by careful selection of the distribution of the alkyl chain (optimum is at C14).

2. *Textiles*. The main volume application is the use of quaternaries with two long alkyl chains as textile softeners for home use as the final rinse in the washing machine. They impart a soft fluffy feel to fabrics by adsorbing on to them with the hydrophobic groups oriented away from the fibre. The main product was bis (hydrogenated tallow alkyl) dimethylammonium chloride, but there is now, doubt on its environ-

254   HANDBOOK OF SURFACTANTS

metal acceptability and it is being replaced by 'ester-based' quaternaries still with two long alkyl chains) or imidazoline-based quaternaries. Antistatic finish for synthetic fibres; dye retardant and dye leveller by competing for positive dye sites on the fibre (e.g benzyl trimethylammonium chloride).

3. *Hair care.* Products chemically very similar to the textile softeners are used as a rinse after shampooing, since they adsorb on the hair giving softness and antistatic properties.
4. *Emulsifiers.* N-Alkyltrimethylammonium chlorides and N-alkyl imidazoline chlorides are used as emulsifiers where adsorption of the emulsifying agent on to the strata is desirable (e.g. bitumen treatment of damp roads, insecticide emulsions). Used for the emulsification of polar compounds (e.g. fatty acids and amines) in O/W emulsions.
5. *Metal working.* Addition to acid (hydrochloric and sulphuric) used in the cleaning and pickling of steel to prevent hydrogen corrosion.
6. *Road building.* Quaternary fatty ammonium and imidazoline salts are used to make bitumen emulsions which can be used to repair roads in wet weather.
7. *Bentonite treatment.* Bentonite can be treated with quaternary ammonium salts to convert the normally hydrophilic bentonite into a product with hydrophobic properties. These products can be used as thickening agents in organic systems, e.g. paint or greases.
8. *Oilfields.* Alkyl trimethyl ammonium chloride is used as a germicide for sulphur-producing bacteria which cause corrosion and hence is known as a corrosion inhibitor.
9. *Antistatic in polymers.* E.g. in pvc belting for coal mines.

Specification

The quaternary compounds cover such a wide range of products that no figures are quoted but the following tests will give significant information with most quaternaries.

Active content
Solvent, usually aqueous based but look for alcohols (e.g. isopropanol)
Flash point, may be low (due to alcohol solvent)
Free amine, should be low
Free alkyl chloride or alkyl sulphate

8.3   Amine and imidazoline salts

Nomenclature

See section 7.13 describing the parent alkyl amines.

CATIONICS 255

Generic
   Salts with hydrochloric acid
   Diamine hydrochlorides
   Imidazoline hydrochlorides
      Alkylimidazoline hydroxyethylamine hydrochlorides
      Alkylimidazoline ethylenediamine hydrochlorides
   Polyamine hydrochlorides
   Primary amine hydrochlorides
   Secondary amine hydrochlorides
   Tertiary amine hydrochlorides
Example
   Dodecyl dimethylamine hydrochloride (or dodecyl dimethylamine
   ammonium chloride), $C_{12}H_{25}N(CH_3)_2H^+Cl^-$

Description

The usual method of forming salts is the neutralisation of the amine with an
acid in aqueous solution, e.g. when hydrochloric acid is used amine
hydrochlorides are formed. If less than the theoretical amount of acid is
used the amine salt can often act as an emulsifier for the unreacted amine
in aqueous solution. If the amine salt is insoluble in water the salt can be
formed by double decomposition. Of the inorganic salts, those with
hydrochloric acid, the hydrochlorides, are made most frequently although
they have poor solubility in cold water. Salts obtained using organic acids
are usually more water-soluble than those from inorganic acids, but care
must be taken to avoid amide formation with carboxylic acids by keeping
temperatures well below 100°C. The acetates are the most commonly made
organic salts. The parent amines and imidazolines have already been
described in section 7.13. However, some comments are needed with
reference to the imidazolines. The imidazoline ring, shown in Figure 8.5, is

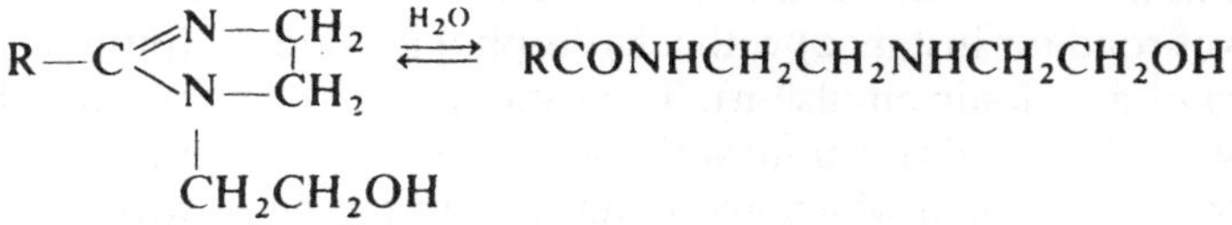

$$R-C{\overset{N-CH_2}{\underset{N-CH_2}{<}}} \overset{H_2O}{\rightleftharpoons} RCONHCH_2CH_2NHCH_2CH_2OH$$
$$\underset{CH_2CH_2OH}{|}$$

Figure 8.5 Hydrolysis of imidazolines.

definitely formed in the manufacture. However there is now considerable
evidence to show that this ring breaks down in aqueous solution by
hydrolysis back to the amide (see Figure 8.5). Thus if salts of imidazolines
are made in aqueous solution the ring may disappear on storage. On the
other hand, if a salt was made with an organic acid in an organic medium
the ring may well be stable for a considerable length of time.

256 HANDBOOK OF SURFACTANTS

General properties

1. *General.* The positively charged ion adsorbs strongly on metal and fibre surfaces.
2. *Solubility.* Inorganic salts have poor solubility in cold water, particularly sulphates, phosphates and silicates. The salts of amidosulphonic acids are more soluble. Organic salts of acetic acid show good aqueous solubility for coco amine but not higher amines, e.g. tallow amine. Salts of hydroxycarboxy acids (e.g. lactic acid, glycollic acid) are readily soluble in cold water. If high molecular weight carboxylic acids are used (e.g. oleic acid, stearic acid), the salts are insoluble in water but readily soluble in fats and oils.
3. *Compatibility with other surfactants.* Form water insoluble products with anionic surfactants.
4. *Chemical stability.* The salts have lower thermal stability than the parent amine, e.g. amine acetates will break down to substituted amides in only a few hours at 100°C.
5. *Surface active properties.* The CMC of a C12 primary amine neutralised with HCl is $1.3–1.5 \times 10^{-2}$M. Surface tension of acid salts of *N,N*-dimethyl-*N*-C3-alkylamidopropylamines can go as low as 26 mN/m at 0.2% (Muzyczko, 1968).
6. *Germicidal activity.* Salts of fatty amines with a chain length 12–16 carbon atoms are the most effective. Salts with acetic acid, glycollic acid, lactic acid and benzoic acid have all proved successful. To increase fungicidal action, use acids or phenols for neutralisation which are effective against fungi, e.g. salicylic acid, *o*-chlorobenzoic acid, *o*-phenylphenol as well as chlorinated phenols (For fatty amines and derivatives, see Hoechst data sheets).
7. *Emulsifying properties.* Cationic emulsions of the oil (or wax) in water type have the property of allowing a normally hydrophobic oil droplet to wet and deposit on to a wide variety of surfaces, both polar and non-polar. They can even be used to treat a water wet surface (a wet road for instance) with a hydrophobic material (bitumen in the form of a cationic emulsion). The cationic surfactant acts as a bridge between the solid substrate which has the polar head attached and the hydrophobic chain which gives adhesion to the oil or wax. An excess of cationic surfactant can give multilayers on the substrate and loss of adhesion.

Applications

1. *Emulsifiers.* Primary amine salts are cationic emulsifying agents below pH 7. Imidazolines can function as oil-soluble emulsifiers producing cationic O/W emulsions. If they are neutralised below pH 8 they become hydrophilic and can act as emulsifiers for polar organic

CATIONICS 257

solvents, e.g. toluene, pine oil or triglycerides. Acetate salts are used for emulsifying waxes although nonionics need to be added in hard water systems. The cationic wax emulsions so formed can give antistatic and water repellent coatings.

2. *Lubricants and metal working.* Corrosion inhibitors for fuels, greases and lubricating oils, and for metal surfaces using long-chain carboxylic acids for neutralisation to give solubility in oils. Short-chain carboxylic acids (acetic, propionic and benzoic) are used in aqueous systems.

3. *Mining.* Stearylamine and tallow fatty amine salts have been used in potassium flotation. Fatty amine acetates are used as collectors for zinc ores. The fatty amines and salts can be used for both collectors and frothers for many silicate ores, although frothing agents are generally needed as well.

4. *Fertilisers.* The free amine (coconut, stearylamine or tallow) or the acetate salt can be used to treat fertilisers to prevent caking together, particularly potassium salts.

5. *Road repairing.* Fatty amines as free bases are used as adhesive agents for hot bitumen. Fatty amine salts can be used to make cationic bitumen emulsions which can be used in wet weather.

6. *Treatment of pigments.* Pigments used in organic coatings need to be quite dry. Water-soluble fatty amine salts will displace the water and produce a water-repellent effect, such that the pigment can disperse readily in an organic medium. Fatty amine oleates are particularly useful for pigments in paints. The fatty amine oleate can be made *in situ* by treatment first with a fatty amine water soluble salt (an acetate) and then treatment with sodium oleate. The dioleate salt of *N*-tallow trimethylene diamine is one specific product used for dispersing inorganic pigments (0.5% on weight of pigment) and organic pigments (2% on weight of pigment).

7. *Textiles.* Antistatic treatment

Specification

Neutralisation (or non amine), 98–100%
p-, *sec*- and *tert*-amine, see section 7.13
Water content: for many organic salts, e.g. acetates and oleates, the water content is low (typically 1%)

Reference

Muzyczko, T.M. *et al.* (1968) *J. Am. Chem. Soc.* **45**, 720–725.

9 Amphoterics

9.1 Amphoterics (general)

Nomenclature

The word 'amphoteric' is derived from the Greek *amphi* meaning both and used to describe surfactants which have both a positive (cationic) and a negative (anionic) group. In acidic solutions they form cations, in alkaline solutions they form anions, and in a middle pH range 'zwitterions' are formed, i.e. molecules with two ionic groups of opposite charges. Sometimes the term ampholyte is used. The nomenclature of some amphoterics has been confused in the past, partly due to mistakes in the original chemical structures, and in addition, more recently, due to retention of the word 'betaine' for products that are strictly not amphoteric. The main categories of amphoterics have been classified as follows.

N-*Alkyl betaines.* The word betaine originally described the compound trimethyl glycine $(CH_3)_3N^+CH_2COOH$, which has a quaternary nitrogen atom. The word was then extended to *N*-trialkyl derivatives of amino acids. In the scientific literature it now means the internal salt of a quaternary ammonium, oxonium or sulphonium ion. They are formed when chloracetic acid reacts with a tertiary nitrogen compound. Note that a quaternary nitrogen group is formed by this reaction, but is not formed in the case of the glycinate and aminopropionate. Strictly speaking, the betaines are not amphoterics because betaines are never anionic. They are more like a quaternary ammonium compound. Nevertheless, by common usage they have been included in amphoterics. Sulphobetaines have been included in this section.

N-*Alkyl glycinates.* These are specific alkyl amino acids with the alkyl group attached to the nitrogen atom of the amine. They are derivatives of glycine, NH_2CH_2COOH, called glycinates and can be formed from chloracetic acid and an alkyl amine:

$$R–NH_2 + ClCH_2COOH \rightarrow R–NHCH_2COOH, \text{ an alkyl glycinate}$$

Products made with R-containing secondary amino groups are included in this section. Such products are known as alkyl polyamino carboxylates or polyamphocarboxy glycinates or polycarboxyglycinates. All these products can form cations, anions or zwitterions.

AMPHOTERICS 259

N-*alkyl aminopropionates or* N-*alkyl iminodipropionates*. The reaction of amines with chlorpropionic acid or acrylic acid will give the following products:

$$R-NH_2 + ClCH_2CH_2COOH \rightarrow R-NHCH_2CH_2COOH,$$
an aminopropionate

$$R-NH_2 + CH_2=CH_2COOH \rightarrow R-NHCH_2CH_2COOH,$$
an aminopropionate

$$R-NH_2 + 2CH_2=CH_2COOH \rightarrow R-N(CH_2CH_2COOH)_2,$$
an iminodipropionate

In the case of the aminopropionate the NH group is still reactive and will react with a further molecule of acrylic acid giving an iminodipropionate, unlike the glycinate. All these products can form cations, anions or zwitterions.

The imidazoline based amphoterics. So far all seems clear, but confusion arose because the early commercial amphoterics were made by reacting an imidazoline with chloracetic acid. The imidazoline is made by reacting aminoethylethanolamine, $NH_2CH_2CH_2NHCH_2CH_2OH$, with a fatty acid (or methyl ester) to form the amide; following this, there is ring cyclisation to form the imidazoline ring as described in section 7.1 and (Figure 7.26). When chloracetic acid reacts with the imidazoline ring it was thought that the ring stayèd intact and that the chloracetic acid reacted with the tertiary amine, and such products were known as mono-carboxy alkyl imidazoline betaines due to the similarity with the betaine structure. However the imidazoline ring breaks down (by hydrolysis, see section 8.3, Figure 8.5) during the reaction with chloracetic acid and derivatives of glycine are formed (See Figure 9.1).

The products of the reaction between an imidazoline and chloracetic acid are as follows.

With one mole of chloracetic acid the products are known as alkyl glycinates or alkylamphoglycinates or sometimes monocarboxylated glycinates. This is structure I in Figure 9.1. It is formed by the hydrolysis of the imidazoline ring and then reaction of the secondary amine with chloracetic acid.

With two molecules of chloracetic acid, the products (as shown in Figure 9.1) are known as alkylcarboxyglycinates or alkylamphocarboxyglycinates or sometimes dicarboxylated glycinates. The composition of such products is complex and they generally contain a mixture of monocarboxylated and dicarboxylated substances. Figure 9.1 only represents the main active constituents; there are other chemical entities in the final product. Very little information seems to be given by most producers on the exact constitution of their products. If reaction with the imidazoline ring occurs

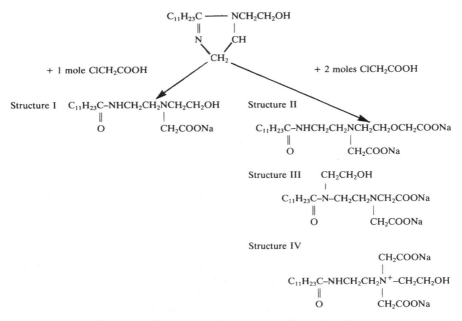

Figure 9.1 Glycinates derived from an alkyl imidazoline.

after hydrolysis, then the imidazoline ring can hydrolyse in two ways; (a) at the C–N bond, when structure II is formed after reaction with ClCH$_2$COOH; or (b) at the C=N bond, when structure III is formed after reaction with ClCH$_2$COOH (see Richardson 1991). The reaction between ClCH$_2$COOH and a secondary amine will proceed in preference to the reaction with a primary hydroxyl group, particularly in aqueous solution. The products of the reaction depend considerably on the reaction conditions, particularly the hydrolysis giving varying amounts of Structures I, II and III (see Lomax, 1991). The quaternary nitrogen (Structure IV) shown in Figure 9.1 is claimed by some authorities but not by others, and it is doubtful if it exists.

The reaction of acrylic acid with imidazolines in aqueous solution follows a similar path in that a mixture of products is obtained depending upon the mode of hydrolysis of the starting imidazoline. The products of reaction, propionates, are shown in Figure 9.2.

Under certain conditions the structures shown in Figure 9.3 also exist; they are stable and can be manufactured. However, such products seem to have no application advantage over the linear structures and they are not usually commercially available.

The situation is further complicated as there are by-products formed during the formation of the original imidazoline ring, the principal products being known as dialkyl amides and dialkylamido esters. The latter

```
    R C—N CH₂CH₂OH
    ‖  |
    N  CH₂              +
     \ /
     CH₂                    CH₂ = CH–COOH

                        (Alkali : NaOH)

    R–C–NHCH₂CH₂N–CH₂CH₂COONa      (Structure 1)
    ‖            |
    O            CH₂CH₂OH

                    +

          ╱ CH₂CH₂NHCH₂CH₂COONa
    RCN ─┤
     ‖    ╲
     O      CH₂CH₂OH                (Structure 2)
```

Figure 9.2 Reaction of hydrolysed imidazolines with acrylic acid.

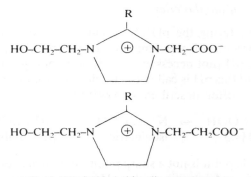

Figure 9.3 Stable imidazoline amphoterics.

will react, with chloracetic acid. There is little point in going into further detail; the important point is that the reaction of an alkyl imidazoline with either one, or more particularly two, molecules of chloracetic acid gives a complex mixture of products, most of them surface-active to a degree. The relative amounts of the mixture will depend very much on the reaction conditions. Most of the products behave as derivatives of glycine. For further details of the possible composition of glycinates derived via imidazolines, see Lomax (1994).

If the name includes the word 'ampho', e.g. cocoamphocarboxy glycinate, then a number of manufacturers have used this description to indicate that the product is a derivative of an imidazoline. Unfortunately this phraseology has not been followed universally, so no particular significance should be attached to the inclusion of this word unless indicated by the manufacturer.

262 HANDBOOK OF SURFACTANTS

Products based on starting materials with more than two nitrogen atoms with a reactive hydrogen group. The products described so far have been made from a raw material with two nitrogen atoms carrying a reactive hydrogen. In the case of the betaines the starting amine would be a tertiary amine. Raw materials are available with more than two nitrogen atoms carrying a reactive hydrogen atom and connected to a hydrophobic group. These can be reacted with chloracetic acid to give glycinates or with propionic acid to give propionates. The main products available are made from alkyl polyamines, $RNH(CH_2CH_2NH)_nH$, and reacted with one or more moles of chloracetic acid. The products formed are clearly defined and described in section 9.2.

Amine oxides. Amine oxides show amphoteric properties, but in neutral and alkaline conditions they are essentially nonionic, and they are included in section 7.2.4.

General properties of amphoterics

1. *General.* By altering the pH of an aqueous solution the anionic or cationic character of the amphoteric can be changed. At some intermediate pH (not necessarily pH 7) both ionic groups show equal ionisation and this pH is called the isoelectric point or area. This type of molecule is often described as a **zwitterion**.

   $$N^+.....COOH \quad \leftrightarrow \quad N^+.....COO^- \quad \leftrightarrow \quad NH.....COO^-$$
   acid pH<3 $\qquad$ isoelectric $\qquad$ pH>6 alkaline

   The isoelectric point is not a sharp point but depends upon the nature of the anionic and cationic groups. The most common anionic group is the carboxyl group (COOH) and the most common cationic group is the amine group (NH). At the isoelectric point amphoterics generally have minimum solubility so there may be a loss in surfactant properties. An amphoteric which is soluble at the isoelectric point is soluble across the whole pH range. The ionic nature of an amphoteric is rarely wholly anionic or cationic above and below the isoelectric area. The betaines differ from the glycinates and propionates in being unaffected in alkali (see Table 9.1).

Table 9.1 Effect of pH on betaines compared to glycinates or propionates

pH	Glycinates or propionates	Betaines
Acid	$N^+-(CH_2)_nCOOH$	N^+-CH_2COOH
Zwitterion	$NH(CH_2)_nCOOH$ and $N^+-(CH_2)_nCOO^-$	$N^+-CH_2COO^-$
Alkali	$NH(CH_2)_nCOO^-$	$N^+-CH_2COO^-$

AMPHOTERICS   263

The isoelectric point would be expected to be of considerable importance but it is not easy to measure in commercial products that are mixtures. The isoelectric points (more strictly areas) are very difficult to measure in the propionates but easier in imidazoline-derived products.

2. *Solubility*. Excellent solubility in aqueous solution but at a minimum in the isoelectric area, although this does not apply to betaines.

3. *Compatibility with aqueous ions*. Excellent.

4. *Compatibility with other surfactants*. Amphoterics show excellent compatibility with other surfactants. Mixed micelles are frequently formed and the mixtures often have functional properties not found in either of the components. For instance, skin irritation can be reduced below the level of either of the ingredients. Biocidal activity sometimes found in amphoterics can be reduced when stable mixed micelles are formed. The effect of adding an amphoteric to an anionic is to reduce the eye and skin irritation, increase the viscosity in the presence of electrolyte, improve foam stability and improve detergency. The effect of adding amphoterics to nonionics is more specific to the amphoteric than in the case of anionics. Most amphoterics will act as hydrotropes in solubilising nonionics in high electrolyte concentrations, particularly at high temperatures. This ability to solubilise depends upon the structure of the amphoteric. Amphoterics with polycarboxy groups and nonionics can show significant synergism in detergency, depending upon the ratio of nonionic to amphoteric (see Amphoterics International, 1987)

5. *Chemical stability*. Practically all amphoterics are resistant to both acids and alkalis. They do not undergo hydrolytic decomposition readily; even those products containing the amide group (amido-propyl betaines, glycinates) are resistant to hydrolysis in cold conditions. The sulphobetaines are particularly suitable for use in strong acids.

6. *Surface-active properties*. The surface activity of amphoterics varies very widely but depends upon two main factors: firstly, the distance apart of the two charges. The CMC passes through a sharp maximum when four methylene groups separate the charges (see Clint, 1992). The maximum is due to a balance between two opposing effects. Increasing the distance between the two polar groups increases the dipole moment, thus making it more hydrophilic, but increasing the length of the chain also introduces more hydrophobic groups, which tend to reduce the CMC. Secondly, as already mentioned, if the pH of the system is near the isoelectric point then the two charges are at their minimum ionisation and the solubility is at a minimum, so the surface activity should be at a maximum, and thus the CMC will also depend upon pH.

264   HANDBOOK OF SURFACTANTS

7. *Functional properties*. Excellent foaming agents, particularly in the presence of high electrolyte concentration and oil. Good, but not outstanding wetting agents, but not as efficient as some other surfactants (e.g. short-chain anionics). Emulsifying properties not particularly good, as the sodium salts.
8. *Biocidal activity*. Amphoterics possess biocidal activity but this is weak for the simpler amphoterics and only becomes pronounced when the number of amine groups (particularly secondary) increases. Some amphoterics show synergism with betaines. $C_{12}H_{25}NHCH_2$ $CH_2NHCH_2CH_2NHCH_2COOH$ is ten times more effective than $C_{12}H_{25}NHCH_2COOH$ (see Sykes, 1965).

Amphoterics have the advantage over quaternary compounds of being unaffected by hard water and alcohol.

Applications

Amphoterics are used in amphoteric/anionic mixtures in shampoos, foam baths, shower gels, liquid soaps, hand cleaners, hand laundry and hand dishwashing. The major products used are: cocoamidopropyl betaine, cocoamphocarboxy glycinate (imidazoline base), cocodimethyl betaine, and tallow polyamino carboxylates. Amphoterics will detoxify anionics in shampoo formulations due to the formation of mixed micelles. However, these mixed micelles may inhibit biocidal behaviour.
Amphoterics can also be effective detergents at extremes of pH.
Amphoterics tend to be less deactivated by protein than quaternary ammonium compounds, so this should make them suitable for use in dairy cleaning. They have the added advantage over quaternaries of being more easily removed from metals, whereas quaternaries are more tenacious and not so easily removed.
The most important property of the amphoterics in general is the extreme mildness on skin and mucous membranes compared with other types of surfactants.

9.2 Betaines

Nomenclature

Generic
   Alkyl amido betaines (nearly always the dimethylamine derivatives)
   Alkyl amidopropyl betaines (nearly always the dimethylamine derivatives)
   Alkyl amidopropyldimethyl betaines.
   Alkyl amidopropyldimethyl sulphobetaines

AMPHOTERICS   265

Alkyl amidopropyl hydroxysultaines
Alkyl betaines (nearly always the dimethylamine derivatives)
Alkyl bis(2-hydroxyethyl) betaines
Alkyl dimethyl betaines
Sulphoamido betaines
Sulphobetaines
Examples
Lauryldimethyl betaine
Cocoamidopropyl betaine.
Oleyl bis(hydroxyethyl) betaine
3-[(3-Cocoamidopropyl)dimethylamino] 2-hyroxypropanesulphonate

Description

The betaines are made by reaction of chloracetic acid with a tertiary amine
to form a quaternary N atom and an ionised COO group (see Figure 9.4)

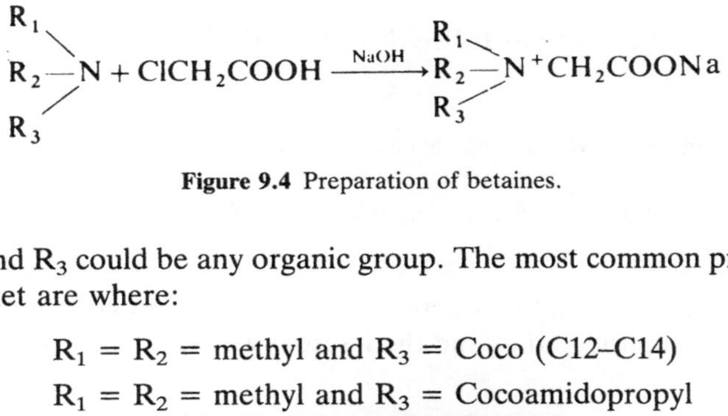

$$R_2 \overset{\displaystyle R_1}{\underset{\displaystyle R_3}{\rule{0pt}{0pt}}} N + ClCH_2COOH \xrightarrow{\text{NaOH}} R_2 \overset{\displaystyle R_1}{\underset{\displaystyle R_3}{\rule{0pt}{0pt}}} N^+ CH_2COONa$$

Figure 9.4 Preparation of betaines.

R_1, R_2 and R_3 could be any organic group. The most common products on
the market are where:

$R_1 = R_2 =$ methyl and $R_3 =$ Coco (C12–C14)

$R_1 = R_2 =$ methyl and $R_3 =$ Cocoamidopropyl
$\quad$ ($C_{11}H_{23}CONHCH_2CH_2CH_2$)

$R_1 = R_2 =$ ethylhydroxy (CH_2CH_2OH) and R_3
$\quad$ = fatty alkyl chain (C12–C18)

The reaction is usually carried out in aqueous solution with caustic soda to
neutralise the hydrochloric acid. The considerable amount of sodium
chloride formed is usually, but not always, left in the product.

N-*Alkyl sulphobetaines* are the only commercial products available
where the anion is not a carboxyl. They have the general formula shown in
Figure 9.5. The sulphobetaines are produced by quite different methods
compared to the normal betaines. At one time propane sultone was used as
an alkylating agent but there are doubts concerning its possible carcino-
genic properties. A better route is by reaction with chlorohydroxypropane
sulphonic acid (see Figure 9.6). There are various other routes, none of

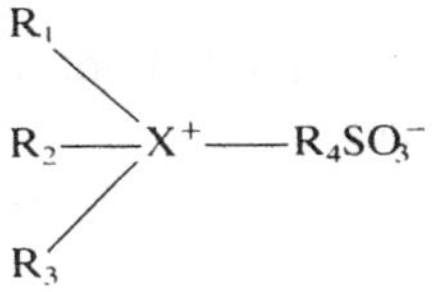

HANDBOOK OF SURFACTANTS

$$R_2\!-\!\!\!\overset{\displaystyle R_1}{\underset{\displaystyle R_3}{X^+}}\!\!\!-\!\!R_4SO_3^-$$

X=N : ammonium sulphobetaine
X=P : phosphonium betaine

Figure 9.5 Sulphobetaines.

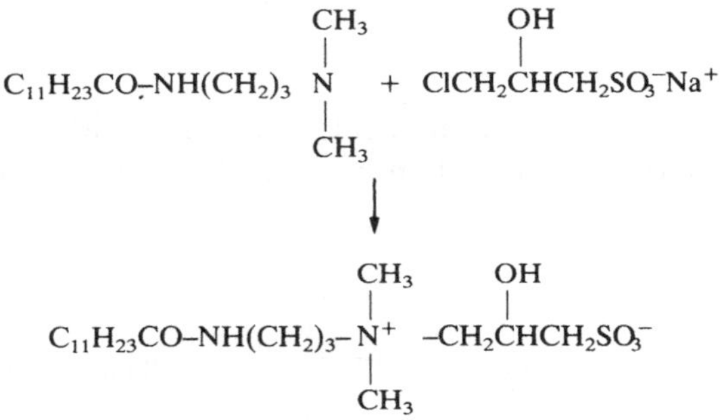

$$C_{11}H_{23}CO\!-\!NH(CH_2)_3\ \overset{\displaystyle CH_3}{\underset{\displaystyle CH_3}{N}}\ +\ ClCH_2\overset{\displaystyle OH}{C}HCH_2SO_3^-Na^+$$

$$\downarrow$$

$$C_{11}H_{23}CO\!-\!NH(CH_2)_3\!-\!\overset{\displaystyle CH_3}{\underset{\displaystyle CH_3}{N^+}}\ -\!CH_2\overset{\displaystyle OH}{C}HCH_2SO_3^-$$

N(N-dodecylamidopropyl)N,N-dimethyl-ammonium-N-(2-hydroxypropyl)sulphonate

Figure 9.6 Preparation of hydroxysulphobetaines.

them easy or economically attractive. But the two most common types of products, the sulphobetaines and the hydroxysulphobetaines are shown in Figure 9.7.

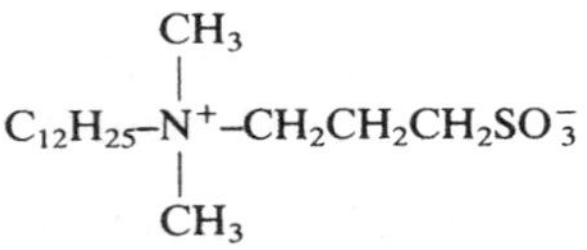

$$C_{12}H_{25}\!-\!\overset{\displaystyle CH_3}{\underset{\displaystyle CH_3}{N^+}}\!-\!CH_2CH_2CH_2SO_3^-$$

N(dodecyl)N,N-dimethyl-ammonium-N-propyl sulphonate

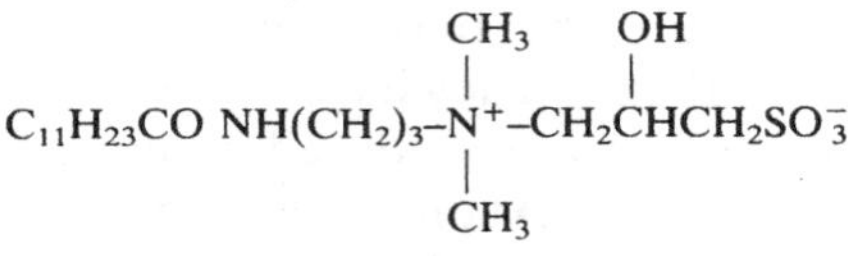

$$C_{11}H_{23}CO\ NH(CH_2)_3\!-\!\overset{\displaystyle CH_3}{\underset{\displaystyle CH_3}{N^+}}\!-\!CH_2\overset{\displaystyle OH}{C}HCH_2SO_3^-$$

N(N-dodecylamidopropyl)N,N-dimethyl-ammonium-N-(2-hydroxypropyl)sulphonate

Figure 9.7 Common sulphobetaines.

AMPHOTERICS 267

General properties

1. *General.* Not really amphoterics, because they cannot donate H^+ in alkaline solution and therefore are never anionic. They are internal quaternary ammonium compounds but do show zwitterionic properties at pH at and above their isoelectric point (neutral and alkaline pHs). They thus show cationic properties similar to quaternary ammonium compounds below their isoelectric point (acid pH).
2. *Solubility.* Soluble in water, insoluble in mineral oil, white spirits, aromatic solvents and perchlorethylene; retain good solubility at and near their isoelectric area (unlike the amino propionates and imino dipropionates)
3. *Compatibility with aqueous ions.* Acid and neutral aqueous solutions are compatible with alkaline earth and other metallic ions (aluminium, chromium, copper, nickel and zinc). Betaines can act as hydrotropes and solubilise alkali in nonionics, but the other amphoterics are superior. Whether they are cost-effective compared to other hydrotropes will depend upon the overall formulation and relative cost, but 0.7 g of sodium cocodimethylbetaine can solubilise 2g of NP+9EO in 3% caustic soda at room temperature. Good lime soap dispersants.
4. *Compatibility with other surfactants.* They are compatible with all classes of surfactants except at low pH with anionics, where they give a precipitate. They thicken anionics in the presence of salt. A ringing gel can be made with triethanolamine lauryl sulphate. At slightly acid pH the cationic nature of the betaine is neutralised by the anionic, and a neutral salt results. This is also the point of minimum solubility, which is the point of maximum viscosity — hence the good thickening properties with anionics. Alkyl amido betaines will complex and solubilise quaternary surfactants into anionic formulations (see Scher, 1983).
5. *Chemical stability.* Excellent chemical stability against oxidising agents, e.g. hypochlorite bleach. Resistant to hydrolysis, even the amidopropyldimethyl betaines, and stable at very low and high pH.
6. *Surface-active properties.* CMC of lauryl dimethyl betaine $= 2 \times 10^{-3}$ M = 0.06%. Surface tension of 1% active lauryl dimethyl betaine = 33 mN/m. CMC of C16 dimethyl betaine $= 2 \times 10^{-5}$M = 0.0007%. Surface tension of 0.1% *N*-coco-*N*,*N*-dimethyl ammonium *N*-propyl sulphonate = 34 mN/m at 25 °C.
7. *Functional properties.* Good foaming agents, but not as good as alkyl sulphates or ether sulphates, although better and more stable at alkaline pH. Alkyl amido betaines give larger foam volumes than alkyl betaines, but dimethyl betaines are better than amidopropyl betaines in soft water in the presence of sebum. Increasing the chain

268 HANDBOOK OF SURFACTANTS

length of the alkyl group reduces foaming properties (optimum at C12); best foaming properties at alkaline pH, but good over a wide pH range (3–11); unaffected by hard water. Good wetting agents with lower molecular weight (C8 alkyl chain) products in acid solution. Good detergent properties, but not as good as anionics used alone; increasing chain length improves detergent properties. Best detergency at alkaline pH; unaffected by hard water. Emulsification properties, good for slightly polar materials but not good for paraffins. When added to anionics give viscosity increase with salt; dimethyl betaines better than amidopropyl betaines. Give excellent foam stability when added to ether sulphate solutions, particularly the sulphobetaines. Detoxifying of anionics is quoted in the literature but the imidazoline-based glycinates and the newer polycarboxy glycinates are superior. The sulphobetaines give excellent wetting and excellent stability against electrolytes, coupled with very low irritancy.

Applications

1. *Shampoos*. Mild characteristics with anionics (low eye and skin irritation), antistatic properties to hair, good conditioning, foam boost and stabilisation, viscosity increase with anionics (at low pH or with salt). Good foaming and detergency on their own. Typically utilised at 3–10% in a formulation, either as a partial or a total replacement for conventional foam boosters (alkanolamides). Coco-amidopropyldimethyl betaine is becoming the main betaine used in shampoos, but in conjunction with ether sulphates and sometimes half ester sulphosuccinates. The product claimed in a baby shampoo patent (Verdicchio and Walts, 1976) is an equimolar concentration of amidobetaine and sulphate with a polysorbate (or EO sorbate ester + EO/PO copolymer). This is analogous to the imidazoline/sulphate system, which is supposed to be tear-free.
2. *Foam baths, liquid soaps, shower gels and hand cleaners*. Foam boost and stabilisation with anionics. Viscosity increase with anionics. Good lime soap dispersant. Good foaming and detergency on its own.
3. *Household products*. Hand textile washing and dishwashing products — mild characteristics with anionics and excellent foam stability and increase in viscosity (at lower pH). pH is usually reduced by using citric acid. Car cleaners (high sequestrant levels). Wetting agent for domestic bleach (hypochlorite)
4. *Industrial detergents*. In foamed alkaline detergents to give added 'cling' to surfaces prior to rinsing.

AMPHOTERICS 269

5. *Textiles*. Cocoamidopropyl betaine is used as antistatic agent in spin finishes. Also used as detergents.
6. *Oilfield applications*. Foaming agents in foam drilling, resistant to high electrolyte concentration.
7. *Foaming agents for gypsum or lightweight concrete.*

Specification

The analysis of amphoterics gives significant problems particularly in the interpretation of the data.

Solids*, 20–55%
Sodium chloride, 5–15%
pH, 6–9
Sodium glycollate*, 1–4%
Free amine, typical 0.25% for cocodimethyl betaine; typical 1.5% for cocoamidopropyl betaine
* Part of the solids may not be surface-active (possibly sodium glycollate derived from the hydrolysis of sodium chloroacetate)

Chemical tests give more indication of the composition of betaines than the other amphoterics. However, it is still recommended that any quality control should include some simple functional tests.

9.3 Glycinates

Nomenclature

See section 9.1

Generic
    Alkyl glycinates
    Alkylamino carboxylic acids
    Alkylamphomonoacetates
    Alkylamphodiacetates
    Alkylcarboxyglycinate
    Alkylamphopolycarboxyglycinates
    Alkyliminodiglycinate
    Alkyl polyaminocarboxylates (APAC)
    Aminoalkanoates
    Amphoglycinates
    Amphocarboxyglycinates
    Alpha-*N*-alkylamino acetic acids
    Carboxy glycinates

270 HANDBOOK OF SURFACTANTS

Dicarboxyl alkyl imidazoline betaines = amphocarboxyglycinates
Glycinates
Hydroxyalkyl alkylamidoethyl glycinates
Imidazoline derivatives
Imidazoline carboxylates
Monocarboxy alkyl imidazoline betaines = amphoglycinates.
Example
N-Lauric acid amidoethyl *N*-2-hydroxyethylglycinate (see Figure 9.8)

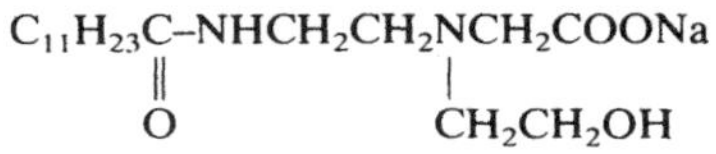

$$C_{11}H_{23}C\text{--}NHCH_2CH_2NCH_2COONa$$
$$\underset{O}{\overset{\parallel}{}} \qquad \underset{CH_2CH_2OH}{\overset{|}{}}$$

Figure 9.8 *N*-Lauric acid aminoethyl *N*-2-hydroxyethylglycinate.

Carboxy glycinates. This name indicates that there are two carboxyl groups present by formation with two molecules of chloracetic acid. It does not define the structure, and the products are nearly always mixtures.

The glycinates which are made from imidazoline and chloracetic acid. These are complex mixtures (see section 9.1). The chemical description of the products differs from one manufacturer to another and also differs in the scientific literature. These names usually give no indication of the actual chemical structure.

Alkyl polyamino carboxylates (APAC) or polyamino polycarboxy glycinates or polyamphocarboxy glycinates or polycarboxyglycinates. These are products where there is more than one nitrogen atom capable of reacting with chloracetic acid. Very often the structure of the parent polyamine is not clear. The abbreviation APAC is now becoming increasingly used. 'Coco APAC' implies that the alkyl group is derived from coconut oil but does not give any idea of the number of nitrogen groups or the number of carboxyl groups. A typical product is shown in Figure 9.9.

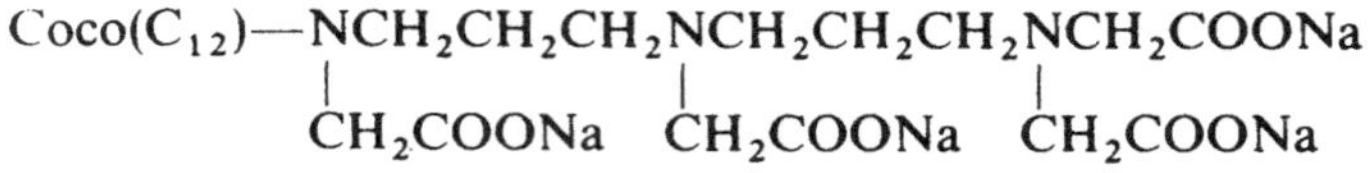

$$Coco(C_{12})\text{--}NCH_2CH_2CH_2NCH_2CH_2CH_2NCH_2COONa$$
$$\underset{CH_2COONa}{\overset{|}{}} \quad \underset{CH_2COONa}{\overset{|}{}} \quad \underset{CH_2COONa}{\overset{|}{}}$$

Figure 9.9 An alkyl polyamine carboxylate.

This product could probably be called sodium carboxymethyl coco polyaminopropionate. At the present time there does not seem to be any agreed nomenclature between manufacturers, nor is there likely to be one while the exact compositions are confidential.

Description

Although the potential for a wide variety of glycinates exists, the actual commercial products available are:

1. *The imidazoline-based glycinates that are based on fatty acids.* These give the complex mixtures already described in section 9.1. They all contain considerable quantities of sodium chloride. They are often known as the alkylamphoglycinates and alkyl amphocarboxyglycinates. The word ampho has no specific chemical significance as such, but is widely used in these type of compounds.

2. *The polyamino polycarboxy products (APAC) based upon fatty amines reacted with acrylonitrile and hydrogenated (Figure 9.9).* These products are much better defined than those made via imidazolines. They all contain sodium chloride (if reacted with chloracetic acid) unless specially purified.

General properties

1. *Solubility.* Products with more than one carboxy group are more soluble in water than the monocarboxy products. They are very soluble in strong acids and alkalis (including in the presence of electrolyte). Solubility is low in most organic solvents (including ethanol). The isoelectric area for the tallow-based APAC derivative with four glycinate groups is quoted as around pH 5 by Palicka (1988).

2. *Compatibility with aqueous ions.* Excellent with hard water, i.e. calcium and magnesium ions. Solubilises phenols and polyphosphates.

3. *Compatibility with other surfactants.* Solubilises quaternary ammonium salts. All products can solubilise nonionics in alkaline solution but efficiency increases as the number of carboxyl groups increases, (at the same hydrophobe chain length).

4. *Chemical stability.* Stable to hydrolysis by acid and alkali.

5. *Surface-active properties.* Surface tension of caprylamphocarboxy glycinate is 29 mN/m at 1%. CMC of 0.3×10^{-5} mol/l. Lowest surface tension for the tallow-based APAC derivative with four glycinate groups is quoted as 44 mN/m, (see Palicka, 1988).

6. *Functional properties.* Substantive to surfaces to give antistatic effects. Good foaming agents which are not affected by change in pH, although the APAC products show excellent foaming at high pH but decreased foaming at lower pH. Effective emulsifying agents for long-chain alcohols and slightly polar compounds, but not for paraffinic oils. Detergency rises with increasing pH; products with several carboxyl groups have synergistic properties, e.g. detergency, particularly with nonionics.

272 HANDBOOK OF SURFACTANTS

7. *Eye and skin irritation.* These are low compared to other surfactants (including betaines and propionates).

Applications

1. *Personal products.* Shampoos, bath additives, etc. Polycarboxy products are used with anionics to decrease skin and eye irritation and give conditioning properties.
2. *Household and industrial detergents.* These products give good solubility in high built liquids, combined with synergism (with nonionics) in detergent properties. They are used in liquid and powder laundry detergents. The polycarboxyglycinates are especially useful in detergents, as they give excellent detergency, high sequestration of heavy metal ions and excellent dispersability of solids, combined with the ability to stabilise enzymes in the detergent.
3. *Shampoo and bath additives.* Dicarboxylated glycinate derivatives are usually used with anionics. Baby shampoos (low eye irritation) have used coconut oil-based products formulated with lauryl sulphates, ether sulphates and ethoxylated sorbitan monoesters. Conditioning shampoos can be formulated with tallow-based glycinates.
4. *Electro-plating.*
5. *Fire fighting.* Foaming agent for foams used in fire fighting.

Specification

Solids[a] 30–40%
Active[a] 20–40%
Sodium glycollate[b], 2–9%
Sodium chloride, 7–12% from glycinates
[a] Active has been often assumed to be equal to solids minus sodium chloride concentration. This is wrong if the sodium glycollate concentration is high.
[b] Formed by hydrolysis of unreacted chloracetic acid, so only found in glycinates. The high levels (6–8%) are found when the glycinates have been prepared and when excess chloracetic has been used and then subsequently hydrolysed. Simple chemical tests give very little indication of the composition of glycinates. Any quality control should include some simple functional tests.

9.4 Amino propionates

Nomenclature

This section describes aminopropionates and iminopropionates.

Generic
  Alkylamino propionates
  Alkylamphopolycarboxy propionates
  Ampho propionates
  Amphocarboxy propionates
  Alkylimino dipropionates
  Aminoalkanoates
  Beta-*N*-alkylalanines
  Beta-*N*-alkylamino propionates = alkylamino propionates
  Beta-*N*-alkylimino propionates = alkylimino dipropionates
  Iminodialkanoates
  Propionates
Example
  Disodium cocoiminodipropionate

Description

The major group of products of commercial importance is made from primary or secondary amines by reaction with acrylic acid (see section 9.1). Unlike in the reaction with chloracetic acid, no sodium chloride is formed as a by-product.

Some commercial products are made from an imidazoline by reaction with with acrylic acid, as already described in section 9.1. Very often their actual structure is not given and they are known as the salt-free imidazoline amphoterics.

General properties

1. *General.* The propionates do not have salt present, in contrast with the betaines and glycinates.
2. *Solubility.* Products with more than one carboxy group are more soluble in water than the monocarboxy products. These products are very soluble in strong acids and alkalis (including in the presence of electrolyte). Solubility is low in most organic solvents (including ethanol).
3. *Compatibility with aqueous ions.* Excellent with hard water, i.e. calcium and magnesium ions. Solubilises phenols and polyphosphates.
4. *Compatibility with other surfactants.* Solubilises quaternary ammonium salts. The coco iminodipropionates are compatible with high concentrations of sulphates, ether sulphates and coco alkanolamides. All products can solubilise nonionics in alkaline solution, but efficiency increases as the number of carboxyl groups increases at the same hydrophobe chain length).

274 HANDBOOK OF SURFACTANTS

5. *Chemical stability*. Stable to hydrolysis, and particularly to acid and alkali.
6. *Surfactant properties*. N-Lauryl aminopropionic acid: CMC = 2×10^{-3} M = 0.052%; N-lauryl iminodipropionic acid: CMC = 1.2×10^{-3} M = 0.04%; surface tension of C12–15 dipropionate = 28 mN/m.
7. *Functional properties*. Substantive to surfaces to give antistatic effects. Excellent wetting at low (C8–12) chain lengths but not very good at C18. Good foaming agent with alkyl chain at C12–C14, but not at C8–C10. Effective emulsifying agents for long-chain alcohols and slightly polar compounds, but not for paraffinic oils. Detergency rises with increasing pH; products with several carboxyl groups have synergistic properties, e.g. detergency, particularly with nonionics. Decreased eye and skin irritation with anionics.

Applications

1. *Personal products*. Shampoos, bath additives, etc. Used in place of other amphoterics where the presence of sodium chloride must be avoided.
2. *Household products*. Have been quoted as used in liquid and powder laundry detergents, but probably not widely.
3. *Industrial detergents*. Addition to acid or alkaline detergents gives improved wetting; use C12 dipropionate for high foam, C8 mono-propionate for low foam. Caprylic acid-based products are used in metal cleaning as they are low foamers.
4. *Electroplating*. The dipropionates give excellent wetting properties in electroplating baths at low pH.
5. *Fire-fighting*. Foaming agent for foams used in fire fighting.
6. *Corrosion inhibitors*. Short alkyl chain dipropionates are used in conjunction with other cationics to give improved wetting and low foam.

Specification

Solids[a], 30–50%
Sodium chloride 0.05% typical
Free amine, —
[a] Unlike the glycinates, the solids should give a good indication of the actives as there is very little sodium chloride and no glycollate.

Simple chemical tests give very little indication of the composition of amphopropionates or amphocarboxypropionates, i.e. made via an imidazoline. Any quality control should include some simple functional tests.

References

Amphoterics International (1987) EP 0,214, 868.
Clint, J. H. (1992) *Surfactant Aggregation*. Blackie, Glasgow and London, p.107.
Lomax, E. (1991) *Recent Developments in the Analysis of Surfactants*, ed. M. R. Porter. Elsevier, London, 109–137.
Lomax, E. (1994) *Amphoterics*. Marcel Dekker, New York, in the press.
Palicka, J. (1988) *Proc. 2nd World Surfactants Congress*, Paris, Vol. III, p.449.
Richardson, F. B. (1991) In *Industrial Applied Surfactants III*, Special Publication no. 107. The Royal Society of Chemistry, Cambridge, pp. 191–213.
Scher, A. (1983) *The Chemistry and Applications of Amido-Amines*, presented to the Society of Cosmetic Chemists Annual Seminar, Cincinatti, Ohio, USA, 5 May 1993.
Sykes, G. (1965). *Disinfection and Sterilisation*, 2nd edn, E. and F.N. Spon, London, pp.377–378.
Verdicchio R. J. and Walts, J. M. (1976) USP 3,950,417.

10 Speciality surfactants

10.1 General

The description of surfactants given in chapters 5–9 concentrated on the surfactants that are commercially available from many producers in volume throughout the world. They have been grouped together on the basis of a common chemical structure. Within these groupings there are hundreds of 'speciality' surfactants. What is a 'speciality' surfactant? The author's definition of a 'speciality surfactant' is one that is different in properties and use to the major volume surfactants used for domestic detergents. Such surfactants are characterised by being significantly higher in price and only available from a limited number of manufacturers, and the exact compositions are often not known, at least by the user. Phosphate esters, most of the amphoterics, ethane sulphonates and sulphosuccinates would come under this heading. However, it is not a definition that is clear and distinct, and a product may become a speciality if its volume declines e.g. alkyl naphthalene sulphonates.

The speciality surfactants described in this chapter are quite different from those defined above. This chapter describes a number of different types of surfactants that either have quite different hydrophobic groups or have specific properties compared with the major product types in chapter 5–9. Fluorinated surfactants and silicone surfactants have different chemical properties in the hydrophobic part of the molecule from those of the normal hydrocarbon chain that constitutes the hydrophobe in all the surfactants in chapters 5–9. A minimum surface tension in the region of 25–27 mN/m is obtained by the incorporation of a suitable polar group at the end of a paraffinic chain 10–18 carbon atoms long. This minimum surface tension is adequate for most applications involving detergency. There are some applications, however, where a lower surface tension in aqueous solution is required, e.g. wetting a polythene film or spreading an aqueous foam on top of a petrol fire. The fluorinated and silicone surfactants were found to possess the ability to depress the surface tension of aqueous solutions down as low as 18 mN/m. However, the fluorine or silicone based surfactants possess properties other than that of depressing surface tension. Many fluorinated and silicone surfactants have been made and tested in the market. Very few are used in any volume, principally due to their high price. It is unlikely that any will achieve major use due to the hydrophobic chain being resistant to biodegradation (see Appendix II).

SPECIALITY SURFACTANTS

However, both the silicone surfactants and the fluorinated surfactants are capable of many variations and tailoring by grafting of organofunctional groups for specific end uses.

Also included in this chapter are surfactants with a normal paraffinic hydrophobic group but possessing some functional property and/or chemical characteristics not normally encountered. Polymerisable surfactants, labile surfactants, bolaform surfactants and Gemini surfactants are described. The technical and patent literature describes many other novel surfactants, but only those which have achieved some commercial success are included here.

10.2 Silicone surfactants

Nomenclature

Generic
    Dimethicone copolyol (CTFA designation)
    Dimethylsiloxane glycol copolymers
    Ionic organo-polysiloxanes
    Organo-polysiloxane copolymers
    Polyether polysiloxanes copolymers
    Polysiloxane glycol copolymers
    Silicone surfactants
Examples
    Polysiloxane polyorganobetaine copolymer

Description

Organosilicones described hereafter are those with a polydimethylsiloxane backbone. Unmodified they are insoluble in water. The incorporation of a water-soluble or hydrophilic group into the silicone structure can give products that exhibit surface-active properties in water. In addition, incorporating organic groups of an organophilic character can give products that exhibit surface-active properties in organic solvents.

A very large number of silicone surfactants are now available commercially, many of them with an inadequate description of their constitution by the manufacturers. The following description is an attempt to simplify the complexity of the product range.

The hydrophobic effect in a silicone surfactant can be obtained either by a polysiloxane chain, which itself can vary in length and degree of branching, or by the addition of a paraffinic hydrophobic at the end of or along the polysiloxane chain. Some possibilities are shown in Figure 10.1.

The hydrophobic effect is obtained by a short siloxane chain of three

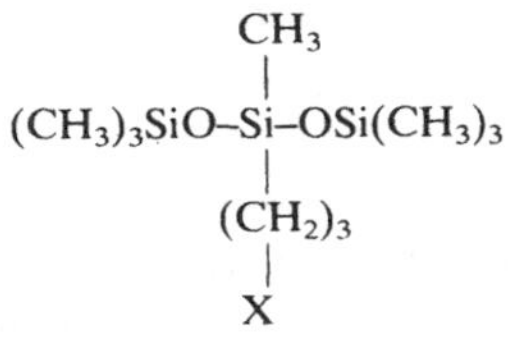

HANDBOOK OF SURFACTANTS

$$CH_3$$
$$(CH_3)_3SiO-Si-OSi(CH_3)_3$$
$$(CH_2)_3$$
$$X$$

X = an organic hydrophilic group, e.g.:
amino, NH_2
polyether, $(EO)_n$
sulphonate, SO_3^-

Figure 10.1 Silicone monomer surfactants.

silicon atoms surrounded by seven methyl groups and one short hydrocarbon chain, to which is attached a hydrophilic group, which can be anionic, nonionic, cationic or amphoteric, identical to those descibed in chapters 6–9. The hydrophobic effect of this short siloxane chain is equivalent to a hydrocarbon chain of C14–C16 chain length. As there is only one hydrophilic group and one hydrophobic group, these surfactants could be described as 'monomeric' silicone surfactants.

Polymeric silicone surfactants. If the siloxane chain in Figure 10.1 is extended to one with a number of polysiloxane links, then thousands of possible structures can be constructed. The hydrophilic group can be situated at the end of a siloxane chain or along the chain; there may be more than one hydrophilic group and these multifunctional groups can be situated close together or randomly along a polysiloxane chain. ABA 'polymeric' or 'polyfunctional' silicone surfactants with the hydrophilic groups at the end of the chain are shown in Figure 10.2.

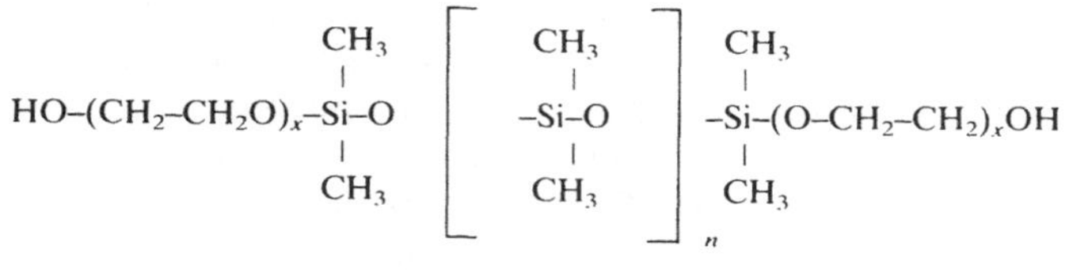

$$HO-(CH_2-CH_2O)_x-\underset{\underset{CH_3}{|}}{\overset{\overset{CH_3}{|}}{Si}}-O\left[\underset{\underset{CH_3}{|}}{\overset{\overset{CH_3}{|}}{-Si}}-O\right]_n\underset{\underset{CH_3}{|}}{\overset{\overset{CH_3}{|}}{-Si}}-(O-CH_2-CH_2)_x OH$$

ABA type polyether copolymer

Figure 10.2 ABA type silicone 'polymeric' surfactants.

These products show a striking similarity in structure to the EO/PO copolymers already described in section 7.9, with the polyoxypropylene chain being replaced by the polydimethylsiloxane chain. Such products show some similar properties. Note that the connection between the polysiloxane chain and the polyoxyethylene chain is by way of a Si–O–C

SPECIALITY SURFACTANTS 279

link, which is hydrolytically unstable. In Figure 10.1, the connection between the polysiloxane chain and the hydrophilic group is by way of a Si–C bond, which is hydrolytically stable.

A multiplicity of hydrophilic groups can be attached along the length of a polysiloxane chain, and such products are known as comb polymers. These are shown in Figure 10.3 where $q = 0$.

$$
CH_3-\overset{\overset{\displaystyle CH_3}{|}}{\underset{\underset{\displaystyle CH_3}{|}}{Si}}-O
\left[\overset{\overset{\displaystyle CH_3}{|}}{\underset{\underset{\displaystyle CH_3}{|}}{-Si}}-O\right]_n
\left[\overset{\overset{\displaystyle CH_3}{|}}{\underset{\underset{\displaystyle (CH_2)_p}{\underset{|}{\displaystyle CH_3}}}{-Si}}-O\right]_q
\left[\overset{\overset{\displaystyle CH_3}{|}}{\underset{\underset{\displaystyle (CH_2)_3}{\underset{|}{\displaystyle (EO)_x}}}{-Si}}-O\right]_m
\overset{\overset{\displaystyle CH_3}{|}}{\underset{\underset{\displaystyle CH_3}{|}}{-Si}}-CH_3
$$

$$(PO)_y$$
$$OH$$

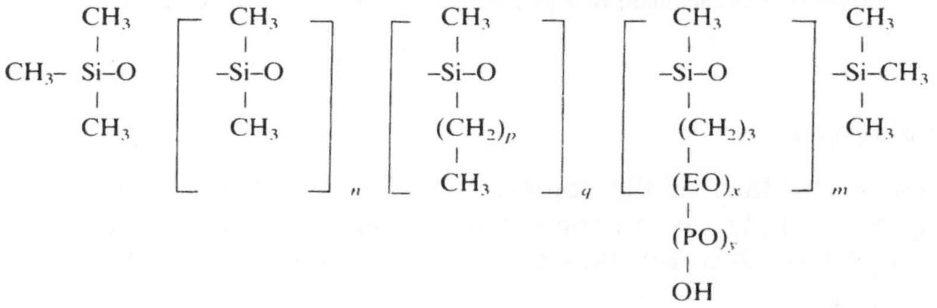

Comb copolymers with (EO)(PO) side groups

Figure 10.3 Comb silicone surfactants.

Dimethicone copolyol (CFTA name) is the structure shown in Figure 10.3 with $q = 0$. Long-chain hydrocarbon side chains may also be attached along the polysiloxane chain, as shown in Figure 10.3 where $q > 0$. In addition, end groups of hydrophilic and/or hydrophobic nature can also be attached to the siloxane chain.

The most common silicone surfactants, which have been used for many years, are those where the hydrophilic group is either a polyoxyethylene chain or a mixed ethylene oxide and propylene oxide copolymer. The alkylene oxide copolymer can be attached to the polysiloxane chain, as shown in Figure 10.3, but in any one product type a very large number of variations are possible, the main ones being: the amount of ethylene oxide; the amount of propylene oxide; whether block or random; and whether end blocked or not. These have been described in section 7.9.

There are excellent possibilities, therefore, of tailoring molecules to give specific properties due to the very large number of variations possible in chemical structure, and hence there is proliferation of silicone surfactants on the commercial market. A very large number of them could be described as polymeric surfactants where there are several hydrophilic groups and several hydrophobic groups in the same molecule. The general features of polymeric surfactants are described in chapter 11.

The methods of synthesis are a combination of silane and silicone chemistry, which have been well described. One common route which is used to add a hydrophilic group to a polysiloxane chain is to have the hydrophilic group attached to an allyl group. The allyl group will react with

280 HANDBOOK OF SURFACTANTS

the H on a Si–H group, as shown in Figure 10.4, and a typical product is shown in the same figure.

$$-\overset{|}{\underset{|}{Si}}-H + CH_2 = CHCH_2R \longrightarrow -\overset{|}{\underset{|}{Si}}-CH_2CH_2CH_2R$$

Figure 10.4 Attachment of a hydrophilic group to a polysiloxane chain.

General properties

1. *General.* Many of the comments in section 7.9 (on EO/PO copolymers) apply, as products can be made of very widely varying properties. However, only a few products have achieved commercial success.
2. *Solubility.* Ability to vary water and organic solvent solubility over a very wide range; silicone surfactants with EO/PO copolymers show inverse solubility with temperature (cloud points) in a similar manner to nonionics (see section 7.1.2).
3. *Chemical stability.* The polydimethylsiloxanes are very stable compounds which are distinguished by chemical inertness to heat and oxygen. When an organic polymer, e.g. a polyoxyethleneglycol, is attached to the siloxane chain, then the chemical stability will depend on (a) the link between the silicone and the non-silicone polymer, and also (b) the chemical stability of the organic polymer. The link between silicone and non-silicone polymer can be (i) an Si–O–C link, where the silicon atom is linked to the carbon atom of a EO/PO copolymer via an oxygen atom (see figure 10.2). Such products are unstable to hydrolysis although they may have enough stability for their end use. The other linkage is (ii) an Si–C link where the silicon atom is directly linked to the carbon atom of a EO/PO copolymer (see Figures 10.1 and 10.3). Such products should be stable to hydrolysis. However, the product shown in Figure 10.1 hydrolyses slowly in aqueous solution at room temperature, but products of higher molecular weight with a number of side chains, shown in Figure 10.3, are much more stable.
4. *Cloud point.* Most of the copolymers with EO/PO groups show a cloud point very similar to other ethoxylates. Above the cloud point, the products are generally low foam and can act as defoamers.
5. *Surface-active properties.* Low molecular weight silicone surfactants can decrease the surface tension of water down to 22 mN/m. A product as shown in Figure 10.1 with X=a polyoxyethylene chain of 16 EO units end-blocked with a methyl group has a CMC of 0.037%

SPECIALITY SURFACTANTS

(Gradzielski *et al*. 1990) and a surface tension of 24 mN/m at the CMC. The higher molecular weight polymeric surfactants have surface tensions somewhat higher (25–35 mN/m) and higher CMCs. However, the published data on CMCs of silicone EO/PO co-polymers are not consistent.

6. *Functional properties*. The functional properties will depend upon the chemical structure e.g. high or low foam products may be produced. The following remarks therefore apply to specific products and are not general. Excellent wetting properties on low energy surfaces are obtained, but low equilibrium surface tension is not necessarily a measure of wetting (see chapter 4). Foaming can be controlled to give very low or zero foam, and silicones with a low EO/PO content are excellent antifoams for aqueous systems. EO-based silicone co-polymers with greater than 60% EO give good foaming agents below their cloud point. The unmodified polydimethylsiloxanes are good antifoams for some solvents and hydrocarbons. By appropriate choice of organic groups and water-soluble group a wide range of products can be emulsified.

7. *Disadvantages*. The polydimethylsiloxanes are stable against bio-degradation and end up unchanged in the sludge fraction of sewage treatment plants; however, the water-soluble copolymers may well be degraded (see Frye, 1983). There is a lack of data on this subject in most manufacturers' literature.

Applications

1. *Foam control in polyurethane foams*. Polysiloxane EO/PO co-polymers are the major cell control additive for flexible and rigid polyurethane foam systems.

2. *Paint additives*. Antifloating of pigments and prevention of surface defects such as crater formation and orange peeling in water – and solvent-based paints; unmodified silicones and EO/PO copolymer modified are both used.

3. *Cosmetic*. Products as shown in Figure 10.3 with C16 alkyl side chains can emulsify fatty acid monoesters and triglycerides in the form of W/O emulsions, which are used as light creams or lotions by the cosmetics industry (Th. Goldschmidt, 1990). In shampoos, polysilox-ane polyether copolymers improve combing properties whilst diqua-ternary polydimethylsiloxanes give conditioning effects (Th. Goldschmidt, 1991).

4. *Textiles*. Mineral oil surface tension can be decreased so that it will wet polypropylene for use in lubricants by the use of a difunctional type 1 where X = the nonylphenyl group; gives soft handle and reduced friction on fabrics. Epoxy and amino functional silicones are

282 HANDBOOK OF SURFACTANTS

becoming increasingly used on cotton and polyster cotton blends to give a very soft handle and improved physical properties (see Joyner, 1989). Such products are applied in the form of microemulsions where the siloxane can be copolymerised and the microemulsion is made at the same time (see Toray Silicone Co. 1987).

5. *Agriculture*. Excellent wetting agents with low foam for herbicides and increased efficiency of nutrient sprays.

Specification

Hydroxyl number, EO and PO content can all be easily measured and are often quoted as a specification, but are misleading and give no indication of structure and/or performance. Cloud point is often used as measure of consistency of molecular structure. It is suggested that a functional test (surface tension) or, more preferably, a simulated performance test is carried out for quality control. Chemical analysis data are difficult to interpret.

10.3 Fluorocarbons

Nomenclature

Generic
Fluorinated surfactants
Fluorochemical surfactants
Fluorosurfactants
Organofluorine surfactants
Perfluoroalkyl surfactants
Perfluoropolyether surfactants
Examples
Amino polyfluorosulphonate
Ethoxylated polyfluoroalcohol
Fluorinated alkyl polyoxyethylene ethanol
Perfluoro polymethylisopropyl ether
Polyfluoroalkyl betaine
Polyfluoropyridinium salt
Polyfluorosulphonic acid salt
Potassium fluorinated alkyl carbonylate

Description

There are two basic types commercially available. The older products (Type 1) are:

$$CF_3-(CF_2)_n-\text{hydrocarbon group}-X$$

Where X = the hydrophilic group and n is usually 6–10. The hydrocarbon group can be alkyl, pyridine group, amidopropyl, etc., and is a link between the fluorinated chain and the hydrophilic group. The hydrophilic group can be $-COOH$, SO_3, $(CH_2-O)_x$, cationic or amphoteric. These types of products are supplied by 3M, ATO Chemie and Asahi Glass Co.

More recent products (Type 2) developed by Montedison are known as the perfluoropolyether surfactants, as they are the fluorinated analogues of ethylene oxide and propylene oxide polymers:

$$CF_3-(O-CF_2-CF(CF_3))_n-(O-CF_2)_m-CF_3$$

where n and m = 20–40

There are also modifications to the above formula with ionic groups at the end of the chain, e.g COOH.

General properties

The fluorochemical based products are very similar to conventional surfactants in having a hydrophobic and hydrophilic group. The special properties are:

1. *Solubility*. Possible immiscibility with both polar and nonpolar solvents. This means that a solvent-phobic group is available in the common solvents and in minerals oils. In Type 1 (above) if group $X = H$, then a surfactant for oils or solvents can be produced. To get the necessary miscibility with water, X must be a water-soluble group for surfactants suitable for aqueous systems.
2. *Compatibility with aqueous ions*. Perfluorocarboxlic acids are more ionised than the corresponding fatty acids, and therefore products with a carboxyl group adjacent to a fluorinated methyl group are unaffected in aqueous solution by mineral acids and polyvalent cations.
3. *Thermal and chemical stability*. The C–F bond is very strong and the size of a fluorine atom is approximately 1.35 Å (hydrogen is 1.2 Å). This means that the fluorine atoms are closely packed around a carbon atom, giving considerably better thermal and chemical stability than hydrocarbons. The sulphonate group has good thermal and oxidative stability.
4. *Surface-active properties*. The perfluorocarbons have very low intermolecular forces between molecules, which gives very low surface energies in the liquid phase and therefore very low surface tensions (as low as 18 mN/m) in aqueous systems. Using different products they can also reduce the surface tension of nonaqueous systems

284   HANDBOOK OF SURFACTANTS

(esters, alcohols, ethers, epoxies, polyesters, urethanes, acrylics) to about 20 mN/m.

5. *Functional properties.* Excellent wetting particularly on low energy surfaces, e.g. polythene. Good foaming and excellent foam stability in very strong acids and alkalis. Defoaming properties in non aqueous systems. Type 1 are poor emulsifying agents for oil and water but excellent for fluorinated monomers in the production of PTFE. Type 2 are excellent O/W emulsifiers.
6. *Disadvantages.* Very expensive; use only if the properties are really required; due to their exceptional stability, there must be doubts on their ability to biodegrade.

Applications

Due to the high price of fluorinated surfactants the applications have been confined to problems which conventional (i.e. lower priced) surfactants cannot solve. These are generally related to wetting and spreading phenomena where the conventional surfactants have a limit of approximately 26 mN/m in aqueous solution (see chapter 4). Difficult-to-wet surfaces are plastic films of polythene and polypropylene, etc. and therefore adhesives, coatings, paints, inks and polishes need to have surface tensions lowered to wet these surfaces. Other difficult-to-wet surfaces are surfaces on plants, insects and the very small fissures in rocks.

Specific applications are as follows:

1. *Fire fighting.* Foaming agents for aqueous-based foam which will spread on burning hydrocarbons to cut off air.
2. *Paint.* Reducing orange peel, cissing, cratering and edge crawling in non-aqueous paints. Improves wetting of low-energy surfaces.
3. *Inkjet printing inks.* Improves wetting of water/glycol inks for plain paper printing without drying of the inkjet nozzles.
4. *Polishes.* To achieve improved levelling, spreading and gloss.
5. *Emulsion polymerisation.* Emulsifiers in the emulsion polymerisation of PTFE.
6. *Petroleum production.* Used in acid fracturing to reduce surface tension for foam generation to open fissures and improve penetration in oil and gas stimulation; good thermal stability and low absorption.
7. *Electroplating.* Gives antifoam effect in very strong acids, e.g. chromic acid.
8. *Cosmetics and barrier creams.* Very stable O/W emulsions.

Specification

The manufacturers, with very few exceptions, do not reveal the exact chemical constitution of their products. Easily measured chemical properties are of very little value in assessing the reproducibility of most fluorochemical products. A simple functional test such as the measurement of the surface tension of a solution would be more suitable. However, such tests are not so easy to carry out. A simple performance test would seem to be the most satisfactory. For example, if a fluorosurfactant was added to a paint to reduce cissing then a cissing test would be most appropriate.

10.4 Miscellaneous specialities

There are a number of types of surfactant that do not easily fit into the chemical classification adopted in this book. They are all 'specialities' in that they do not have large-scale use and are expensive, and there are few manufacturers. The following descriptions are brief and only intended to point out particular characteristics and specific end uses.

10.4.1 Bolaform surfactants

These should not be confused with Gemini (see later). Bolaform surfactants are also known as bola-amphiphiles, bolaphiles or α, ω-surfactants, and the names refer to surfactants where the alkyl chain (or a more complex hydrophobic moiety) is terminated **at both ends** by a polar group, which may be ionic, nonionic or amphoteric (see Figure 10.5). One could call the Pluronic® EO/PO copolymers (see chapter 11) bolaform surfactants.

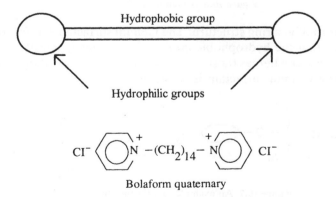

Figure 10.5 Bolaform surfactants.

Bolaform surfactants have a number of properties which distinguish them from conventional surfactants:

1. More than 12 methylene groups in the hydrophobic chain are necessary for bolaform surfactants to aggregate.
2. There is some controversy on the structure of the micelles formed, but it is believed that the two ionic groups locate at the micellar surface via a loop-type structure.
3. The counterion dissociation from the micelle is relatively large.
4. Their CMCs are smaller than would be expected by comparison with a surfactant with half the length of the hydrophobic chain and one polar group.
5. The aggregation number is small compared with usual micelles.

The above findings would suggest a much looser form of packing in the micelles compared with normal surfactants. It is also worth mentioning that the structure of the micelles formed by the Pluronic EO/PO copolymer with the hydrophobic chain in the middle is still uncertain.

10.4.2 Gemini surfactants

Gemini surfactants possess in sequence a long hydrocarbon chain, an ionic group, a rigid spacer, a second ionic group, another hydrocarbon long chain (see Figure 10.6). The spacer is an organic group, which can be linear

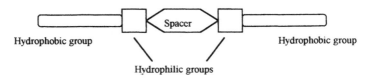

Figure 10.6 A Gemini surfactant.

or, more normally, a ring structure. The Gemini surfactants differ from the bolaform in that the hydrophobic groups in an aqueous micelle point out towards the water whereas those in the bolaform do the exact opposite. An example of a Gemini surfactant is shown in Figure 10.7.

Figure 10.7 An example of a Gemini surfactant.

Quite surprisingly, many of these compounds are soluble in water (see Menger and Lirttau, 1991). Measurements of surface tension of the phosphate esters shown in Figure 10.7 at two temperatures gave interesting results (see Table 10.1).

SPECIALITY SURFACTANTS

Table 10.1 Surface tension of Gemini phosphate esters

Hydrocarbon chain	Temperature (°C)	Lowest surface tension (mN/m)	CMC (M)
$C_{12}H_{25}$	23	39	0.0006
$C_{16}H_{33}$	23	53	0.01
$C_{20}H_{42}$	23	60	Not detectable
$C_{12}H_{25}$	50	38	0.0001
$C_{16}H_{33}$	50	38	0.006
$C_{20}H_{42}$	50	42	0.003

The discovery of clear CMC values was evidence of aggregation. Different methods were examined to find evidence of aggregation by dye colour change methods. Such methods gave CMCs below the values obtained by surface-tension measurements. Note that the CMC increases with increasing chain length and that the lowest surface tension is obtained with the shortest hydrophobic chains, both effects being contrary to those found with normal surfactants.

Similar products were made by substituting the quaternary ammonium group for the phosphate group, but these products tended to be much less water-soluble than the phosphate derivatives. Studies were made on a film balance suggesting that the molecules lie **flat** on the surface.

Zana and Talmon (1993) describe two alkylquaternary ammonium species linked by a hydrocarbon chain. The surfactants with a short-chain spacer (two or three carbon atoms) form long thread-like micelles even at very low concentrations, while the corresponding monomeric quaternary surfactants only show spherical micelles. When the chain length of the spacer is four or more, normal spherical micelles are formed. Thus, in Figure 10.6, either the spacer must be a ring or other rigid chain or, if flexible, must be very short.

10.4.3 Labile surfactants

Labile surfactants are those surfactants that carry out their surface-active function efficiently but, when that function is no longer needed, lose their surfactant properties. One might well classify biodegradable surfactants as 'labile', as they lose surfactant properties and degrade to inorganic compounds when their function, e.g. detergency, is no longer needed. It is now a requirement for all surfactants that are used as detergents to be biodegradable (see appendix II).

However, the products included in this chapter are where the surfactant properties need to be changed deliberately. A specific example is in emulsion polymerisation, where the surfactant is needed to carry out the emulsion polymerisation process and yet the final polymer emulsion may

288   HANDBOOK OF SURFACTANTS

be used as a dry film to repel water. The presence of the surfactant in the dry film will reduce the water-repelling properties of that film, and therefore it would be advantageous if the surfactant properties could be lost **after** the polymerisation process is complete. In practice, this is very difficult to achieve; one method is to use a polymerisable surfactant. Aqueous-based car polishes require a surfactant for emulsifying the wax and for cleaning, but its presence in the dry film reduces water repellency. Using volatile amine salts of long-chain fatty acids will give reasonable emulsions, and on drying, particularly in hot sunny conditions, the amine volatilises off leaving the free fatty acid which is water-repellent. This technique is probably the easiest and most practical method of reducing surface activity.

By changing the pH the surfactant properties of amphoterics and also of some other surfactants, can be changed significantly. There are two types of surfactants that become unstable reasonably quickly in alkali, particularly at temperatures say above 60°C. These are the isethionates (section 6.3) and the silicone–polyglycol copolymers where the two polymers are joined by a Si–O–C link (see section 10.2). Also, all of the glycerides (see Section 7.12) and polyglyceryl esters contain ester groups, which will hydrolyse to give soaps, which can then be deactivated by calcium ions. Thus a combination of alkali, heat and calcium is needed, but much higher pH and temperatures will be needed compared with the isethionates or silicone copolymers.

Many papers and patents have been published on the synthesis of specific groups susceptible to pH or chemical action. One recent example is the synthesis of surfactants with the 1,3-dioxalane ring, which is readily decomposed at low pH (see Ono 1993).

10.4.4 Polymerisable surfactants

There has been considerable interest in polymerisable surfactants for many years. These are products that have surfactant properties in a monomeric form but are capable of either homo- or co-polymerisation with other monomers. The major uses are:

1. *Emulsion polymerisation.* The polymerisable surfactant can be used as one of the surfactants used in polymerisation in water. Normally, the surfactant will remain in the polymer emulsion and may enhance, detract from, or have no effect on the properties of the emulsion. In general it will have some effect, but there will be some applications which may need its effect to be eliminated (see labile surfactants) or if it is not eliminated, then the surfactant needs to be incorporated in the polymer. An example of the latter is where the polymer is being used as a wound dressing and the surfactant must not be allowed to

SPECIALITY SURFACTANTS

diffuse into the patient's system. Another example is based on the need to reduce the wetting effect in surface coatings (e.g. outdoor paints), in particular the tendency of surfactants to migrate to the surface of the paint film, rendering this more hydrophilic and sensitive to moisture. As nearly all surfactants are biodegradable they will provide a source of food for bacteria and algae in and on the paint film. A number of surfactants have been proposed for use in emulsion polymers which will copolymerise with the polymer system. The type of reactive group must be similar to that of the polymer system in which it is to be incorporated. Some examples are as follows:

Ethoxylated fatty acid alkanolamides. The polyunsaturated linseed (or soybean) amide ethoxylates can be used as polymerisable surfactants in, e.g., alkyd emulsions. The unsaturated grouping reacts with the alkyd binder in the presence of catalysts (see Brink *et al.*, 1992). *Allyl derivatives of nonylphenol ethoxylates, sulphates and phosphates.* Nonylphenol ethoxylates, nonylphenol ethersulphates and nonylphenolethoxyphosphates are used extensively in the emulsion polymerisation of vinyl-type monomers. Yokata *et al.* (1991) described the synthesis and use of allyl derivatives in the emulsion polymerisation of ethylacrylate and butylacrylate/styrene. Typical surfactants are shown in Figure 10.8. These monomers were surfactants and gave the surface tension values shown in Table 10.2, which are taken from Yokota *et al*, (1991).

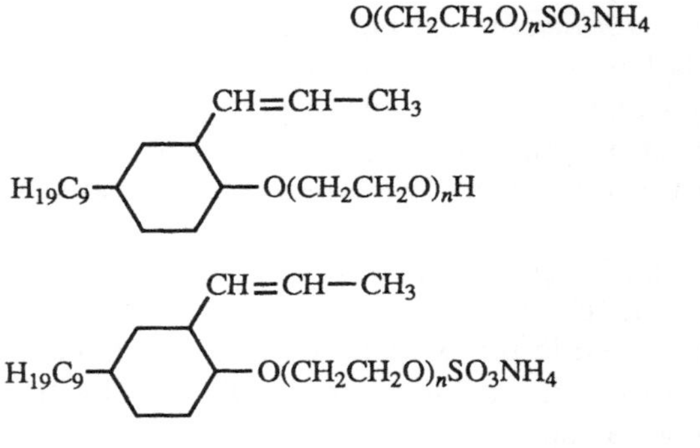

Structure I

Structure II

Structure III

Structure IV

Figure 10.8 Polymerisable nonylphenol ethoxylate derivatives.

HANDBOOK OF SURFACTANTS

Table 10.2 Surface tension of 0.1% solution of polymerisable surfactants shown in Figure 10.8

Structure	No. of moles of EO (n)	Surface tension (mN/m) at 25°C
Ethoxylate – I	20	39.9
Ether sulphate – II	10	41.2
Ethoxylate – III	20	38.3
Ether sulphate – IV	10	40.5
NP+21EO	21	39
NP+6EO sulphated	6	39

Thus, the surfactant monomers with allyl groups as side chains gave very similar surface tensions to conventional surfactants of similar structure. The phosphate esters prepared by reacting structures I and III with phosporic anhydride (P_2O_5 gave very slightly lower surface tension than the sulphates. All the monomers gave a high level of foam only slightly lower than similar products without the unsaturated groups.

The polymerisable monomers were then used as surfactants in the emulsion polymerisation of ethylacrylate and butyl acrylate. It was shown that 70–80% was incorporated into the polymer, with surfactants containing 20 moles of EO, and 80–85% with surfactants containing 10 moles of EO. The resulting polymers had similar properties to those made with conventional surfactants, except that they all gave films that had much improved water resistance and a very high contact angle with water. Those emulsions made with polymerisable surfactants with 10 EO content also had very low foaming properties.

It was therefore postulated that polymerisable surfactants were distributed uniformly through a polymer film, while a conventional surfactant gave a very high distribution of surfactant at the surface of a film (see Figure 10.9).

Methacrylate esters with betaine or sulphonate groups Kurze (1991) described a sulphobetaine and a sulphonate both containing a methacylate group capable of functioning wholly or partly as surfactants in the emulsion polymerisation process to give polymer films with high initial water resistance. Two typical products are shown in Figure 10.10.

Maleate esters and sulphosuccinate derivatives. Maleic anhydride will react readily with long-chain alcohols to give half esters of maleic acid; this reaction is used in the preparation of the sulphosuccinates (see section 6.7). However, the half ester of maleic acid with a long-chain alcohol is a surfactant with a carboxyl group and will copolymerise with vinyl monomers. The copolymerisation does not readily

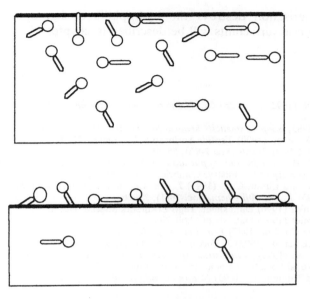

Figure 10.9 Surfactant distribution in a polymer film: (a) polymerisable surfactant; (b) conventional surfactant.

$$CH_2 = \underset{\underset{CH_3}{|}}{\overset{}{C}} - COOCH_2CH_2\underset{\underset{CH_3}{|}}{\overset{\overset{CH_3}{|}}{N^+}} - CH_2CH_2CH_2SO_3^-$$

$$CH_2 = \underset{\underset{CH}{|}}{\overset{}{C}} - COOCH_2CH_2CH_2SO_3K$$

Figure 10.10 Polymerisable surfactants with methacrylate groups.

take place, neither are the soaps very good surfactants in emulsion polymerisation, and this type of monomer is rarely used.

Urquiola *et al.* (1992) described the preparation of sodium dodecyl allyl sulphosuccinate, which is used to polymerise vinyl acetate and was compared with its non-polymerisable counterpart (sodium dodecyl propyl sulphosuccinate).

2. *To make polymeric surfactants.* The polymerisation of a 'polymerisable' surfactant will give a polymer with a number of hydrophilic groups and also a number of hydrophobic groups, unless the mode of

292 HANDBOOK OF SURFACTANTS

polymerisation destroys the hydrophilic or hydrophobic groups. Polymeric surfactants will be described in chapter 11.

References

Brink, C. *et al.* (1992) *Proc. 3rd CESIOInternational Surfactants Congress*, London, vol. D, pp. 211–216.
Frye, C.L. (1983) *Soap/Cosmetics/Chemical Specialities*, August, 32–42.
Gradzielski, M. *et al.* (1990) *Tenside Surfactants Detergents*, **27**, 366–378.
Joyner, M.M. (1989) *Textile Asia* **20**(6), 55–58.
Kurze, W. (1991) In *Industrial Applications of Surfactants III*, Special Publication no. 107. The Royal Society of Chemistry, Cambridge, pp. 238–253.
Menger, F.M. and Littau, C.A. (1991) *J. Am. Chem. Soc.* **113**(4), 1451–52.
Ono, D. (1993) *J. Am. Oil Chem. Soc.* **70**(1), 29–36.
Th. Goldschmidt (1990) Data sheet 'Tego Cosmetics Special' No. 3.
Th. Goldschmidt (1991) Data sheet 'Abil Silicones'.
Toray Silicone Co. Ltd (1987) Eur. Pat. Appl. 0,291,213.
Urquiola, M.B. et al. (1992) *J. Polym. Sci A*, **30**(12), 2619–29.
Yokota, K. *et al.* (1991) In *Industrial Applications of Surfactants III*, Special Publication no. 107. The Royal Society of Chemistry, Cambridge, pp. 29–48.
Zana, R. and Talmon, Y. (1993) *Nature (London)*, **362**(6417), 228–30.

11 Polymeric surfactants

In recent years there has been considerable interest in polymeric surfactants, but there is no clear definition of a polymeric surfactant. One group of products, exemplified by polyacrylic acid of low molecular weight ($< 15\ 000$), condensates of naphthalene sulphonate with formaldehyde, and sulphonated styrene–maleic anhydride copolymers, is often included in descriptions of polymeric surfactants. Natural products such as alginates, pectins, and protein-based products could all be looked upon as polymeric surfactants. Most of these products show only weak surfactant properties in respect of surface tension reduction in aqueous solution and the formation of micelles. All these products are better classified as polymeric electrolytes (or polyelectrolytes) as they have a number of hydrophilic groups but no specific well-defined hydrophobic groups, and therefore are not included in this chapter.

The word **polymeric** would suggest high molecular weight, with a large number of repeating molecular units. Most polymeric surfactants which have appreciable surface-active properties have only a few repeating units and the term oligomeric would seem to more descriptive. However the word polymeric has now been firmly established and so will be used in this book. The products that will be described in this chapter are those where there is a number of hydrophilic groups (generally more than two) and a number of hydrophobic groups (generally more than two), with both types of groups linked in the same molecule by covalent bonds

Most polymeric surfactants can be placed into two main groups: comb polymers and block copolymers.

Comb polymers

The comb polymers are ones where both the hydrophilic and the hydrophobic groups are linked to a polymeric regularly repeating unit such as a vinyl polymer chain, a siloxane chain or a phenol formaldehyde polymer chain. A diagrammatic view of a comb polymer is shown in Figure 11.1. Such a product is known as a **comb** polymeric surfactant as the hydrophobic groups and/or the hydrophilic groups are like the teeth of a comb sticking out from the backbone of the comb. A typical example would be an alkyl phenol that has been polymerised with formaldehyde and then reacted with ethylene oxide, as shown in Figure 11.2.

Such a system has an enormous number of variations with very few raw

294 HANDBOOK OF SURFACTANTS

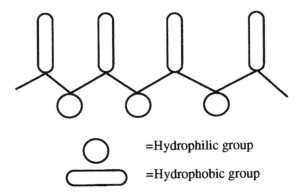

Figure 11.1 Typical comb polymer.

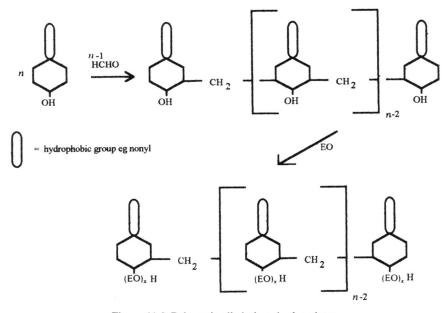

Figure 11.2 Polymeric alkyl phenol ethoxylates.

materials. The size of the hydrophobic group, the length of the backbone chain (the value of n), and the amount of EO (the value of x) are all easily varied. Not only that, but the polyoxethylene chain $(EO)_x$ could be varied by incorporating propylene oxide in a random or block manner, and the end group–H could be alkylated or end-blocked with a variety of end groups. The major use of this type of product was in crude oil demulsification, where the hydrophobic groups are C4–C8 hydrocarbons. It is surprising that so few products of this type have found their way on to the

POLYMERIC SURFACTANTS 295

market for other uses. It may be the suspect stability of the methylene bridge between the aromatics, coupled with concern over the environmental properties of the alkyl phenol group, has restricted their use.

An investigation on the micellar properties of the polymeric alkylphenol ethoxylates was carried out by Ishigami (1978), who measured the CMC of products from nonylphenol and $x = 11$ (NF-11) and 20 (NF-20). There were no values given for n. Using light-scattering, he stated that micelles were formed of aggregation number approximately 10, with CMC values of 10^{-6}M. As the molecular weight was not stated, this was difficult to compare with other products. Solubilisation of water in benzene and 2-ethyl-1-hexanol was better than with the corresponding NP + 11EO or NP + 20EO, but in hexane it was lower. Ishigami gave the following data:

Product	CMC (M)	Surface excesss conc. $\times 10^{-10}$ (mol/cm^2)	Surface area (Å^2/mole)
NF–11	2×10^{-6}	5.00	55
NF–20	2×10^{-6}	2.58	64
NP+18E	150×10^{-6}	2.21	75

The puzzling feature is that the surface area per mole is similar for the nonylphenol + 18EO (NP+18E) and for the polymers.

Two examples of comb polymeric surfactants with a siloxane chain are shown in Figure 11.3. The first example has a polysiloxane chain of $m+n$ repeat units with m side chains, each side chain having a sulphonate group, giving an anionic polymeric surfactant. There are $n + 2$ hydrophobic groups represented by the two $Si(CH_3)_3$ groups and the n $Si(CH_3)_2$ groups.

The second example shows an EO/PO copolymer in place of the sulphonate group, giving a nonionic polymeric surfactant. Once again a wide range of surfactants can be made. Variations in HLB values are produced merely by varying the ratio of n to m in either molecule. The overall molecular weight $(m+n)$ will also affect the properties of the surfactant. In the second example, further variation is possible by varying the ratio of x to y and also the value of $x+y$.

Block copolymers

The other main group of polymeric electrolytes is block copolymers, where there are blocks of hydrophobic groups (B) and block of hydrophilic groups (A), the simplest being in an A-B-A configuration. EO/PO copolymers of the **Pluronic®** type could be looked upon as polymeric surfactants (see section 7.9). The EO/PO copolymers and derivatives have

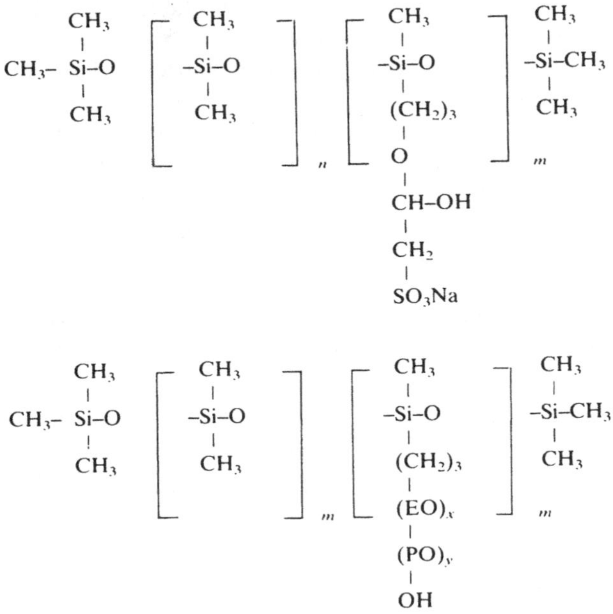

The structures in Figure 11.3:

$$
CH_3-\underset{\underset{CH_3}{|}}{\overset{\overset{CH_3}{|}}{Si}}-O\left[\underset{\underset{CH_3}{|}}{\overset{\overset{CH_3}{|}}{-Si}}-O\right]_n\left[\underset{\underset{\underset{\underset{\underset{\underset{SO_3Na}{|}}{CH_2}}{|}}{CH-OH}}{\overset{\overset{\overset{\overset{CH_3}{|}}{-Si}-O}{|}}{(CH_2)_3}}}{}\right]_m \underset{\underset{CH_3}{|}}{\overset{\overset{CH_3}{|}}{-Si}-CH_3}
$$

$$
CH_3-\underset{\underset{CH_3}{|}}{\overset{\overset{CH_3}{|}}{Si}}-O\left[\underset{\underset{CH_3}{|}}{\overset{\overset{CH_3}{|}}{-Si}}-O\right]_m\left[\underset{\underset{\underset{\underset{OH}{|}}{(PO)_y}}{\overset{(EO)_x}{|}}}{\overset{\overset{\overset{CH_3}{|}}{-Si}-O}{|}}{(CH_2)_3}}\right]_m \underset{\underset{CH_3}{|}}{\overset{\overset{CH_3}{|}}{-Si}-CH_3}
$$

Figure 11.3 Silicone polymeric surfactants.

blocks of water-soluble groups and blocks of hydrophobic groups. The simplest EO/PO copolymers (the ABA type) have only one hydrophobic group but two hydrophilic groups, the former being a polymer of propylene oxide and the latter being a polymer of ethylene oxide:

$$(EO)_x\,(PO)_y\,(EO)_x$$

Such products show properties which are common to other to polymeric surfactants such as good wetting and dispersing, but low foam.

However, on this simple basis a large number of variations can be made by replacing the starting (PO) with a molecule of greater functionality than two. A good example using a tetrafunctional amine derivative is the complex EO/PO polymeric surfactant shown in Figure 11.4, which is used in the dyeing of fibres (see Eisenlauer and Horn, 1984).

Block copolymers made from polyesters. Considerable efforts have been made to use the formation of polyesters for the formation of block copolymers. The block copolymers so far described have polyoxyethylene as the hydrophilic block and polyoxypropylene glycol as the hydrophobic block. The latter group may be replaced by poly(12-hydroxystearic acid) as the hydrophobic block, but products of type BAB, i.e. the hydrophobic

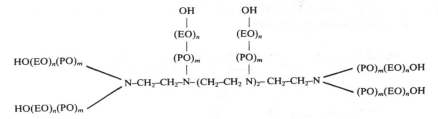

Figure 11.4 A complex EO/PO polymeric surfactant.

groups on the outside of the molecule, have found a practical use. A simplified structure is shown in Figure 11.5.

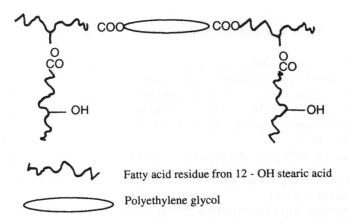

Figure 11.5 BAB type polyester formed from condensation of 1 mole PEG with 4 moles 12-OH stearic acid.

Aston et al. (1985) have described properties of well-defined products of polyethylene glycol (PEG) and poly(12-hydroxystearic acid) in organic solvents. The products described were BAB block copolymers, where B= 5–6 units of poly(12-hydroxystearic acid) (PHSA). A1 = PEG 1500, A2 = PEG 4000, A3 = 50% PEG 1500 and 50% PEG 4000. All are waxy solids, soluble in aliphatic hydrocarbons. A1 is a W/O emulsifier with a calculated HLB = 6.

Much more complex structures are commercially available with a 'random structure' prepared by reacting aliphatic carboxylic acids, aliphatic and/or aromatic polycarboxylic acids or anhydrides, polyalkylene glycols (usually PEG) and polyols. The products are statistical mixtures with a wide molecular weight distribution, containing loops of hydrophilic moieties (PEG) bound through ester linkages to the lipophilic moieties in a random tridimensional network.

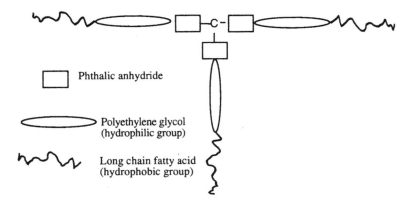

Figure 11.6 Simplified 'random structure' polyester surfactant.

There are no published structures of the **random structure** polyesters, but the author has attempted to give a very simplified structure, and this is shown in Figure 11.6. It must be emphasised that the actual surfactant molecules are probably very different in shape, but this Figure does indicate the type of hydrophilic and hydrophobic blocks and how they are joined together.

Bognolo (1992) gave the principal uses of these type of compounds as:

1. Emulsion explosives (90% saturated solution of nitrates + 5–7% fuel and emulsifier)
2. Acrylamide polymerisation through the 'inverse emulsion' process
3. W/O/W emulsions: for primary W/O, use BAB block copolymers of polyhydroxystearic acid and PEG, and then for final O/W use AB or ABA EO/PO.

Another type of polyester polymeric surfactant is that based on alkenyl succinic anhydrides. An alkenyl succinic anhydride is reacted with a PEG, a polyol (pentaerythritol) and optionally a monocarboxylic acid. The length of the alkenyl chain can be quite long: the molecular weight of the alkenyl succinic anhydride is 1000 to give good solubility in mineral oil and to act as a good emulsifier. The polyesters are poor emulsifiers if the chain length of the alkenylsuccinic anhydride is C18 or less. However, if blended with polyesters made with long alkenyl chains, they give good emulsifiers showing synergy but only by leaving for weeks at room temperature or for 4h at 125°C. The long-chain alkenylsuccinic anhydrides are prepared from polyisobutene and are commonly used as oil additives. PEG (30%) reacted with polyisobutenylsuccinic anhydride (molecular weight=1000)

POLYMERIC SURFACTANTS 299

and small amounts of glycerol, neopentyl alcohol, and monocarboxylic fatty acids are soluble in paraffins and used as O/W emulsifiers (HLB=8) (see Baker, 1979).

Polymeric silicone surfactants

The siloxane chain offers considerable scope for producing polymeric surfactants. Mono-, di-, tri- and tetra-functional siloxane chains are available. The lengths of the chains are easily controlled. By using a condensation polymerisation, very short chain lengths are readily made, and a variety of functional monomers, which can be incorporated into the chain, are commercially available. This chapter will not attempt to describe the various methods of producing the silicone polymeric surfactants, but will concentrate on the principal types available and their main properties. The siloxane chain is illustrated in Figure 11.7.

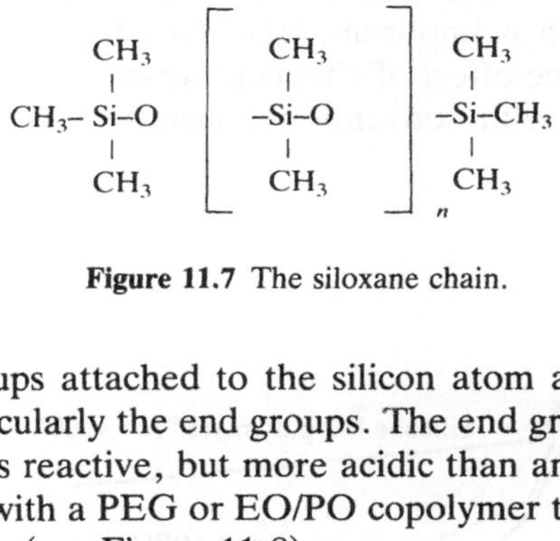

$$CH_3\text{--}\underset{\underset{CH_3}{|}}{\overset{\overset{CH_3}{|}}{Si}}\text{--}O \left[\underset{\underset{CH_3}{|}}{\overset{\overset{CH_3}{|}}{-Si}}\text{--}O \right] \underset{\underset{CH_3}{|}}{\overset{\overset{CH_3}{|}}{-Si}}\text{--}CH_3$$

Figure 11.7 The siloxane chain.

The methyl groups attached to the silicon atom are stable and highly hydrophobic, particularly the end groups. The end groups can be replaced by Si–OH, which is reactive, but more acidic than an alcohol group. This will form an ester with a PEG or EO/PO copolymer to form an ABA type of block copolymer (see Figure 11.8).

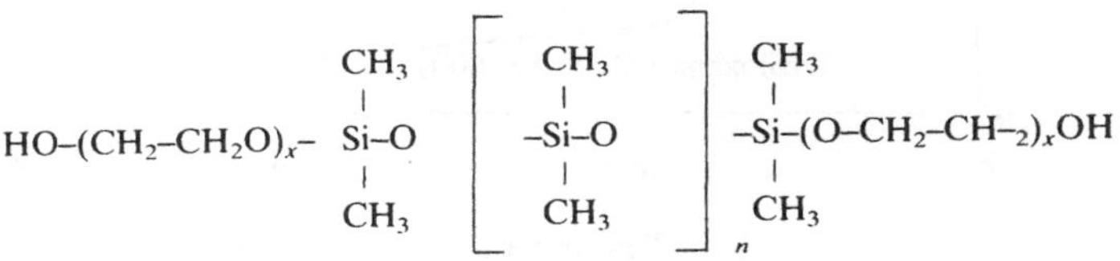

$$HO\text{--}(CH_2\text{--}CH_2O)_x\text{--}\underset{\underset{CH_3}{|}}{\overset{\overset{CH_3}{|}}{Si}}\text{--}O \left[\underset{\underset{CH_3}{|}}{\overset{\overset{CH_3}{|}}{-Si}}\text{--}O \right]_n \underset{\underset{CH_3}{|}}{\overset{\overset{CH_3}{|}}{-Si}}\text{--}(O\text{--}CH_2\text{--}CH\text{--}_2)_x OH$$

Figure 11.8 ABA type siloxane EO/PO copolymer.

Hydrophilic side groups such as polyoxyethylene or sulphonate in comb type polymers have already been shown in Figure 11.3. Other functional groups such as carboxyl, quaternary ammonium, betaine and epoxy groups may be incorporated. Thus anionic, nonionic, cationic and amphoteric polymeric silicone surfactants can be prepared.

Properties of polymeric surfactants

The structures of the majority of polymeric surfactants have not been defined, either because of the method of synthesis and/or because the manufacturers do not reveal the chemical structure of their products. As a result, there is very little published information on the relationship between chemical structure and surfactant properties. Due to the relatively large molecular weight of the polymeric surfactants, they are seldom used for their foaming, wetting or detergent properties. Instead, the principal applications are as demulsifiers (condensed alkyl phenol alkoxylates), antifoams (the modified EO/PO copolymers) and dispersing agents (polyesters).

Properties of polymeric silicone surfactants. The dynamic surface tension is the rate at which surface tension falls, and it measures how quickly the surfactant lowers surface tension. This can be important in many practical applications such as wetting, where it is not only the final surface tension that is achieved which is important (see also chapter 4). Union Carbide (1992) have studied the effect of chemical structure of silicone surfactants and compared them with conventional surfactants. Selected results are shown in Figure 11.9.

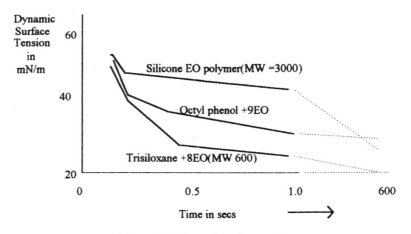

Figure 11.9 Dynamic surface tension.

The influence of molecular weight is clear, in that the higher molecular weight silicone EO polymer is very much slower to achieve a low surface tension, particularly if compared with the octyl phenol ethoxylate, which achieves its lowest surface tension very rapidly. However, the silicone EO polymer eventually achieves a lower surface tension than the octyl phenol

POLYMERIC SURFACTANTS 301

+ 9EO. The trisiloxane has values of $n =0$, $m =1$, $x =0$ and $y =8$ in the second example in Figure 11.3; thus it is closer to a conventional nonionic surfactant. The trisiloxane ethoxylates seem to have the ideal characteristic of a rapid drop in surface tension combined with a very low ultimate surface tension. Unfortunately, they tend to be hydrolytically unstable. The dynamic rate for silicone polymeric surfactants can be improved by incorporating PO in the hydrophilic group, but there is a loss in the final surface tension and it is difficult to achieve an equilibrium surface tension of less than 30 mN/m and at the same time obtain a high dynamic rate.

The dimethylsiloxy chain is not only hydrophobic but also lipophobic above a certain chain length, and will not dissolve in many organic media such as mineral oils. If some of the methyl groups are replaced by more lipophilic groups (e.g. long-chain hydrocarbon groups) then the modified silicone oil becomes partially oil soluble and surface-active in organic media.

A polymeric silicone surfactant with alkyl side groups and z usually in the range 12–18 is shown in Figure 11.10.

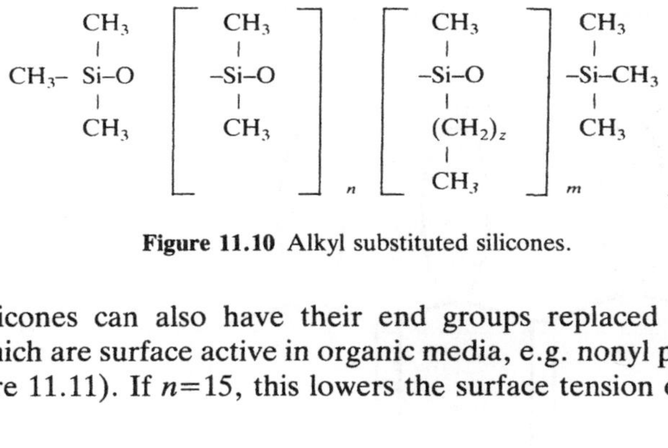

Figure 11.10 Alkyl substituted silicones.

The silicones can also have their end groups replaced with organic groups which are surface active in organic media, e.g. nonyl phenol groups (see Figure 11.11). If $n=15$, this lowers the surface tension of mineral oil

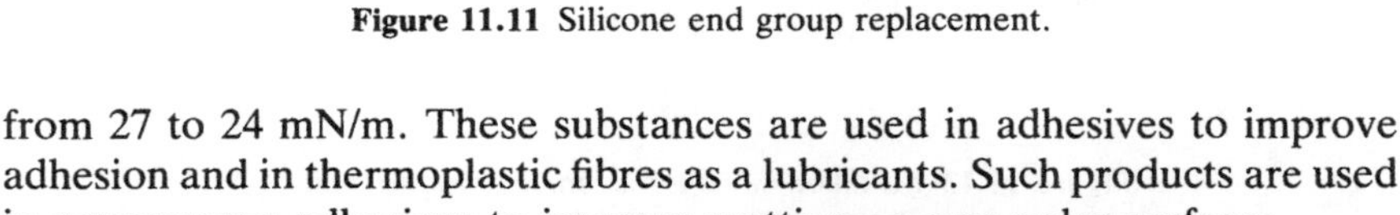

Figure 11.11 Silicone end group replacement.

from 27 to 24 mN/m. These substances are used in adhesives to improve adhesion and in thermoplastic fibres as a lubricants. Such products are used in nonaqueous adhesives to improve wetting on non-polar surfaces

In all the silicone products shown so far, the dimethyl and alkyl groups along the silicone chain have provided the hydrophobic and/or lipophobic effect. By putting a hydrophilic group on the siloxane chain (e.g. a poly(EO) chain) as well as a hydrophobic side chain (e.g. a C_{18} alkyl group), products with quite novel properties can be obtained. An example is shown in Figure 11.12.

$$CH_3-\underset{\underset{CH_3}{|}}{\overset{\overset{CH_3}{|}}{Si}}-O\left[\underset{\underset{CH_3}{|}}{\overset{\overset{CH_3}{|}}{-Si}}-O\right]_n\left[\underset{\underset{\underset{CH_3}{|}}{(CH_2)_p}}{\overset{\overset{CH_3}{|}}{-Si}}-O\right]_q\left[\underset{\underset{\underset{\underset{OH}{|}}{(PO)_y}}{\underset{|}{(EO)_x}}}{\overset{\overset{CH_3}{|}}{-Si}}-O\right]_m\underset{\underset{CH_3}{|}}{\overset{\overset{CH_3}{|}}{-Si}}-CH_3$$

Figure 11.12 Silicones modified with alkyl and alkoxy side groups.

Such products can give extremely good W/O emulsions with the hydrophobic groups in the oil phase, the hydrophilic groups in the water phase and the siloxane chain at the interface. This is shown diagramatically in Figure 11.13.

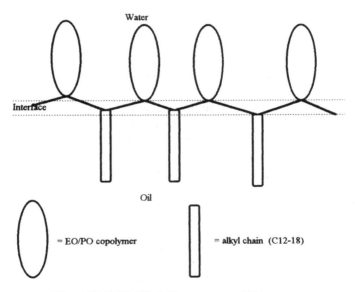

Figure 11.13 Modified silicones as emulsifying agents.

CMC and the formation of micelles in solution. There is very little published on the CMCs of well defined polymeric surfactants but the results which are published are contradictory. Data obtained from a paper entitled 'The aggregation behaviour of silicone surfactants in aqueous solutions' by Gradzielski et al. (1990) and one on a very similar system by Ohno et al. (1992) showed contradictory results and are illustrated in Figure 11.14. Gradzielski et al. (1990) show an ill defined CMC at about

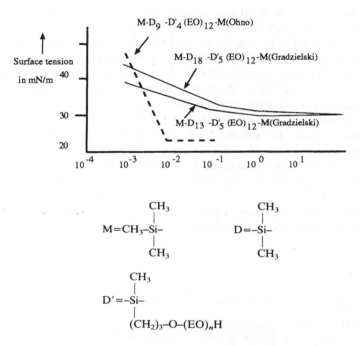

Figure 11.14 Polymeric silicone surfactants.

5g/l whilst Ohno et al. (1992) show a sharp CMC at about 0.01 g/l. A lower surface tension was obtained by Ohno et al. as well; this would be expected with a more hydrophobic surfactant. However, the very significant change in behaviour is surprising. There are two possible explanations:

1. The surface tension–concentration behaviour changes markedly as D_n goes from 9 to 13.
2. The molecules, although appearing very similar, were in fact quite different in structure. There were various ways that they could differ:
   1. The distribution of the EO, i.e. there would be EO contents greater than and less than 12.
   2. The distribution of the Si–O as the products (i.e. the molecular weights) are only averages.
   3. The distribution of the $(EO)_{12}$ group along the siloxane chain.

304 HANDBOOK OF SURFACTANTS

With the present evidence it is difficult to decide where the differences lie. However, it may show that the actual arrangement of hydrophobic and hydrophilic groups along the chain can also influence surfactant behaviour.

General surfactant properties of polymeric surfactants and how they differ from normal surfactants. The literature of ionic polymeric surfactants shows general agreement in the surface-active properties of ionic polymeric surfactants. They show general similarities to normal surfactants. They do form micelles or aggregate, and CMCs can be determined with a reasonable degree of confidence. However, the aggregates are not well defined, and they are generally smaller than the polyelectrolytes of similar structure, but without the hydrophobic group. For instance viscosities are of the order of 10–100 times smaller. The surface activity is lower than for the monomeric surfactants and they take more time to achieve their performance. Rates of wetting and attainment of equilibrium surface tension are often slower.

Literature on nonionic polymeric surfactants with polyethoxylates as the hydrophilic groups shows differences in behaviour which, at present, cannot be explained.

Polymeric surfactants show superior practical advantages as emulsifying agents, foaming agents or defoamers and as dispersing agents.

References

Aston, M.S., Bowden, C.J., Herrington, T.M. and Sahi, S.S. (1985) *J. Am. Oil Chem. Soc.* **62**, 1705–1709
Baker, A.S. (1979) Polymeric surfactants for industrial applications, presented at SCI Colloid and Surface Chemistry Group Symposium on *Surface Active Agents*, University of Nottingham, 26–28 September.
Bognolo, G. (1992) *Proc. 3rd CESIO International Surfactants Congress*, London, vol. C, p. 117.
Eisenlauer, J. and Horn, D. (1984) *Proc. World Surfactants Congress*, Mumich, vol. IV, p. 260–266.
Gradzielski, M. *et al.* (1990) *Tenside Surfactants Detergents*, **27**, 366–378.
Ishigami, Y. (1978) *Yukagaku*, **9**, 565–571.
Ohno, M. *et al.* (1992) *J. Am. Oil Chem. Soc.* **69**(1), 80–84.
Union Carbide (1992) Data Sheet 'Silwet Surfactants'.

Appendix 1

Names of hydrophobes

Carbon chain	Hydrocarbon	Systematic name	Common name of acid
C1	CH_4	methane	formic
C2	C_2H_6	ethane	acetic
n-C3	C_3H_8	propane	propionic
n-C4	C_4H_{10}	butane	butyric
n-C5	C_5H_{12}	pentane	valeric
n-C6	C_6H_{14}	hexane	caproic
n-C7	C_7H_{16}	heptane	heptoic or oenathic
n-C8	C_8H_{18}	octane	caprylic
n-C9	C_9H_{20}	nonane	pelargonic
n-C10	$C_{10}H_{22}$	decane	capric
n-C11	$C_{11}H_{24}$	undecane	undecylic
n-C12	$C_{12}H_{26}$	dodecane	lauric
n-C13	$C_{13}H_{28}$	tridecane	tridecylic
n-C14	$C_{14}H_{30}$	tetradecane	myristic
n-C15	$C_{15}H_{32}$	pentadecane	pentadecylic
n-C16	$C_{16}H_{34}$	hexadecane	palmitic (cetyl)
n-C17	$C_{17}H_{36}$	heptadecane	margaric
n-C18	$C_{18}H_{38}$	octadecane	stearic
cis-C18:1(d9)	$C_{18}H_{36}$	octadecane	oleic
trans-C18:1	$C_{18}H_{36}$	—	elaidic
C18:2	$C_{18}H_{34}$	—	linoleic
C18:3	$C_{18}H_{32}$	—	linolenic
OH-C18:1	$C_{18}H_{37}OH$	—	ricinoleic
n-C20	$C_{20}H_{42}$	eicosane	arachidonic
n-C22	$C_{22}H_{46}$	docosane	behenic
cis-C22:1	$C_{22}H_{44}$	—	erucic
n-C24	$C_{24}H_{50}$	tetracosane	lignoceric
n-C26	$C_{26}H_{54}$	hexacosane	cerotic
C28	$C_{28}H_{58}$	octacosane	montanic

Average composition of fats and oils

Oil	Carbon chain							
	C8–10	C12	C14	C16	C18	C18:1	C18:2	C18:3
Coconut	15	48	18	8	2	6	2	—
Corn	—	—	—	12	2	28	57	1
Olive (California)	—	—	8	2	83	6	—	
Palm kernel	6	50	18	9	2	13	1	—
Safflower	—	—	—	7	3	13	77	—
Soyabean	—	—	11	3	23	53	8	
Tallow (beef)	—	—	6	29	19	44	2	—
Tallow (mutton)	—	—	5	25	30	36	4	—

Appendix 2   Ecological and toxicity requirements

All industrialised countries have developed legislation and guidelines for the use and disposal of chemicals. In Europe the EC has issued a Directive 67/548/EEC entitled 'Classification, Packaging and Labelling of Dangerous Substances' which lays down criteria for packaging and labelling of chemical products that are deemed to be dangerous on the basis of various tests and standards. Included, amongst other criteria, were toxicity and danger to the environment. Other countries, e.g. the USA, Japan and Switzerland, have similar legislation and/or guidelines, but the detailed test methods and standards differ from one regulatory authority to another. The situation on legislation is continually changing, albeit slowly, and as described in this Appendix it will probably be out of date on publication. Nevertheless, the principles on test methods do not change and a description of the evolution of the EC Directives relating to biodegradation will be valuable background to understanding the concern on the environmental safety of surfactants.

Biodegradation

At the time of writing, there is a heightened concern world-wide about environmental issues. Although concern for surfactants in the environment has been enshrined in European legislation for nearly 20 years, only more recently has there been a more general concern over the fate of all synthetic chemicals in the environment. The pattern of use of the large volume surfactants is well known and it is possible to calculate the probable environmental concentration (PEC) in the various parts of the aquatic environment. Toxicity testing on fish and algae can be carried out, and the concentration having 'no observed effect' (NOEC) can be measured. If PEC is lower than NOEC by 1–2 orders of magnitude or more the chemical may be considered safe in the environment. However, if the PEC is more than or of the same order of magnitude as NOEC then the chemical could pose a problem. Most of the surfactants used in households and in industry go into the sewers. Surfactants have a relatively high PEC in sewage. The PEC figure will give an annual volume that enters the environment but gives no indication of what happens thereafter. It was hoped that measurement of biodegradation would give a measure of the persistence in the environment.

ECOLOGICAL AND TOXICITY REQUIREMENTS

Biodegradation is a process carried out by bacteria that are present in nature. Bacteria are micro-organisms that can metabolise an organic chemical and convert it into less ·complex chemicals by a series of enzymatic reactions. Finally, the end products are reached: carbon dioxide, water and oxides of the other elements Thus, if an organic chemical undergoes biodegradation it is transformed into its inorganic substituents and in effect disappears from the environment. If a product does not undergo natural biodegradation then it is stable and persists in the environment. The study of a surfactant's biodegradation is therefore important as a measure of its life in the natural environment.

Water, oxygen, mineral salts, growth factors and the appropriate species of bacteria have to be available before biodegradation takes place. Biodegradation takes time, since the molecules change one step at a time in the biochemical reactions. Biological reactions in aqueous media are also heavily dependent upon concentration, pH and in particular temperature. The rate of bacterial breakdown of chemicals can vary from a half-life (under optimum conditions) of 1–2 hours for fatty acids and sugars, 1–2 days for linear alkyl benzene sulphonates and months for branched chain alkyl benzene sulphonates. Note that even the latter will degrade given enough time but their rate of degradation is very slow, and hence they could build up in the environment if the rate at which they are added is faster than the rate at which they decay. Conversion of organic carbon into carbon dioxide 100% is not possible because some of the carbon is utilised in the synthesis of new bacterial cells, and soluble intractable organic products are often formed as a small proportion of the original carbon.

Petrochemical based surfactants have probably been studied in greater detail than any other group of chemicals and it would be expected that there would be no problems by now. Such is not the case, but the reason is complex and the explanation is partly technical and partly political. It is wise, therefore, that the surfactant user should have a grasp of the essential reasons behind this concern and should have the required data on the products available in order to answer questions on this subject.

Soap has been used for thousands of years and many synthetic surfactants have been in use over the last 100 years, but it is only during the last 25 years that problems have arisen. During the 1950s persistent foam began to accumulate on many rivers in the USA and Europe. It was found that this foam was due to alkyl benzene sulphonates based on tetrapropylene which seemed to persist in rivers. This surfactant (ABS) was replacing soap in domestic washing powders and was a constituent of the new liquid washing-up liquids. Testing in the presence of bacteria showed that the ABS was relatively stable and only degraded slowly, whilst soap was unstable in the presence of bacteria, its surfactant properties were destroyed and the foam disappeared. Substitution of alkylbenzene sulphonates based on linear alkyl chains (LABS) for ABS eliminated the foam

308 HANDBOOK OF SURFACTANTS

present in rivers. This piece of history is very interesting in that action was taken by all the major surfactant producers and users in advance of legislation. The point at which action was taken was when a significant proportion of the users (the population) were aware of the problem and that the problem (foam) was found to be due to the surfactant. In the case of foam the general population was very aware, as they could see it. But when the foam disappeared the problem was apparently solved. The change over from 'hard' ABS to 'soft' LABS was taken by the surfactant manufacturers before legislation was passed by the EC. Nevertheless, legislation was passed by the EC in 1973 to the effect that detergents must be capable of being degraded by natural bacteria. The first legislation EC Directive 73/404/EEC stated that the detergents (**not** the surfactants) must be capable of at least 90% biodegradation. The definition of a detergent is:

'For the purposes of this Directive, detergent shall mean the composition which has been specially studied with a view to developing its detergent properties, and which is made up of essential constituents (surfactants) and in general, additional constituents (adjuvants, intensifying agents, fillers, additives and other auxiliary constituents)'

This Directive does not give a clear definition of a detergent, and non-detergent uses are not covered by this legislation. However, as detergents are the major use of the 'commodity' surfactants, in practice all large volume surfactants will comply with the requirements of this legislation.

The EC Directive 73/404/EEC did not define the test methods by which the standard of 90% should be achieved. Subsequent directives did define tests methods as follows. Anionics: EC Directive 73/405/EEC and EC Directive 82/243/EEC; nonionics: EC Directive 82/242/EEC. Each of the EC countries have national legislation covering these directives. Any of four test methods can be used:

1. The OECD (Organisation for Economic Co-operation and Development) method published in the OECD Technical Report of 11 June 1976, Proposed Method for the Determination of the Biodegradability of Surfactants used in Synthetic Detergents (anionic and nonionic).
2. The German method established by the 'Verordnung über die Abbaubarkeit anionischer und nichtionischer grenzflachenaktiver Stoffe in Wasch-und Reinigungsmitteln of 30 January 1977, published in the Bundesgesetzblatt 1977, Part 1, page 244, as set out in the Regulation amending that Regulation of 18 June 1980, published in the Bundesgesetzblatt 1980, Part 1, page 706 (anionic and nonionic).
3. The French Method approved by the Decree of 28 December 1977 published in the Journal Officiel de la Republique Française of 18 January 1978, pages 514 and 515, and experimental standard T73–260

ECOLOGICAL AND TOXICITY REQUIREMENTS 309

of June 1981 (for anionic surfactants) or T73–270 of March 1974 (for nonionic surfactants), published by the Association Française de Normalisation (AFNOR).
4. The United Kingdom method called the Porous Pot Test as described in the Technical Report No 70/1978 of the Water Research Centre.

In the case of conflict on test methods an annex to Directive 73/405/EEC provides the confirmatory test method for anionic surfactants, and likewise 82/242/EEC for nonionics. At the present time there is no EC Directive defining the test method for cationics or amphoterics.

The general principles behind all the test methods are:

1. A dilute solution of known concentration of the surfactant is mixed with bacteria and the solution is analysed for **the surfactant** after a particular period of time (T). The fall in concentration of the surfactant after time T is expressed as a percentage and this percentage will be the amount degraded. Thus, if the amount of surfactant remaining after time T is 25% of the original concentration then 75% biodegradation will have occurred in time T.
2. The time T is increased until the biodegradation is a constant value.
3. The method of analysis for the anionic is by titration with methylene blue, which will analyse for an anionic showing surfactant properties.
4. The method of analysis of the nonionic is a colorimetric method known as the Wickbold method, which, however, measures only nonionics that show surfactant properties and is specific to nonionic surfactants containing 6–30 alkylene oxide groups.

The test methods in the EC Directives referred to above actually consist of two methods. Firstly, there is a **screening test** which is relatively easy to carry out. If the surfactant passes the screening test then no other tests are carried out. If the surfactant fails the screening test then the **confirmatory test** is carried out. In each of the four methods referred to above the confirmatory test will differ in detail but in all cases the bacterial concentration is much higher in the confirmatory test than in the screening test. All the major surfactants used in household products in Europe will pass the requirements of these Directives and generally pass the screening tests. The **OECD Screening Test** is one in which a dilute solution of the surfactant is mixed with a bacterial sample from a sewage plant and some inorganic nutrients. The test solution is agitated under carefully controlled conditions for a period of up to 19 days. During the test the change in the concentration of the surfactant is measured by the standard methods quoted in the Directive. The results are compared with those for two control substances, one of high biodegradability and one of low bio-degradability. If the concentration of surfactant is reduced by 80% or more then the surfactant passes the screening test. If not then the **OECD**

310   HANDBOOK OF SURFACTANTS

Confirmatory Test is carried out. This test is intended to provide a laboratory simulation of an activated sludge plant. It gives much more opportunity than the OECD Screening Test for the bacteria to adapt to the surfactant. The surfactant is mixed with synthetic sewage and fed continuously at a known concentration into an aeration vessel. There is a running-in period of 42 days to enable the bacteria to adapt to the surfactant, and then the removal of surfactant is measured over a 21-day period using the same analytical techniques as for the screening test. The same criterion i.e. 80% or more removal, then applies for the surfactant to pass the EC Directives using the OECD Confirmatory Test.

However, these test procedures for biodegradability have several unsatisfactory features which have led to growing criticism of the results. The tests were intended to show whether a surfactant was degraded by bacteria in natural waters and/or a sewage plant. They have been shown to be very good tests in showing whether or not a product persists in giving foam on rivers. They do not show what actually happens to the surfactant, only that it loses its surfactant (foaming) properties. The reason is that the analytical method used to estimate the amount of surfactant remaining in the solution measures the amount of the original anionic or nonionic surfactant remaining. Once the test substance has lost its surfactant properties it can no longer be measured by the method of analysis, and therefore has apparently disappeared and been biodegraded. Also, the analytical methods that have been developed are difficult to apply in practice. Biodegradation only occurs in dilute solution and the starting concentration of surfactant in biodegradation tests is typically 15–40 ppm of surfactant. In order to measure biodegradation to an accuracy of say 5%, then the actual concentration of surfactant must be measured to an accuracy of less than 1 ppm. There are only a few types of surfactants where there are methods available that can be relied upon to give such accuracy. There are no methods available for many surfactants, particularly many nonionics. In addition, there are no Directives specifying test methods and standards for cationics and amphoterics.

This measurement of biodegradability is now known as **primary** biodegradation. The material having lost its surfactant properties generally goes on biodegrading until all the carbon, hydrogen, nitrogen and oxygen atoms in the original molecule have been transformed into carbon dioxide, nitrogen gas and water, and this is called **ultimate** biodegradation. Considerable work has been carried out to show that, in the EC tests as outlined above, ultimate biodegradation does in fact take place but is not measured. Thus, the test procedure detailed above does not measure the ultimate biodegradation, only the primary biodegradation.

These test procedures entail practical problems in giving reproducible results, such as the number of bacteria, the source of bacteria, bacterial evolution and change, the conditions of degradation and the analytical

ECOLOGICAL AND TOXICITY REQUIREMENTS 311

method. Nevertheless, it is the probelm of measuring ultimate biodegradation that is the major concern at the time of writing. What are these problems?

Ultimate biodegradation

In the EC Directive 67/548/EEC (Classification, Packaging and Labelling of Dangerous Substances) the criteria for labelling a product 'dangerous to the environment' are based on toxicity tests on fish, daphnia and algae, partitioning behaviour in organic solvents mixtures, and the concept of **ready biodegradability**. Tests for ready biodegradability are not the same as the biodegradation tests described under primary biodegradation for the following reasons:

1. The need for test methods that are widely applicable to all surfactants.
2. The primary biodegradation tests only determine loss of surfactant properties and not the breakdown to inorganic compounds.

The problem of testing for biodegradation for all surfactants, or for that matter all chemicals, had been apparent long before the particular needs of ultimate biodegradability were addressed. The major problem is one of measuring the breakdown of the organic molecule (surfactant from now on). The analytical methods for assessing primary biodegradation (see above) measured the concentration of the starting material that remained after exposure to the bacteria. This had the drawback that the method of analysis must differentiate between the original material and its subsequent breakdown products and also be accurate at very low concentrations. Thus each type of surfactant needed a different test method. There were, however, some methods available that had been used for a considerable time for measuring the potential for an organic compound to biodegrade. They all depend upon the organic carbon in the surfactant being converted into carbon dioxide. These methods are shown in Table A.1.

Table A.1 Analytical methods of assessing ultimate biodegradability

Analytical method	Expected values for ready biodegradability
BOD/COD ratio (see the text)	$BOD_5/COD > 50\%$ or $BOD_{28}/COD > 60\%$
DOC (dissolved organic carbon)	DOC removal $> 70\%$
Carbon dioxide evolution	Depends upon test apparatus and conditions
Oxygen consumption	Depends upon test apparatus and conditions

BOD is the biological oxygen demand and is measured by the number of grams of oxygen consumed during biodegradation over 5 or 28 days (BOD_5 or BOD_{28}). The COD is the chemical oxygen demand and is measured by heating with chromic acid and measuring the amount of chromic acid

312   HANDBOOK OF SURFACTANTS

consumed. The BOD measures the amount of oxygen consumed by the bacteria in converting the organic carbon in the surfactant to carbon dioxide. The COD measures the total theoretical amount of oxygen needed to convert the organic carbon into carbon dioxide. BOD is always less than COD because some of the organic carbon from the surfactant is converted into cellular material. DOC (dissolved organic carbon) is measured by removing solids by filtration and then analysing the filtrate for organic carbon by standard methods of analysis.

The analytical tests in Table A.1 can be applied to any surfactant and the chemical structure need not be known. However, there is still the problem of the practical conditions of biodegradation involving what type of bacteria, how many bacteria, how long for adaptation, what temperature, whether aerated or not, what nutrients and how much. Thus a number of practical test methods have evolved, but these have now been generally placed into three categories:

1. *Ready biodegradability.* These are test conditions where the conditions of biodegradation are relatively poor, e.g. low concentrations of bacteria. If surfactants will degrade under these conditions, then it is assumed that they will biodegrade rapidly and completely in a wide range of aerobic environmental conditions, e.g. streams. The OECD Screening Test for primary biodegradation has been modified as an example. Other tests are the modified AFNOR, STURM and MITI(I) tests (see Table A.2). The tests are intended as simple, economical, fail-safe screening tests. If the surfactant fails to biodegrade sufficiently it does not mean that it is non-biodegradable.

2. *Inherent biodegradability.* These are tests where the conditions for biodegradation are much more favourable than in the ready biodegradability tests, e.g. a much higher concentration of bacteria, additional organic nutrients, etc. Examples of such tests are the modified semicontinous activated sludge or modified Zahn–Wellens tests (see Table A.2). These tests are intended to determine whether a surfactant is biodegradable under the most favourable conditions.

3. *Simulation tests.* These are tests carried out under conditions as close as possible to those in the environment. These are not normally considered necessary, particularly if field data are available. However, they are carried out when the results of ready and inherent biodegradability tests are inconclusive. The OECD confirmatory test for primary biodegradation is in fact a type of simulation test. By appropriate changes in analysis it can be applied to any surfactant. Other simulation tests are being proposed.

Table A.2 is included to show the complexity of the present situation. It may become simpler, but in the author's opinion this is very unlikely in the next few years.

ECOLOGICAL AND TOXICITY REQUIREMENTS

Table A.2 Ultimate biodegradation tests — proposed EC methods (as at 1992)

Test method	Method of anaylsis
Ready biodegradability	
Modified AFNOR	DOC
Modified Sturm	Carbon dioxide
Modified MITI(I)	BOD
Modified OECD Screening	DOC
Closed bottle	BOD
Inherent biodegradability	
Modified semicontinuous activated sludge	DOC
Modified Zahn–Wellens	DOC or COD
Simulation tests	
Activated sludge simulation tests	DOC

Is a particular surfactant biodegradable? This question is often asked. Does this mean primary, or is ready (ultimate) biodegradation required? Must it pass any specific legislation e.g. Directive 73/404 EEC? Is any particular standard required? In which country is the answer required? Only when one has the answer to these subsidiary questions can one give a reasonable answer. However, for the majority of commodity surfactants used in industry today, one may safely state that they will be readily biodegradable within the meaning of that phrase described in this appendix.

Toxicity

Most of the commodity surfactants have been thoroughly investigated in respect of their possible toxic effects on humans. Considerable information exists on the testing of surfactants, and much of it has been published. The majority of this information has been collected by tests on laboratory animals, but there is also a considerable amount of information, not always published, which the large detergent companies collect on accidental ingestion by the public. It would be safe to say that the present surfactants that have been used on a large scale for a number of years are as safe to use as natural soaps. However, there are a number of exceptions to this statement of which users should be aware.

1. Synthetic surfactants can alter in composition by raw material changes and/or process changes by the manufacturer.
2. A surfactant that has only been used in small volumes suddenly finds a new use and is produced and used in considerable volume in consumer markets. Long-term testing is very seldom carried out until after a surfactant has been produced in large volume.

314 HANDBOOK OF SURFACTANTS

3. Most surfactants for industrial use do not undergo the extent of testing carried out on surfactants used in consumer markets.

Surfactants are subject to all the regulations governing the general use of chemicals and there is no intention in this book to cover the general safety of chemicals. It will focus on the description of those features of toxicity more relevant to the use of surfactants. The general application of surfactants in industry is similar to that of other chemicals. However, surfactants are different to most other synthetic organic chemicals when they are used in personal and household products. Such products are regularly applied by millions of people to their hair and skin. Accidental ingestion of small quantities is impossible to avoid, e.g. in toothpaste and dish-washing products. Thus, the emphasis is on evaluation of the short- and long-term effects of ingestion and the short- and long-term effects of exposure of the skin and mucous membranes to the products.

Considerable research has been carried out and is still continuing on the effects of surfactants on biological membranes, due to their ability to adsorb at interfaces. This research generally uses cultured cells *in vitro*, and the morphological structure of the cell can be changed. Anionic and cationic surfactants can adsorb on to proteins, and induce physicochemical changes that result in a loss in the biological activity of the protein. There has also been considerable research on the adsorption of surfactants by the skin. Surfactants defat the skin by emulsification of the lipids and thereby remove some of the protection offered by the skin against aqueous systems; this can also lead to a loss of water by the skin. However, the lipid regenerating factor of the skin is considerable, and it is continuous exposure of skin to surfactants that results in roughness, scaling and dry skin. Fortunately, these effects have been recognised by users for many years in the use of soap in washing, and most people take sensible precautions.

The practical problem facing users and formulators with surfactants is the various labelling regulations now appearing, which place upon the manufacturer or supplier the need to label the product if it is dangerous to human beings or the environment. This account cannot be up to date on this subject, but the manufacturer or supplier should have the necessary data available on which a decision can be made. The exact interpretation of toxicity data should be left to a competent expert but the collection of the appropriate data usually falls on the formulator/seller. Therefore, it is becoming more important for a user of surfactants to understand the various tests that are carried out for labelling purposes. The rest of this appendix is therefore devoted to a simple description of those tests used for labelling purposes.

There are a number of labelling regulations but the two principal regulations are:

ECOLOGICAL AND TOXICITY REQUIREMENTS

- EC Directive 67/548, The classification, packaging and labelling of dangerous substances
- The Federal Hazardous Substances Act (USA) — Code of Federal Regulations Title 16, Parts 1500.3, 1500.4, and 1500.41, National Archives of the United States, Washington, DC (1973)

There have been a considerable number of amendments and additions to these regulations. EC member countries have national regulations to implement Directive 67/548. Non-EC member countries in Europe have similar legislation in respect of the toxicity tests required but can differ in the standards required.

The regulations give standards of results on animal tests whereby the product may be then classified as dangerous and is given an appropriate label. The protocol for the animal tests is standardised. Protocol means the exact way in which the test is carried out. There is no intention to describe the tests in detail, but rather to give the principal features of the tests.

1. *Acute toxicity oral.* Normally carried out by feeding a number of animals with varying doses of the surfactant and estimating the dose at which 50% die. This is known as the LD_{50} value and is usually expressed in milligrams of surfactant/kilogram of body weight. The higher the figure the less toxic is the substance. The LD_{50} value of most common surfactants is high, alkyl sulphates giving figures of 1000–10000 mg/kg. Tests for labelling only may be made at only one level, i.e. that greater than the maximum allowed for a particular safety label, and therefore if < 50% die the product does not require the safety label. A sufficient number of animals must be used in this type of test, so always ascertain how many animals have been used.
2. *Acute toxicity percutaneous* (skin toxicity or dermal toxicity). Similar to oral tests, but the test substance injected under the skin.
3. *Acute toxicity by inhalation.* Generally not applicable, as surfactants are very rarely in powder or spray form.
4. *Chronic toxicity.* These are long-term tests carried out in conjunction with tests for carcinogenicity. Surfactant is added to the food or drinking water of a group of laboratory animals at a level below the LD_{50} over periods of up to 2 years. The effects on the test animals are compared to those on a similar group of animals with no added surfactant. If surviving, at the end of the test period all the animals are killed and post-mortem examinations carried out. One criticism of these tests is the very high level of surfactants fed to the animals if the LD_{50} value is high; this may not represent anywhere near the levels of surfactants normally ingested by humans over a period of time.
5. *Primary skin irritation.* Surfactant is applied to the bare skin of animals and the skin irritancy is measured by the swelling (oedema)

316   HANDBOOK OF SURFACTANTS

and the redness (erythema) of the skin. This is done visually on a scale (0–4 in the case of the original Draize test). Variations are possible in this test, such as the concentration of the surfactant and the time of contact with the skin.

6. *Secondary irritation or skin sensitisation.* Chemical compounds can give allergies to certain individuals. An allergy is severe skin irritation and possibly effects on the respiratory system (e.g. bronchial asthma) when exposed to minute quantities of a chemical compound. Sensitisation is the development of an allergy after renewed exposure to the compound. The big problem in evaluation is that not everyone is subject to an allergy produced by a compound and not everyone is subject to sensitisation. The surfactant industry is now extremely cautious on possible effects of sensitisation, due to the well documented effects of very small traces of sultones in washing-up liquids in Norway in 1965–66. There are a number of tests used in evaluation; the Magnusson–Kligman and Buehler tests on guinea-pigs are the best known. These consist of injection and/or application of the test substance to the guinea-pig's skin and then application of the test substance, similar to the primary skin irritation test. Tests on humans are preferable to form the best evaluation, and the best known test procedure is the repeated insult patch test by Marzulli and Maibach (1977)

7. *Corrosiveness.* In the EC Directive a substance is considered to be corrosive if, when it is applied to healthy intact animal skin, it produces full thickness destruction of skin tissue. Thus the tests are basically primary skin irritation tests.

8. *Eye irritation.* Accidental exposure to the eye is very common, e.g. with shampoos. Tests on the eyes, however, also represent the effects on mucous membrane. Surfactant is injected into the eye of a rabbit and irritation first causes reddening through increased blood pressure in the conjunctivae. This can further lead to destruction of the cell walls followed by bleeding. Damage can also occur to the cornea, which in the case of severe attack can lead to irreversible clouding, and to the iris. The most common test is the Draize test on the rabbit (see Draize *et al* 1944). There is considerable research taking place into replacing animals in toxicity tests by *in vitro* methods. Fentern and Balls (1992) summarised the various alternatives to the Draize eye irritation test.

9. *Carcinogenicity.* Carried out in a similar manner to the acute toxicity tests; the organs of the test animals are compared to those of the control animals at the end of the test period. It is almost certain in large studies that some of the control animals will have abnormal organs. The effect of the surfactant can only be determined by the

ECOLOGICAL AND TOXICITY REQUIREMENTS 317

difference between the test animals and the control animals and whether this difference is statistically significant.

10. *Mutagenicity*. Mutagenicity is the danger of genetic damage being passed on to the offspring by mutations affecting the ovum before fertilisation. The most common test is the Ames test (see Ames *et al.*, 1975), which is easy and relatively cheap to carry out. However, there is a problem in interpreting the Ames test, as a negative result is meaningful but a positive result does not necessarily mean that the product tested causes mutagenicity.

11. *Embryotoxicity (teratogenicity)*, Embryotoxicity is the effect of chemicals on the mother during pregnancy causing death or a change in the embryo. Such testing is difficult, particularly in interpretation, and there have been reports, particularly from Japan, of surfactants causing some changes, but these have not been reproduced elsewhere.

References

Ames, B.N., McCann, J. and Yamasaki, E. (1975) *Mutation Res.* **31**, 347–364.
Draize, J.H, Woodward, G. and Calvery, H. (1944) *J. Pharmacol. Exp. Ther.* **82**, 377.
Fentern, J. and Balls, M. (1992) *Chem. Ind.* **6**, 207–211.
Marzulli, F.N and Maibach, H.I. (1977) *Dermatotoxicology and Pharmacology*. Wiley, New York.

Index

acetylenic diols 187
acetylenic glycols 187
acetylenic surfactants 187
N-acyl derivatives of
  sarcosine 109
acyl oxyalkane
  sulphonates 111
adhesives with poleric
  surfactants 301
adsorption 27, 29, 31, 46
AE *see* alcohol ethoxylates
AES *see* ether sulphates
aggregation number of
  mixtures 90
alcohol ether sulphates (AES)
  see ether sulphates
alcohol ether (or ethoxy)
  sulphonates 134
alcohol ethoxylate in ether
  sulphates 122
alcohol ethoxylates 114, 189
alcohol in alkyl sulphates 119
alcohol sulphates (AS) 118
  in ether sulphates 122
  with ether sulphates 123
  with half ester
    sulphosuccinate 163
alkane sulphonates 135
alkanolamide ethorylates 210
alkanolamides 194
  with alkyl sulphates 119
  with alpha-olefin
    sulphonates 154
  with cationics 251
  with ether sulphates 124
alkene sulphonates in alpha-
  olefin sulphonates 153
alkene sulphonates/hydroxy
  alkane sulphonates 151
alkenyl alkyl polyglycol ether
  sulphonates 134
alkenyl succinic
  anhydrides 298
alkoxyalkanoic acids 105
alkoxyalkanols 189
alkyl(poly-1-oxapropene)
  oxaethane carboxylic
  acids 105
alkyl acid phosphates 113
alkyl amido betaines 264
alkyl amidodimethyl
  propylamines 238
alkyl amidopropyl
  betaines 264
alkyl amidopropyl
  dimethylamine. 236
N-alkyl amidopropyl-dimethyl
  amine oxides 198, 200
alkyl amidopropyldimethyl
  betaines 264
alkyl amidopropyldimethyl
  sulphobetaines 264
alkyl amidopropyl
  hydroxysultaines 265

alkyl amines 236
alkyl amino carboxylic
  acids 269
alkyl amino propionates 273
N-alkyl amino propionates
  259
alkyl amphocarboxyglycinate
  259, 271
alkyl amphodiacetates 269
alkyl amphoglycinates 259,
  271
alkyl amphomonoacetates
  269
alkyl amphopolycarboxy-
  glycinates 269
alkyl amphopolycarboxy
  propionates 273
alkyl benzene bottoms
  sulphonates 155
alkyl benzene
  sulphonates 138
alkyl betaines 265
N-alkyl betaines 258
N-alkyl bis(2-hydroxyethyl)
  amine oxides 198
alkyl bis(2-hydroxyethyl)
  betaines 265
alkyl carboxyglycinates 259,
  269
alkyl carboxyl substituted
  cyclic pentanes 102
n-cocoyl-N-methyl
  taurine 166
alkyl diethylene
  triamines 236
N-alkyldimethyl amine
  oxides 198
alkyl dimethylamines 238
alkyl dimethyl benzyl
  annnonium chloride 251
alkyl dimethyl betaines 265
alkyl ether phosphates 113
alykl ethylene diamines 236
N-alkylglucamine 206
alkyl glucosides 202
alkyl glycinates 259, 269
N-alkyl glycinates 258
N-alkyl imidazoline
  chlorides 254
alkyl imidazoline
  ethylenediamine 236
alkyl imidazoline
  ethylenediamine
  hydrochlorides 255
alkyl imidazoline
  hydroxyethylamine
  hydrochlorides 255
alkyl imidazoline
  hydroxyethylamines 236
alkyl imidazolines 236
alkyl iminodiglycinate 269
alkyl imino dipropionates
  273

N-alkyl imino
  diproprionates 259
N-lauric acid amidoethyl-N-2-
  hydroxyethylglycinate 270
alkyl N-methyl
  glucosamates 202
alkyl monoethanolamide
  ethoxylates 210
alkyl naphthalene
  sulphonates 134, 148, 276
3-alkyloxypropylamines 236
alkyl phenol ether
  sulphates 129
alkyl phenol ether (or ethoxy)
  sulphonates 134
alkyl phenol ethoxylates 243
alkyl phenol polyglycol ether
  carboxylic acids 105
alkyl phenol polyglycol
  ethers 243
alkyl phosphates 113
N-alkylphthalamates 109
alkyl polyamine
  ethoxylates 212
alkyl polyamines 236, 262
alkyl polyaminocarboxylates
  258, 269, 270
alkyl polyglycol ether
  carboxylic acids 105
alkyl polyglycosides 170,
  202
alkyl polyoxyethylene
  amines 212
alkyl polyoxyethylene
  glycols 189
alkyl primary amines 236
alkyl propanediamines 236
alkyl secondary amines 236
alkyl substituted
  silicones 301
alkyl sulphates 118
  similar properties to
    Isethionates 112
  with fatty ester
    sulphonates 147
  with sarcosinates 111
N-alkyl sulphobetaines 265
alkyl tertiary amines 236
alkyl tetraethylene
  pentamines 236
alkyl triethylene
  tetramines 236
N-alkyltrimethylammonium
  chlorides 254
allyl derivatives of nonyl
  phenol derivatives 289
alph-N-alkylamino acetic
  acids 269
alpha-olefin sulphonates
  (AOS) 64, 151
  with taurates 186
alpha-olefin sulphonates with

INDEX

half ester
sulphosuccinate 163
alpha sulphonated fatty acids
(or esters) 144
ALS (ammonium lauryl
sulphate) 118
amide carboxylates 109
amide ethoxylates 210
amides of methyl taurine 112
amine oxides 169, 198, 262
with alpha-olefin
sulphonates 154
with ether sulphates 124
with paraffin
sulphonates 137
with sarcosinates 111
aminoalkanoates 269, 173
aminoethyl imidazoline 238
amino functional
silicones 281
amino
polyfluorosulphonate 282
aminopropionates 272
amphocarboxyglycinates 269,
270
amphocarboxy
propionates 273
amphoglycinates 269, 270
ampholyte 258
amphopropionates 273
amphoteric surfactants 18,
276
amphoterics, changing pH 43
analysis 18
anhydrohexitol esters 226
anionic classification 99
anionic polymeric
surfactant 295
anionic surfactants 17
anionics with amine
oxides 201
biodegradation 309
cations and solubility 99
chemical stability 100
wetting 64
with alkanolamides 197
with amphoterics 263
with amino
propionates 274
with amphoterics 264
with betaines 267, 268
with cationics 249
with monoethanolamides
197
with nonionics 183
with polycarboxyglycinates
272
with sarcosinates 111
anisotropic phases 52
anisotropic properties 52
anticaking with alkyl
naphthalene
sulphonates 151
antifoams 69
for anionic surfactants 116
with polymeric
surfactants 300
AOS *see* alpha-olefin
sulphonates
APAC 270
applications 19
applications directory 19
AS *see* alcohol sulphates

baby shampoo 268

baby shampoos 272
bactericidal action of
cationics 249
bactericide 252
bacteriostat 252
Ballestra reactor 132
barium dinonylnaphthalene
sulphonate 155
bending modulus of a
bilayer 45
benzalkonium chloride 250,
251
benzene sulphonates 134
benzyl trimethylammonium
chloride 254
Berol reactor 132
beta-*N*-alkylalanines 273
beta-*N*-alkylamino
propionates 273
beta-*N*-alkylimino
propionates 273
betaines 258
with alpha-olefin
sulphonates 154
with ether sulphates 124
bilayers 40
biocidal activity with
amphoterics 263, 264
biocides 253
biodegradability of nonyl
phenol ethoxylates 244
biodegradable
surfactants 287
biodegradation 19, 220, 276
and silicone
surfactants 281
bis(hydrogenated tallow alkyl)
dimethylammonium
chloride 249
bis(2-hydroxyethyl) amine
oxides 200
bis(2-hydroxyethyl)
dodecylamine 212
block copolymers 293, 295
BOD 311
bola-amphiphiles 285
bolaform surfactants 285
bolaphiles 285
borate esters 198
broad cut acid 138
builders 86
building viscosity 197
butyl naphthalene
sulphonate 150
by-products of
ethoxylation 174

calcium petroleum
sulphonate 155
carbon chain distribution 12
carboxy glycinates 269
carboxylic acid salts 101
carboxymethylated
alcohols 105
carcinogenicity 316
carpet backing with
sulphosuccinamates 166
castor oil ethoxylate 243
catalyst residues from
ethoxylation 175
cationic softeners 43
cationic surfactants 18
CD-ROM records 22
CESIO 23
charge-fluctuation forces 80

Chemical Abstracts 20, 21,
22
Chemithon reactor 132
choice of surfactant 1, 7
chronic toxicity 315
Classification Packaging and
Labelling of Dangerous
Substances EEC
Directive 306, 311
cloud point of ethoxylated
alcohols 50
cloud point of
ethoxylates 175, 176
cloud point of nonionics 43,
49
CMC 27, 31, 36–37, 52, 68
amphoterics 38
detergency 87
Krafft point 47
polymeric surfactants 302
sources of 96
surfactant miles 89
cocoamidopropyl betaine 265
with alkyl sulphates 119
with alpha-olefin
sulphonates 154
cocoamidopropyldimethyl-
amine 239
coco amido propyl dimethyl
amine oxide 198
3-[(3-cocamidopropyl)-
dimethylamino] 2-hydroxy-
propanesulphonate 265
cocoamphocarboxy
glycinate 261
coco bis(2-hydroxyethyl)
amine oxide 200
cocodiethanolamide with
alpha-olefin sulphonates 154
cocomonoethanolamide with
alkyl sulphates 120
coconut diethanolamide 194
coconut fatty acid 2-sulpho
ethyl-1-ester sodium salt 112
coconut
monoethanolamide 194
COD 311
comb polymers 293
comb silicone surfactants 279
condensed alkyl phenol
alkoxylates with polymeric
surfactants 300
confirmatory test for
biodegradation 309
corrosion inhibitor 111
corrosive 316
cosmetic chemicals 20
critical micelle concentrations
see CMC
critical solid tension 61
crude oil demulsification 294
cubic liquid-crystalline
phases 52–55
cumene sulphonates 134

decagylcerol tristearate 231
defoamers 65, 69, 220, 221
mixed micelles 72
use of EP/PO
copolymer 183, 220
use of glycol and glyceryl
esters 235
use of soaps 103
demulsifiers with polymeric
surfactants 300

HANDBOOK OF SURFACTANTS

Derwent patent abstracts 22
detergency 11, 84, 85
detergents 18
   control of foam 221
   used in 1:1:1
      sulphonate:soap:nonionic
      mixtures 103
dialkanolamide 194
dialkanolamides with ether
   sulphates 124
dialkyl benzene
   sulphonates 155, 156
dialkyl pyrophosphates 113
dialkylsulphosuccinate 159
diamine hydrochlorides 255
diamines 236
di-carboxyl alkyl imidazoline
   betaines 270
dicarboxylated glycinates 259
diesters of glycerol 233
diether in ethoxylates 181
dihydroxy ethyl piperazine in
   alkanolamides 195
dimethicone copolyol 277
N,N-dimethyl-N-(3-alkyl
   amidopropyl)amines 237
N,N'dimethyl-N-(3-lauryl
   amidopropyl)amine 237
dimethyl octynediol 187
dimethyl octynediol or
   3,6-dimethyl-4-octyne-3,6
   diol 187
dimethylsiloxane glycol
   copolymers 277
dioleate salt of N-tallow
   trimethylene diamine 257
1,3 dioxalane ring 288
1,4 dioxane
   formation during
      sulphation 132
   in alcohol ethoxylates 190
   in ether sulphates 122
   in ethoxylates 175
   in nonylphenol ether
      sulphates 130
diquaternary
   polydimethylsiloxanes 281
directories of surfactant
   manufacturers 19
disodium coco alcohol
   ethoxylate(3)
   monosulphosuccinate 160
disodium
   cocoiminodipropionate 273
disodium coco
   monoethanolamide
   ethoxylate(5)
   monosulphosuccinate 160
dispersing 80
dispersing agents with
   polymeric surfactants 300
dispersing properties of
   nonionics 186
dispersion forces 81
distribution of chain
   lengths 12
disulphonate in fatty ester
   sulphonates 145
disulphonates in alpha-olefin
   sulphonates 153
disulphonic acid 140
DLVO (the Derjaguin Landau
   Verwy and Overbeek)
   theory 81
Dobanes (alkyl benzene) 138

dodecylbenzene sulphonates
   see LABS
dodecyl benzene sulphonic
   acid 138
dodecyl dimethylamine
   ammonium chloride 255
dodecyl dimethylamine
   hydrochloride 255
dodecyl diphenyl oxide
   disulphonate 138
dodecyltrimethyl ammonium
   chloride 248
Draves wetting test 62, 63
dynamic surface tension of
   polymeric surfactants 300

EC Directives 306, 308
electrodynamic forces 80
embryotoxicity 317
emulsification and HLB 185
emulsifier
   for herbicides 115
   for kerosene 129
   for mineral oil 115
   for organic non-water-
      soluble materials 142
   for pine oil and
      creosote 128
   with anticorrosive
      properties 115
emulsifiers and
   demulsifiers 20
emulsion explosives use by
   polymeric surfactants 298
emulsion paint 70
emulsion polymerisation 20,
   125, 130
   with fluorinated
      surfactants 284
   with labile surfactants 287
   with polymerisable
      surfactants 288
emulsion technology 19, 20,
   55
emulsions, 20, 55
   with polymeric
      surfactants 302
end-blocked nonionics 177,
   189
enhanced oil recovery 20,
   137, 158
EO/PO chain in silicone
   surfactants 279
EO/PO copolymers 79, 285,
   295
epoxy functional
   silicones 281
esteramide 206
   in alkanolamides 195
ester based quaternaries 254
ester carboxylates 107
ester sulphonates 111, 144
esters of isethonic acid 112
ethane sulphonates 134, 276
ether carboxylates 105
ether group 169
ether sulphates (AES) 64,
   121
   aggregation number 35
   with amine oxides 202
   with betaines 268
   with cocoamidopropyldi-
      methyl betaine 268
   with diester
      sulphosuccinates 163

with ester carboxylates 108
with ethoxy
   carboxylates 106
with ethoxylated sorbitan
   esters 229
with fatty ester
   sulphonates 147
ether sulphonates 134
ethoxy carboxylates 105
ethoxy sulphates 121
ethoxylated alcohols 161
ethoxylated
   alkanolamides 210
ethoxylated amines 212
ethoxylated castor oil 242
ethoxylated esters 231
ethoxylated ethane
   sulphonates 134
ethoxylated fatty acid
   alkanolamides 289
ethoxylated fatty acids 64
ethoxylated fatty alcohols 64,
   189
ethoxylated lanolin 242
ethoxylated
   monoalkanolamides 161,
   210
ethoxylated nonionics and
   foams 69
ethoxylated
   polyfluoroalcohol 282
ethoxylated polyglyceryl
   esters 231
ethoxylated sorbitan
   derivatives with glyceryl
   esters 235
ethylene glycol esters or glycol
   esters 231
ethylene glycol
   mono-esters 232
ethylene oxide
   in alcohol ethoxylates 190
   unreacted 175
ethylene oxide/propylene
   oxide block copolymers 215,
   216
eye irritation 316
   and mixed micelles 92

fabric softeners for household
   use 248, 251
FAS (fatty alcohol sulphate)
   see alcohol sulphates
fatty acid ethoxylates 222
fatty acid sugar esters 202
fatty acid sulphonates 144
fatty alcohol ether
   sulphosuccinate 160
fatty alcohol sulphates see
   alcohol sulphates
fatty amide polyglycol
   ethers 210
fatty amine ethoxylates 212
fatty amine oleates 257
fatty monoglyceride
   sulphates 127
fatty primary amines 236
fire fighting with fluorinated
   surfactants 284
fluorinated alkyl
   polyoxyethylene ethanol
   282
fluorinated surfactants 276,
   282
foam 65, 66, 69, 307

INDEX

food applications
with glyceryl esters 235
with sorbitan esters 229
formulated products 3, 4, 6, 9, 20
furanosides 205

gel-like behaviour 56
gels and cubic phases 53
Gemini surfactants 286
germicidal activity of fatty amine salts 256
germicidal properties of cationics 252
germicide 252
Gibbs–Marangoni effect and emulsions 75
Gibbs surface elasticity and foams 67
glucosamides 202
glucose esters 202
glucosyl alkyls 202
glycerides 288
glycerol esters — mono- or diglycerides 231
glycerol monolaurate 231
glycerol (or glycerol) monostearate 231
glyceryl esters 231
glyceryl monostearate 235
glycinates 270
glycine 258
glycol esters 232
Government publications 23
green acids 156
gypsum board production 125

half ester sulphosuccinate 160
hard acid 138
heavy alkylate sulphonates 134, 155
hexagonal structure 52, 57
HLB 72, 73
Hostapur SAS60 136
hydrated crystal structure 49
hydroperoxide in ethoxylates 179
hydrophilic effect 28–29
hydrophilic group 11, 36
hydrophilic/lipophilic balance see HLB
hydrophobic effect 28, 29, 277, 278
hydrophobic group 11, 12, 36
hydrotropes
using amphoterics 263
using betaines 267
hydroxyalkane alkyl polyglycol ether sulphonates 134
hydroxy alkane sulphonates in alpha-olefin sulphonates 153
hydroxyalkyl alkylamidoethyl glycinates 270
hydroxyethyl imidazoline 238
hydroxyl group 169
hydroxysulphobetaines 265

Igepon A 12
imidazoline-based amphoterics 259
imidazoline-based glycinates 271

imidazoline-based quaternary compounds 251
imidazoline carboxylates 270
imidazoline derivatives 236, 270
imidazoline hydrochlorides 255
iminodialkanoates 273
iminopropionates 272
indanes in LABS 141
industrial applications of surfactants 17
inherent biodegradability 312
inkjet printing inks with fluorinated surfactants 284
interfacial tension 31, 60, 74, 95
iodophors 220, 221, 246
ionic organo-polysiloxanes 277
isethionates 111, 288
isopropyl naphthalene sulphonate 150

journals and periodicals 20

Krafft point 46–49, 52
addition of electrolytes 47
Kritchevsky alkanolamide 194

labelling regulations 314
labile surfactants 110, 287
LABS (linear alkyl benzene sulphonates) 138
biodegradation 307
substitution by alkyl sulphates 120
with alcohol ethoxylates 183, 192
with amine oxides 137, 201
with EO/PO copolymer 184, 220
with fatty ester sulphonates 147
with nonylphenol ether sulphates 130
with soap 103
lamellar phase 52
lamellar structures 39
lanolin ethoxylate 242
LAS 118
lauric diethanolamide 194
lauric monoethanol-amide 194
lauroaminopropionates with DEA lauryl sulphate 120
lauryl amidopropyl dimethylamine 237
laurylamine 237
laurylamine ethoxylate 212
lauryl dimethyl amine oxide 200
lauryldimethyl betaine 265
lauryl ether sulphate with paraffin sulphonates 136
lauryl polyethyleneglycol phosphate 113
lauryl sarcosine 109
lauryl sulphate
with half ester sulphosuccinate 163
with monoethanol-amide 197

leather processing with paraffin sulphonates 137
legislation 306, 308
LES with ethoxycarboxylates 106
linear alkyl benzene sulphonates see LABS
linseed amide ethoxylates 127
liposomes 39, 43
liquid crystalline phase and detergency 87
liquid crystal phases and structure 43, 50, 52
liquid crystals with ethoxylates 177
liquid explosives with sorbitan esters 230
London forces 79
long chain carboxylic acid esters 222
long chain esters of hydroxy acids 107
lubricant with polymeric surfactants 301
lubricants, greases and lubrication 21
lubricating oil additives with petroleum sulphonates 158
lyophilic solids 83
lyophobic solids 80
lyotropic liquid crystals 52
lyotropic mesomorphs 52
lyotropic mesophases 52

macroemulsions 72
mahogany acids 156
maleate esters as polymerisable surfactants 290
Marangoni effect and foams 67
Mazzoni reactor 132
Meccaniche Moderne reactor 132
Meroxapol 216
metal working with petroleum sulphonates 158
methacrylate esters with betaine or sulphonate groups 290
methyl glucoside esters 202
micelles 34
aggregation 42
alkylpolyglycosides 40
change shape 57
counterion 42
cylindrical 39–42
dynamic nature 39
effect of cosurfactant 42
effect of salt 42
geometrical packing 40
inverted 40
Krafft point 46
lamellar 43, 52, 57
long thin thread 40
mixed 90
monodisperse 41
properties 27
shape of micelles 35
spherical 40
viscosity 41
microemulsions 19, 55, 57, 59
and detergency 87
inversion 61

322 HANDBOOK OF SURFACTANTS

microemulsions *contd*
  with silicone surfactants 282
middle-phase 52
minimum surface tension 276
mixed adsorption layers 90
mixtures of anionics and
  cationics 44
monoalkanolamides 194
monoalkyl phosphates 113
monoalkylpolyethylene glycol
  ethers 189
mono-carboxy alkyl
  imidazoline betaines 270
monocarboxylated
  glycinates 259
mono-, di-, or tri-glyceride
  sulphates 127
monoesters of glycerol 233
1-monolaurin 231
multi-hydroxyl groups 170
multi-hydroxyl products 171
mutagenicity 316

n-acyl-*N*-alkyl-taurate 166
naphthalene sulphonates 134
  condensed with
    formaldehyde 84, 293
naphthenates 102
narrow cut acid 138
narrow EO distribution 173
narrow range alcohol
  ethoxylates 122
narrow range AE used for
  ether sulphates 124
natural petroleum
  sulphonates 156
neat phase 52
nitroso content of
  alkanolamides 199
N-nitrosodiethanolamine 199
nonionic polymeric
  surfactant 295
nonionic surfactants 18
  biodegradation 309
  solubilised with amino
    propionates 273
  with amphoterics 263
  with sarcosinates 111
Nonoxynol 9 43
nonylphenol ether sulphates
  129, 289
nonyl phenol ethoxylates
  186, 243, 289
  detergency 186
nonylphenol
  ethoxyphosphates 289
no observed effect
  (NOEC) 306

oil field chemicals 125
  and acetylenic
    surfactants 188
  and ethane
    sulphonates 135
  and ethoxylated
    amines 215
  and petroleum
    sulphonates 158
  and soaps 104
oil slick dispersants 230
oleamide diethanolamide with
  alpha-olefin sulphonates 154
oleyl bis(hydroxyethyl)
  betaine 265
oligomeric surfactants 293

omega-sulphonated fatty acids
  (or esters) 144
on-line databases 23
organofluorine
  surfactants 282
organo-polysiloxane
  copolymers 277
Ostwald ripening 75
overbased sulphonates 155
oxo alcohols 116, 189

paint additives 21
paraffin sulphonates 135
patents 21
PEG esters 222
perfluoro polymethylisopropyl
  ether 282
perfluoroalkyl
  sulphonates 283
perfluoroalkyl
  surfactants 282
perfluoropolyether
  surfactants 282, 283
petroleum sulphonates 134,
  155
phase inversion of
  nonionics 73
phase inversion
  temperature 73
  and detergency 87
phenol formaldehyde
  polymer 293
phosphate esters 113, 276
phosphated alcohols 113
phosphine oxides 198, 200
phospholipid molecules 44
physical properties of
  surfactants 16
pine oil disinfectants with
  soaps 103
PIT *see* phase inversion
  temperature
plaster board manufacture
  with LABS 142
Pluronics 216
Polaxamer 216
Poloxamine 216
polyacrylic acid 84, 293
polyalkoxylated ether
  glycollates 105
polyalkylene oxide block
  copolymers 216
polyamine
  hydrochlorides 255
polyamines 237
polyamino polycarboxy
  glycinates 270, 271
polyamphocarboxy
  glycinates 258, 270
polycarboxyglycinates 258,
  270
polydimethylsiloxane 277
polyelectrolytes 293
polyesters as polymeric
  surfactants 296, 298
polyether polysiloxanes
  copolymers 277
polythylene glycol esters 222,
  231
polyethylene glycol esters with
  ether sulphates 124
polyethylene glycols in nonyl
  phenol ethoxylate 244
polyfluoroalkyl betaine 282
polyfluoropyridinium salt 282

polyfluorosulphonic acid
  salt 282
polyfunctional silicone
  surfactants 278
polyglyceride 231, 233
polyglyceryl esters 231, 238
polyglycol in PEG esters 223
polyglycols by-products from
  ethoxylation 174
polyglycols in alcohol
  ethoxylates 190
polyhydroxy amides 202
polyhydroxy derivatives 170
poly(12-hydroxy stearic
  acid) 296
polymeric alkyl phenol
  ethoxylates 294
polymeric electrolytes 293
polymeric silicone
  surfactants 221, 278, 299
polymeric surfactants 80, 83,
  84, 304
polymerisable nonylphenol
  ethoxylate derivatives 289
polymerisable surfactant 127,
  288
polyol monoester 231
polyoxyalkylene glyceride
  esters 231, 233
polyoxyalkylene glycol
  esters 231
polyoxyalkylene polyol
  esters 231
polyoxyalkylene propylene
  glycol esters 231
polyoxyethylene
  alkylphenols 243
polyoxyethyleneglycol(400)
  triglycerol monostearate 232
polyoxyethylated
  alkylamides 210
polyoxyethylated
  alkylphenols 243
polyoxyethylated fatty
  alcohols 189
polyoxyethylated fatty
  amines 212
polyoxyethylated
  polyoxypropylene
  glycols 216
polyoxyethylenated straight-
  chain alcohols 189
polyoxyethylene alcohol
  sulphates 121
polyoxyethylene alcohols
  189
polyoxyethylene
  alkylamines 212
polyoxyethylene esters 222
polyoxyethylene fatty acid
  esters 222
polyoxyethylene nonyl phenol
  sulphates 129
polyoxyethylene polyglyceryl
  ester 231
polyoxyethylene sorbitan
  esters 226
polyoxyethylene(20) sorbitan
  monolaurate 226
polyoxypropylated
  polyoxyethylene glycols 216
polysaccharides 202
polysiloxane chain 277
polysiloxane glycol
  copolymers 277

INDEX

polysiloxane polyorganobetaine copolymer 277
polyunsaturated linseed amide ethoxylates 289
potassium fluorinated alkyl carbonylate 282
primary amine hydrochlorides 255
primary biodegradation 310
primary skin irritation 315
probable environmental concentration (PEC) 306
propionates 273
propylene glycol esters 231
propylene glycol monoesters 232
published books 17
pyrosulphonic acid 140

quality control 2
quasi-cationic 200
quaternary ammonium compounds 21, 249
quaternary softeners 44
quaternary surfactants with alkylamido betaines 267
quaternised polyoxyethylene fatty amines 250

rape seed oil sulphates 128
ready biodegradability 312
regulations for surfactants 314
reverse hexagonal 55
reverse micelles 55
reverse Pluronics 216
ricinoleic acid triglyceride 128
'ringing gels' 56
rinse aids in machine dishwashing 221

safety data 16
salt effect 57
salts of alpha-sulphonated fatty esters 144
sarcosinates 109
 with ether sulphates 124
 with LABS 142
screening test for biodegradation 309
SDS *see* sodium dodecyl sulphate
secondary alkane sulphonates 135
secondary alkyl sulphates 117
secondary amine hydrochlorides 255
secondary irritation 316
secondary n-alkane sulphonates 135
shampoos — thickened 57
silicone end groups 301
silicone monomer surfactants 277
silicone–polyglycol copolymers 288
silicone polymeric surfactants 296
silicone surfactants 276, 277
siloxane EO/PO copolymer 299

simulation tests for biodegradation 312
skin sensitisation 316
soap 28, 48, 101
 crystalline phases 102
 defoaming of other surfactants 103
 in alkanolamides 195
 in fatty ester sulphonates 148
 replacement 100
 with alpha-olefin sulphonates 155
 with ether sulphates 168
 with fatty ester sulphonates 147
 with taurates 168
Soap and Detergent Association (USA) 23
sodium carboxymethyl coco polyaminopropionate 270
sodium dibutylnaphthalene sulphonate 148
sodium di(2-ethylhexyl)-sulphosuccinate 160
sodium dilaureth-7 citrate 107
sodium di(lauryl alcohol + 7EO)citrate 107
sodium dodecane/oxyethylene/3 sulphate 121
sodium dodecyl allyl sulphosuccinate 291
sodium dodecylbenzene sulphonate 138
sodium dodecyl propyl sulphosuccinate 291
sodium dodecyl sulphate 117, 118
 with lauryl dimethylbetaine 90
sodium glycollate 272
sodium isethionate 112
sodium isopropylnaphthalene sulphonate 148
sodium laureth-3 sulphate 121
sodium laureth-7 tartrate 107
sodium lauryl alcohol +7EO tartrate 107
sodium lauryl ether sulphate 122
sodium lauryl sulphate 118
 with lauryl dimethylbetaine 90
 with monoisopropanol-amides 197
sodium nonyl phenol 2-mole ethoxylate ethane sulphonate 134
sodium N-octadecyl sulphosuccinamate 164
sodium N-octylphthalamate 109
sodium olefin (C14–16) sulphonate 151
sodium salt of dialkyl (C10–14) benzene sulphonate 155
sodium tallow methyl ester alpha-sulphonate 144
soft acid 138
solubilisation 19, 57, 58
 using nonionics 184
solubility 27, 28, 30
 and liquid crystals 46

behaviour of ethoxylated nonionics 47
 of ionic surfactants 47
sorbitan ester ethoxylates 226
sorbitan esters 226
sorbitan fatty acid esters 226
sorbitan monolaurate 226
sorbitan trioleate 226
Spans 226
Special Training 23
specification 16
spreading coefficient 61
stability of an emulsion 74
Stepan reactor 132
Strecker reaction 135
structure/performance relationships 19
substantivity 248
sucroglycerides 202
sucrose esters 202
sucrose monolaurate 207
sucrose monooleate 208
sucrose monostearate 207
sugar esters 202
sulphated alkanolamide ethoxylates 126
sulphated butyl oleate 129
sulphated ethoxylated alkanolamides 126
sulphated methyl esters 128
sulphated methyl ricinoleate 128
sulphated mono-, di- or triglycerides 127
sulphated monoglycerides 128
 of coconut fatty acids 128
sulphated nonyl ethoxylates 129
sulphated oils 127
sulphated polyoxyethylated aliphatic alcohols 121
sulphated polyoxyethylene amides 126
sulphates, manufacture of 116
sulphation
 using sodium bisulphite 161
 with sulphamic acid 129
sulphoalkyl amides 166
sulphoalkyl esters 112
sulphoamido betaines 265
sulphobetaines 265, 266
sulphoester acid in fatty ester sulphonates 145
sulphonated acids 143
sulphonated oleic acid 143
sulphonated styrene/maleic anhydride copolymers 293
sulphonated unsaturated acids 143, 144
sulphonation
 with chlorsulphonic acid 131
 with gaseous sulphur trioxide 131
 with oleum 131, 149
 with SO_3 131
sulphones 143
sulphosuccinamates 164
sulphosuccinate derivatives as polymerisable surfactants 290

324 HANDBOOK OF SURFACTANTS

sulphosuccinates 160, 276
  with LABS 142
sulphoxides 198, 200, 202
sulphoxidation 136
sultones in alpha-olefin
  sulphonates 153
superamides 194
surface properties of
  solids 249
surface tension 31
  and foams 66
  and wetting 62
  dynamic 63
surfactants
  aggregation 19
  chemical stability 95
  classification 97
  compatability with aqueous
    ions 96
  description 95
  general properties 95
  generalisations 98
  interactions 89
  mixtures 89
  nomenclature 95
  properties of polymeric
    surfactants 304
  safety 97
  solubility 95
  specifications 97

surfactants in consumer
  products 17
Surfynols 187
surgical scrubs 111
synergism 85
  in detergency 263
synthetic long chain alkyl
  benzene sulphonates 155
synthetic petroleum
  sulphonates 155, 156

tall oil soap 101
tallow soap 101
taurates (amide
  sulphonates) 166
teratogenicity 317
tertiary amine
  hydrochorides 255
tetralin in LABS 141
2,4,7,9-tetramethy-5-decyn-
  4,7-diol 187
tetrasodium N-(12-
  dicarboxyethyl)-N-octadecyl
  sulphosuccinamate 165
thickening bleach
  solutions 56
toluene sulphonates 134
triglycerol monostearate
  231

trimethyl glycine 258
Turkey red oil (sulphated
  castor oil) 127
Tweens 226

ultimate biodegradation 310
unilamellar vesicles 43–45
use of surfactants 3

vesicles 39, 43–45
viscosity 52, 55, 56
viscous phase 52

water-based coatings 21
wetting 61
  with polymeric
    surfactants 301
wool fat 242

xylene sulphonates 134

Young's equation 61

Ziegler alcohols 116, 189
zinc pyrithane 201
zwitterions 258, 262

CPSIA information can be obtained at www.ICGtesting.com
Printed in the USA
LVOW01s2252291213

367301LV00018B/84/P